D1346342

NEUROMETHODS □ 21

Animal Models of Neurological Disease, I

NEUROMETHODS

Program Editors: Alan A. Boulton and Glen B. Baker

21. **Animal Models of Neurological Disease, I: Neurodegenerative Diseases**
    Edited by *Alan A. Boulton, Glen B. Baker, and Roger F. Butterworth,* 1992
20. **Intracellular Messengers**
    Edited by *Alan A. Boulton, Glen B. Baker, and Colin W. Taylor,* 1992
19. **Animal Models in Psychiatry, II**
    Edited by *Alan A. Boulton, Glen B. Baker, and Mathew T. Martin-Iverson,* 1991
18. **Animal Models in Psychiatry, I**
    Edited by *Alan A. Boulton, Glen B. Baker, and Mathew T. Martin-Iverson,* 1991
17. **Neuropsychology**
    Edited by *Alan A. Boulton, Glen B. Baker, and Merrill Hiscock,* 1990
16. **Molecular Neurobiological Techniques**
    Edited by *Alan A. Boulton, Glen B. Baker, and Anthony T. Campagnoni,* 1990
15. **Neurophysiological Techniques: Applications to Neural Systems**
    Edited by *Alan A. Boulton, Glen B. Baker, and Case H. Vanderwolf,* 1990
14. **Neurophysiological Techniques: Basic Methods and Concepts**
    Edited by *Alan A. Boulton, Glen B. Baker, and Case H. Vanderwolf,* 1990
13. **Psychopharmacology**
    Edited by *Alan A. Boulton, Glen B. Baker, and Andrew J. Greenshaw,* 1989
12. **Drugs as Tools in Neurotransmitter Research**
    Edited by *Alan A. Boulton, Glen B. Baker, and Augusto V. Juorio,* 1989
11. **Carbohydrates and Energy Metabolism**
    Edited by *Alan A. Boulton, Glen B. Baker, and Roger F. Butterworth,*1989
10. **Analysis of Psychiatric Drugs**
    Edited by *Alan A. Boulton, Glen B. Baker, and Ronald T. Coutts,* 1988
9. **The Neuronal Microenvironment**
    Edited by *Alan A. Boulton, Glen B. Baker, and Wolfgang Walz,* 1988
8. **Imaging and Correlative Physicochemical Techniques**
    Edited by *Alan A. Boulton, Glen B. Baker, and Donald P. Boisvert,* 1988
7. **Lipids and Related Compounds**
    Edited by *Alan A. Boulton, Glen B. Baker, and Lloyd A. Horrocks,* 1988
6. **Peptides**
    Edited by *Alan A. Boulton, Glen B. Baker, and Quentin Pittman,* 1987
5. **Neurotransmitter Enzymes**
    Edited by *Alan A. Boulton, Glen B. Baker, and Peter H. Yu,* 1986
4. **Receptor Binding Techniques**
    Edited by *Alan A. Boulton, Glen B. Baker, and Pavel D. Hrdina,* 1986
3. **Amino Acids**
    Edited by *Alan A. Boulton, Glen B. Baker, and James D. Wood,* 1985
2. **Amines and Their Metabolites**
    Edited by *Alan A. Boulton, Glen B. Baker, and Judith M. Baker,* 1985

NEUROMETHODS ☐ 21

Animal Models of Neurological Disease, I

Neurodegenerative Diseases

Edited by

Alan A. Boulton
University of Saskatchewan, Saskatoon, Canada

Glen B. Baker
University of Alberta, Edmonton, Canada

and

Roger F. Butterworth
University of Montreal, Montreal, Canada

Humana Press • Totowa, New Jersey

999 Riverview Drive, Suite 208
Totowa, New Jersey 07512

Printed in the United States of America.

Library of Congress Cataloging-in-Publication Data

Main entry under title:

Animal models of neurological disease / edited by Alan A. Boulton,
Glen B. Baker, and Roger F. Butterworth.
p. cm. — (Neuromethods ; 21)
Includes bibliographical references and index.
Contents: v. 1. Neurogenerative diseases.
ISBN 0-89603-208-6
1. Nervous system—Diseases—Animal models. I. Boulton, A. A.
(Alan A.) II. Baker, Glen B., 1947– . III. Butterworth, Roger F.
IV. Series.
[DNLM: 1. Disease Models, Animal. 2. Nervous System Diseases.
W1 NE337G v. 21 / WL 100 A5984]
RC346.A54 1992
616.8'0427—dc20
DNLM/DLC
for Library of Congress 91-35401
CIP

Preface to the Series

When the President of Humana Press first suggested that a series on methods in the neurosciences might be useful, one of us (AAB) was quite skeptical; only after discussions with GBB and some searching both of memory and library shelves did it seem that perhaps the publisher was right. Although some excellent methods books have recently appeared, notably in neuroanatomy, it is a fact that there is a dearth in this particular field, a fact attested to by the alacrity and enthusiasm with which most of the contributors to this series accepted our invitations and suggested additional topics and areas. After a somewhat hesitant start, essentially in the neurochemistry section, the series has grown and will encompass neurochemistry, neuropsychiatry, neurology, neuropathology, neurogenetics, neuroethology, molecular neurobiology, animal models of nervous disease, and no doubt many more "neuros." Although we have tried to include adequate methodological detail and in many cases detailed protocols, we have also tried to include wherever possible a short introductory review of the methods and/or related substances, comparisons with other methods, and the relationship of the substances being analyzed to neurological and psychiatric disorders. Recognizing our own limitations, we have invited a guest editor to join with us on most volumes in order to ensure complete coverage of the field. These editors will add their specialized knowledge and competencies. We anticipate that this series will fill a gap; we can only hope that it will be filled appropriately and with the right amount of expertise with respect to each method, substance or group of substances, and area treated.

Alan A. Boulton
Glen B. Baker

Preface to the Animal Models Volumes

This and several other volumes in the Neuromethods series will describe a number of animal models of neuropsychiatric disorders. Because of increasing public concern over the ethical treatment of animals in research, we felt it incumbent upon us to include this general preface to these volumes in order to indicate why we think further research using animals is necessary and why animal models of psychiatric and neurologic disorders, in particular, are so important.

We recognize that animals should only be used when suitable alternatives are not available. We think it self-evident, however, that humans can only be experimented upon in severely proscribed circumstances and alternative procedures using cell or tissue culture are inadequate in any models requiring assessments of behavioral change or of complex in vivo processes. However, when the distress, discomfort, or pain to the animals outweighs the anticipated gains for human welfare, then the research is not ethical and should not be carried out.

It is imperative that each individual researcher examine his/her own research from a critical moral standpoint before engaging in it, taking into consideration the animals' welfare as well as the anticipated gains. Furthermore, once a decision to proceed with research is made, it is the researcher's responsibility to ensure that the animals' welfare is of prime concern in terms of appropriate housing, feeding, and maximum reduction of any uncomfortable or distressing effects of the experimental conditions, *and that these conditions undergo frequent formalized monitoring.*

In the present volume of Neuromethods, we have included a chapter on the ethics of animal models by Dr. E.

Olfert, a veterinarian who also directs a laboratory animal care facility. As indicated in Dr. Olfert's chapter, it is essential to conform to national and local animal welfare regulations, whether codified in law or by self-regulatory bodies. We urge readers who wish to adopt any of the procedures described to follow closely not only the letter of their own national and local regulations, but also the spirit of these guidelines.

The Editors

Preface

Studies of pathophysiologic mechanisms as well as the development of new approaches to the treatment of neurological disorders continue to rely on the use of experimental animal models. The present volume attempts to bring together detailed methodology currently in use in this area. As mentioned in the "Preface to the Animal Model Volumes," the first chapter deals with the issue of the ethics of using animal models. Written by Ernest Olfert, the chapter covers in detail the issues of selection of appropriate animal models, acceptable methods of euthanasia, and other concerns of an ethical nature.

Animal models for use in the study of the three most commonly studied neurodegenerative disorders, namely Alzheimer's Disease (by Gary Wenk), Huntington's Disease (by Dwaine Emerich and Paul Sanberg), and Parkinson's Disease (with a chapter on rodent models by François Jolicoeur and Ribert Rivest followed by a chapter on primate models by Paul Bédard and coauthors) are covered in four chapters. Similarities as well as differences between the human condition and that obtained using animal models are highlighted. Animal models of neurological disorders associated with loss of myelin are covered in two chapters by Gregory Konat and Richard Wiggins (Genetic Dysmyelination models) and Sheldon Miller (Nongenetic models). Animal models of the Cerebellar Ataxias (Roger F. Butterworth) and of Lesch-Nyhan Disease (Roberta Palmour) are covered in two separate chapters.

Chapters treat systematically the use of various approaches from genetic animal models of neurodegenerative disorders to the use of surgical lesions and selective neurotoxins. Where appropriate, example protocols are included in sufficient detail to enable the reader to directly perform the experiments described. Particular methodological or ethical considerations additional to those raised by Dr. Olfert are also included.

It is hoped that the volume may provide an important step toward a rationalization and standardization of experimental protocols for use in studies of animal models of neurodegenerative diseases.

Roger F. Butterworth

Contents

Animal Models of Alzheimer's Disease
Gary L. Wenk

Animal Models of Huntington's Disease
Dwaine F. Emerich and Paul R. Sanberg

Nongenetic Animal Models of Myelin Disorders
Sheldon L. Miller

Contributors

PAUL J. BÈDARD • *Departments of Pharmacology and Neurology, Laval University, Québec, Canada*

PIERRE BLANCHETTE • *Department of Neurology, Laval University, Québec, Canada*

RENÉ BOUCHER • *Department of Anatomy, Laval University, Québec, Canada*

ROGER F. BUTTERWORTH • *Department of Medicine, University of Montréal, Montréal, Québec, Canada*

DWAINE F. EMERICH • *CytoTherapeutics, Inc., Providence, Rhode Island*

BALTAZAR GOMEZ-MANCILLA • *Department of Neurology, Laval University, Québec, Canada*

FRANÇOIS B. JOLICOEUR • *Departments of Psychiatry and Pharmacology, University of Sherbrooke, Québec, Canada*

GREGORY KONAT • *Department of Anatomy, West Virginia University, Morgantown, WV*

SHELDON L. MILLER • *The Wistar Institute, Philadelphia, PA*

ERNEST D. OLFERT • *University of Saskatchewan, Saskatoon, Saskatchewan, Canada*

ROBERTA M. PALMOUR • *Departments of Psychiatry and Biology, McGill University, Montreal, Québec, Canada*

ROBERT RIVEST • *Departments of Psychiatry and Pharmacology, University of Sherbrooke, Québec, Canada*

PAUL R. SANBERG • *Department of Surgery, University of Florida, Tampa, Florida*

GARY L. WENK • *Department of Psychology, Johns Hopkins University, Baltimore, MD*

RICHARD C. WIGGINS • *Department of Anatomy, West Virginia University, Morgantown, WV*

Ethics of Animal Models of Neurological Diseases

Ernest D. Olfert

1. Introduction

1.1. General Remarks

Some people might wonder about the propriety of asking a laboratory animal veterinarian to author a chapter on the ethics of animal use in neuropsychiatric research. After all, she or he is part of the infrastructure that supports biomedical research rather than an independent ethicist or philosopher. On further reflection, however, it makes a lot of sense. Laboratory animal veterinarians are involved in making ethical decisions on a daily basis, decisions that directly affect the well-being of the animals used in biomedical research, teaching, and testing. They are involved (by law in some countries) in the research protocol review process. They are regularly involved in the management of facilities where research animals are housed and used. And in the animal room or laboratory, they are directly involved in ministering (providing veterinary care) to the animals being used, a vital aspect of which is the prevention and relief of pain and suffering. Veterinarians take the role of the "animal's advocate" in this whole process.

Discussions on moral and ethical issues often seem out of place next to scientific articles in scientific journals or books, since they are, for the most part, presentations of opinions and views, rather than interpretations of experimentally derived data and

From: *Neuromethods, Vol. 21: Animal Models of Neurological Disease, I*
Eds: A. Boulton, G. Baker, and R. Butterworth

"facts." Insofar as they are based on critical thought and reasoning, however, they have many similarities. Because ethical discussions do contain opinions and judgments, such discussions inevitably carry with them the perspective of the author (scientist, philosopher, veterinarian, animal welfarist, animal rightist, and so on). Likewise this chapter will have the perspective of the author—a veterinarian involved on a daily basis with the practical aspects of the welfare of laboratory animals—and as such will focus the discussion on the animals rather than on the science of neuropsychiatric research. It will no doubt also have the personal biases each of us carries. Hopefully, the latter will not cloud the principles being put forward.

The intent in this chapter is to discuss in general the ethical principles embodied in the current positions on the use of animals for research, teaching, and testing. These ethical principles can guide us in our use of the animals that are judged to be necessary to advance the knowledge in the neurological sciences. Any set of ethical guidelines for the use of animals in biomedical research should be continuously subject to critical review. The biomedical research community must continue to act responsibly and hold its animal use practices up for such a review.

1.2. Societal Concerns

It is generally accepted that our society wants to continue to derive the benefits of improved health and well-being as a result of scientific and biomedical research. Where animals are used in this pursuit (the pursuit of knowledge, the training of scientists, the safety/toxicity testing of drugs), society deserves the assurances that the animal use is justified, and that the animals used are in fact used in a humane manner and are not subjected to any unnecessary pain and suffering. The public must also be assured that there are adequate controls in the system.

The responsibilities for all these assurances are placed at several levels. In most countries federal agencies set standards for the care and use of research animals, and monitor compliance. In some countries the standards are set by law (US, Britain, other European countries) and in others (Canada, for example) the standards are set by nonregulatory agencies (e.g., the Canadian Council on Animal Care). The burden of ensuring that the standards are observed in the intended manner falls primarily on research institutions and their institutional animal care and use committees. More directly on a day-to-day basis the responsibilities lie with the principal investigators and the laboratory animal veterinary directors. Monitoring compliance may be done via assessments or accreditations (based on guidelines) or inspections (based on legislative regulations).

The medical community, whose ethic contains a commitment to improving the health and well-being of humans, has in the past been less than eager to discuss the ethics of animal use in biomedical research, teaching, and testing. It has been relatively silent. The biomedical research community must enter much more clearly into this public debate if the issues raised by opponents to research are to be answered.

2. Ethics of Animal Experimentation

2.1. General Remarks

It is beyond the scope of this chapter to enter into a discussion of animal rights and the question of the moral status of animals. A brief comment on the subject seems appropriate, however, to provide a framework for the current attention to the issue, particularly as it relates to our discussion of the ethical principles of animal use in neuropsychiatric research.

The past 10–20 years have seen a tremendous number of disputations on the rights of animals (press articles, commentaries, symposia, books, and so on). Discussions on the nature of rights, on the question of the moral status of nonhuman animals, on whether or not animals do (or should have) rights and what these rights should be, and on how these rights should be enforced have all found receptive audiences. At one level this is a philosophical debate with several positions more or less clearly taken and put forward by various philosophers (for example, Regan, Singer, Rollin), with scientists and other interested people only peripherally involved in this debate. These positions range from arguing that animals have (should have) all the same rights as human beings to the position that animals cannot have rights but that we have many responsibilities toward them. At another level this is an emotional public debate, with the various factions (the animal liberationists, the animal rightists, the animal welfarists) vying for the hearts of the public.

There seems to be general agreement that we do have moral responsibilities toward animals. With respect to the use of animals in biomedical research, the following argument is often used. If animals aren't similar to humans, and don't perceive pain and suffering similar to humans, then why are they being used in research? And if animals do have enough similarities to humans to make them useful research models, then we must accept that they be regarded as moral objects, worthy of moral consideration. Although there are widely differing views on what these moral considerations (or responsibilities) are, most would agree that they must include a concern for the health and well-being of the animals, and for the prevention of any unnecessary pain and suffering.

Several articles have been published in the last decade on the ethical considerations of the use of animals in neuropsychiatric and psychological research (Iggo, 1979; Marcuse and Pear, 1979; Bowd, 1980; Bond, 1984; Rollin, 1985; Casey and Dubner, 1989; Franklin and Abbott, 1989). The reader is referred to these articles for additional perspectives on this subject. From these, and from the existing current guidelines, a practical view on ethical guidelines for the humane use of animals in neuropsychiatric research will be presented. After all, one function of ethics is to set down practical guidelines for human conduct.

2.2. The Utilitarian Principle

The ethical principle that best encompasses our current views and practice of using of animals for research, teaching, and testing is utilitarianism. Briefly stated, this principle says that the action being taken (our use of animals in research) must be viewed in the light of the consequences resulting from that action. An action is "good" if there are overall benefits arising from that action. Of course there arise questions about whose benefits are to be considered, and whose costs; these require answers too. The utilitarian ethical approach suggests that for the use of animals in science to be justified there must be some recognizable benefit to humans (or to other animals). This utilitarian principle is, however, difficult to apply in cases of basic biomedical research, where the pursuit of knowledge in and of itself is the objective, one that humankind and society accept as valuable.

The utilitarian principle is implicit in the statements of most organizations involved in biomedical research, or in its regulation. In the Ethics of Animal Experimentation statement of the CCAC (June 1989), the leading sentence reads, "Research involving the use of animals is acceptable *only* if it promises to contribute to understanding of fundamental biological or physiological principles, or to the development of knowledge that can reasonably be expected to benefit a significant proportion of human beings or sentient animals." Similar sentiments are included in many other ethics statements, including those of the American Psychological Association (APA), the International Association for the Study of Pain (IASP), and the Canadian Psychological Association (CPA). It is also implicit in the British laboratory animal welfare legislation, which states, in section 5.4 of the Animals (Scientific Procedures) Act 1986; "In determining whether and on what terms to grant a project licence the Secretary of State shall weigh the likely adverse effects on the animals concerned against the benefit likely to accrue as a result of the programme to be specified in the licence."

Rollin (1981) proposed two principles for ethical guidance in using animals for biomedical research. First, the research should be evaluated on the "Utilitarian Principle" (the benefit to humans and/or animals should clearly outweigh the pain and suffering of the animals). Second, the "Rights Principle" (the research "should be conducted in such a way as to maximize the animal's potential for living its life according to its nature or telos, and certain fundamental rights should be preserved as far as possible, given the logic of the research, regardless of considerations of cost") should be applied.

In the utilitarian ethical approach, all the risks and the (potential) benefits are to be weighed. It is here that we must include, in our judgments, the potential negative effects (the ethical costs) to the animals in terms of pain, suffering, distress, loss of health, deviation from a physiological norm, confinement, disability, and so on of a given research proposal, experiment, or specific technique applied to the animals. Thus, if the potential benefits are expected to be high, a higher ethical cost might be accepted on the part of the animals; for "trivial" research, there should be little or no ethical cost to the animals being used. Drewett and Kani (1981) wrote, "It is a quite widely held principle among the general public that if experiments that injure animals are to be conducted, they should at least be restricted to those that might reasonably be expected to improve medical care."

One of the criticisms of this utilitarian approach is that an investigator might be able to "justify" almost any degree of pain and suffering merely by stating that the benefits were expected to be high. In practical terms, however, there are several levels of controls in place that would not permit this to happen. There are the guidelines of the granting agencies and the policies of the scientific institutions, the ethical reviews of protocols by the institutional animal care and use committees and the veterinary medical judgments of the laboratory animal veterinarians, and the ethical principles of the editorial boards of scientific journals.

2.3. Unacceptable Procedures on Animals

Clearly there are limits to the use of animals in biomedical research. Certain procedures are judged to carry so high an ethical cost to the animals that they are either banned outright, or special review procedures are required if their use is proposed. For example, in the British Animal (Scientific Procedures) Act 1986, there are controls over the reuse of an experimental animal, and over the use of neuromuscular blocking agents. Under the 1989 regulations of the Laboratory Animal Welfare Act (USA), paralytic drugs without accompanying anesthesia are not allowed. The CCAC *Ethics of Animal Exerimentation* statement (June 1989) states "... the following experimental procedures inflict excessive pain and are thus unacceptable: a) utilization of muscle relaxants or paralytics (curare and curare-like) alone, without anesthetics, during surgical procedures; b) traumatizing procedures involving crushing, burning, striking, or beating in unanesthetized animals." Investigators in neuropsychiatric and psychological research should note that identified among these unacceptable procedures are some techniques that have in the past been regularly used in this field of research.

2.4. Other General Ethical Principles

Some other general ethical principles that have often been cited as guiding us to a more humane use of animals in research include the now famous 3-R tenet of Russell and Burch (1959); those of *Replacement* (of animals with other, nonsentient material), *Reduction* (of numbers of animals used), and *Refinement* (of technique, "to reduce to an absolute minimum the amount of distress imposed on those animals that are still used"). Their book—*The Principles of Humane Experimental Technique* (Russell and Burch, 1959)—was a landmark publication, and still deserves to be consulted for its comments on the humane use of animals in the neurosciences.

2.5. Pain and Suffering in Animals

It is part of our understanding of the (higher) animals that they are conscious beings that have feelings, can experience pain and suffering, and so on. We believe that there are differing degrees of these attributes along the phylogenetic order. The renewed interest in this matter over the past 15 years has stimulated philosophical deliberations of this common sense of the nature of animals. Several authors have proposed that it is the sentient animal's ability to experience pain and to suffer (these seem to be the most important but by no means the only attributes of sentient animals that are considered to be morally relevant), that require that we give them moral consideration (Singer, 1975; Rollin, 1981,1989; Regan, 1983; Rowan, 1984).

It is beyond the scope of this section to review the extensive literature on pain in animals, the neurophysiology of pain perception, the experience of pain, or the motivational behavior of pain and suffering in animals. The reader is directed to the recent literature on pain in animals (Yoxall, 1978; Flecknell, 1984; Gibson, 1985; American Veterinary Medical Association, 1987;

Soma, 1987), various veterinary anesthesia texts, and to the broader body of literature on pain in general.

The question of distress in animals and how to define and measure it is still quite perplexing and with no precise answers, although the literature on this subject is quite extensive. A great deal of additional research will be needed before we will be able to provide clear scientific support for our judgments about stress and distress in animals.

2.6. Recognition and Assessment of Pain in Animals

In her book *Animal Suffering—The Science of Animal Welfare,* Dawkins (1980) sets out the basis for assessing pain and suffering in animals. Information on animal pain and suffering can come from physiological and behavioral observations, from the health status (illness or disease) of the animal, from comparisons of the behavior of natural (wild) animals with domesticated (confined or captive) animals, and from animal choice experiments. Dawkins also stresses that a comparison with the same procedure applied to humans is useful when assessing the degree of pain and suffering an animal may be experiencing. The IASP Ethical Guidelines (Zimmermann, 1983) agree; "If possible, the investigator should try the pain stimulus on himself/herself; this principle applies for most noninvasive stimuli causing acute pain." Franklin and Abbott (1989) state a similar view; "Studies in humans confirm that such shocks (electrical shocks of sufficient intensity to produce vocalisation that lasts beyond the time when the stimulus is applied) evoke extremely unpleasant painful sensations. On ethical grounds, therefore, one should not use vocalisation after-discharge as a pain measure unless this type of response is essential to the experiment."

Burghardt (1985) called this "critical anthropomorphism," a term that neatly describes the general premise that there is a Darwinian (evolutionary) continuum from animals to humans, not only with respect to the physical but also with respect to the behavioral and psychological, and that one must approach the interpretation of animal thinking and behavior with caution. This evolutionary continuum has been tacitly accepted by neuropsychiatric and psychology researchers. The words pain, punishment, fear, depression, reward, and suffering are all part of the normal language in their publications on animal-based research, although euphemisms like "nociception" are now appearing much more frequently.

Recently attention has been directed at developing practical guidelines for assessing pain, distress, and suffering in laboratory animals. These guidelines can assist scientists, animal care staff, veterinarians, and institutional animal care and use committees in their evaluation of the ethical costs of research on the animals. It is these guidelines that will be discussed here.

Morton and Griffiths (1985) developed a set of criteria for assessing pain, distress, and discomfort in laboratory animals, based on evaluating five aspects of an animal's condition. The five aspects are:

1. Changes in body weight (including levels of food and water intake);
2. External appearance;
3. Measurable clinical signs (e.g., changes in heart rate, in respiratory rate, and nature);
4. Unprovoked behavior; and
5. Responses to external stimuli.

In each of these categories, a rating of 1 (normal or mild) to 3 (severe changes from normal) is made, the cumulative rating indicating increasing deviation from the normal in the animal. The cumulative rating is interpreted as an indication of increasing pain, distress, and suffering. As indicated below in the section on laws, the new British Act set down three categories of invasiveness, similar to those proposed by Morton and Griffiths (1985). The British Association of Veterinary Teachers and Research Workers (1986) prepared a set of guidelines for the recognition and assessment of pain in animals, providing some specifics on the changes (behavioral and physiological) in the various animal species that may indicate the presence of pain. Soma (1987) presents a discussion of the general clinical signs of pain in animals.

The publications mentioned above (Morton and Griffiths, 1985; AVTRW, 1986; Soma, 1987) serve to focus on an important matter—that of learning to make more objective assessments of the pain, distress, and suffering that may be inflicted on animals in the course of biomedical research. Surely it is our obligation to continue to improve our ability to accurately assess this in animals. Increased observation of the behavior of animals related to pain and suffering, and the publication of these observations, will make a significant contribution to the body of information that is developing.

2.7. Risk/Benefit Analysis

The main question that must be addressed is that of attempting to evaluate the ethical cost to the animal subjects in terms of pain and suffering. The basis for recognizing pain, distress, and suffering in animals was discussed in the previous section *(above)*. This ethical cost must then be balanced with the potential benefits of a research project. "Ethical costs" as used here would include pain, suffering, loss of health, distress, depression or anxiety, inability to cope with the experimental situation, deviations from normal physiological and behavioral parameters, confinement or restraint, and so on.

Several methods of categorizing the techniques used in laboratory animals with respect to the levels of pain, distress, and suffering produced by these techniques have been developed in the last few years. The Scientist Center for Animal Welfare (SCAW) has provided leadership in this area, and developed a five-category table—*Categories of Biomedical Exeriments Based on Increasing Ethical Concerns for Non-human Species* (Anon., 1987), with the recommendation that such a classification system be used in protocol review. The Canadian Council on Animal Care (CCAC) also uses a five-level *Categories of Invasiveness in Animal Experiments* (CCAC, July, 1989) to assess the ethical cost to the animals of research techniques. These categories (A to E) describe experiments of increasing invasiveness, with Category E containing "Procedures that involve inflicting severe pain near, at, or above the pain tolerance threshold of unanesthetized, conscious animals." Examples given for this category that are pertinent to neuropsychiatric and psychological research are "behavioural studies about which the effects of the degree of distress are not known; use of muscle relaxants or paralytic drugs without the use of anesthetics."

In Britain, a categorization scheme has been incorporated into the regulations of their new Animals (Scientific Procedures) Act 1986. These regulations identify a three band categorization of severity of potential animal pain and suffering. The terms *mild, moderate,* and *substantial* define the three bands of potential stress/distress and pain and suffering a laboratory animal might endure in an experimental procedure. The onus is on the licensee (principle investigator) to accurately determine which band the proposed experimental procedures fall into, and to be familiar with the anticipated (abnormal) behavior of the animals that would be perceived as pain or distress-induced behavior.

3. Ethics of Animal Models of Neuropsychiatric Disease

3.1. Animal Models of Pain

Because it is an inherent aspect of studies into the various aspects of pain in humans and animals that some pain must be produced, such experiments raise special ethical concerns that demand careful consideration. This difficulty was recognized early as the study of pain became a more distinct discipline. The ethical statement from the International Association for the Study of Pain (IASP), published by the IASP Committee for Research and Ethical Issues in 1980 and revised in 1983 (Zimmermann, 1983) reflects that concern. The IASP *Ethical Guidelines for Investigations of Experimental Pain in Conscious Animals* (Zimmermann, 1983) contain seven points relating to the justification of the experiment, and to the potential severity and duration of the pain produced in the experimental animal. These seven points are:

1. It is essential that the intended experiments on pain in conscious animals be reviewed beforehand by scientists and laypersons. The potential benefit of such experiment to our understanding of pain mechanism and pain therapy needs to be shown. The investigator should be aware of the ethical need for a continuing justification of his investigations.

2. If possible, the investigator should try the pain stimulus on himself; this principle applies for most noninvasive stimuli causing pain.
3. To make possible the evaluation of the levels of pain, the investigator should give a careful assessment of the animal's deviation from normal behavior. To this end, physiological and behavioral parameters should be measured. The outcome of this assessment should be included in the manuscript.
4. In studies of acute or chronic pain in animal measures should be taken to provide a reasonable assurance that the animal is exposed to the minimal pain necesary for the purposes of the experiment.
5. An animal presumably experiencing chronic pain should be treated for relief of pain, or should be allowed to self-administer analgesic agents or procedures, as long as this will not interfere with the aims of the investigation.
6. Studies of pain in animal paralyzed with a neuromuscular blocking agent should not be performed without general anesthetic or an appropriate surgical procedure that eliminate sensory awareness.
7. The duration of the experiment must be as short as possible and the number of animals involved kept to a minimum.

Franklin and Abbott (1989), in their review of drug effects on pain responses, state that some 50 different pain tests (and variations on these) have been described in the literature. A critic might say that this only shows the tremendous ingenuity of scientists to cause pain to experimental animals. It is clear, however, that the pain response and its perception in the subject (human or animal) is a very complex phenomenon, and one that is not easily or simply measured.

The ethical issues for the scientist conducting pain research are several, as indicated by the IASP statement. Some of these bear reemphasizing. First, *there must be clear scientific justification for the proposed experiment*. Second, and as will be expanded upon below (*see* Section 4. Selecting the Right Animal Model), there must be careful design and conduct of the experiment, so that meaningful results are obtained using the fewest numbers of animals necessary, and so that the pain produced is not wasted. Related to this is the question of the *choice of the pain test*. Given the complexity of the pain experience and the many different pain tests described in the literature, the scientist has an obligation to determine as precisely as possible which pain test will provide the correct information in the experiment under consideration.

The question of reusing the same animals in pain tests must also be considered. As a general principle, *repeated pain tests in the same animal should be avoided*, recognizing that doing so may be part of a given experimental design.

Sessle (1987) suggests a number of questions (to be asked of the investigator) that would serve to focus on the need to minimize the pain in the experimental animals. The kind of questions that a protocol review committee should ask of an investigator proposing to conduct pain tests would include the following. Is the scientific justification acceptable? Are there alternatives to this experiment? Does the protocol conform to the general ethical guidelines of this kind of research? Is the amount of the stimulus the lowest one necessary to achieve the effect? Is the timing of the noxious stimulus the minimum necessary? Are the minimum number of animals proposed? Has the investigator justified the choice of pain test? Has the investigator demonstrated that there are no alternatives to the experiment?

In the following two sections dealing with models of acute and chronic pain, more specific ethical concerns will be discussed appropriate to each of these.

3.1.1. Animal Models of Acute Pain, and Pain Testing

There are several ethical considerations when acute pain is being studied, or when pain testing is being conducted. Franklin and Abbott (1989) deal with this subject with a great deal of sensitivity toward the animals. They state that whatever pain test is being used, it should be one that *"inflicts the briefest and least intense pain that is consistent with the demands of the experiment."*

As noted in the IASP statement (above) it is incumbent on the investigator to be thoroughly familiar with the behavior of the animal species being used, and to document the behavioral and physiological changes being observed in animals subjected to stimuli causing pain. This is also important so that unexpected severe levels of pain are recognized if they occur. Some of the tests for acute pain have been extensively used on humans, and those experiences should guide investigators regarding the nature of the pain produced.

Many pain tests use a threshold pain level, one that triggers some avoidance or withdrawal behavior in the subject. If the stimulus intensity is suprathreshold, special care must be taken in its application. *Threshold levels of pain stimuli rather than suprathreshold levels should be used whenever possible.* Franklin and Abbott (1989) indicate that in many instances suprathreshold pain stimuli are unnecessary.

In some of these tests the animals can terminate the painful stimulus by their action (e.g., tail flick test, and the hot plate test where the animal's first response is taken as the endpoint, and the animal removed from the stimulus). In other tests (e.g., the acute tissue injury-inducing tests, which include the writhing test and the formalin test) the painful stimulus presumably goes on longer than the test period, and the animals cannot escape from it. *Using avoidance tests would be preferable to ones where the pain continues after the results are obtained.* If models of acute pain, or acute pain tests, are being used where the pain is not terminated by the animal's reaction, but may extend beyond the time necessary to obtain results, then *the pain should be terminated as quickly as possible.* This may require humanely killing the animals as soon as the test is completed. The investigator should also consider the use of analgesic drugs in these cases.

The use of bradykinin injections (often given intra-arterially) must be approached with particular caution, since bradykinin injections produce "severe pain" in humans (Lineberry, 1981).

3.1.2. Animal Models of Chronic Pain

Animal models of chronic pain are of particular concern, since the ethical cost to animals where chronic pain is induced (adjuvant-induced arthritis in rats, for example) can be very high. It is well known that such disease states in humans are often accompanied with severe, unremitting pain and suffering. Unless particular attention is paid to the animal models used in such a study, the pain and suffering could easily extend beyond that necessary for the purposes of the research.

Franklin and Abbott (1989) suggest that the pain produced in adjuvant-induced arthritis in rats approaches an intensity such that, based on the rating proposed by Morton and Griffiths (1985), relief (analgesics) should be given. The IASP guidelines *(above)* agree, recommending that analgesia be provided, "as long as this will not interfere with the aims of the investigation." In my view this should be strengthened by stating that *"animal models presumably experiencing chronic pain should be provided with adequate analgesia at all times; exception to this should be restricted to those times justified to the institutional animal care and use committee by the investigator."* The principal investigator, the institutional laboratory animal veterinarian, and the institutional animal care committee should ensure that these animals are afforded every consideration for limiting or easing their discomfort and pain.

All animals with chronic pain should receive special attention relating not only to the relief of their pain, but also *with respect to their husbandry and housing*. A great deal can be done, beyond the routine care given normal laboratory animals, to make these special animal models comfortable. The expertise of the laboratory animal veterinarian and the animal health technicians should be consulted in this matter. This extra care should be provided regardless of cost. A further contribution would be the description of such special animal care procedures in publications of the research.

An editorial by Casey and Dubner in the journal *Pain* (1989) raises the issue of the ethics of searching for new animal models of chronic pain. They suggest that certain animal models (e.g., models of nerve injury) should be developed, because these could contribute to new approaches for controlling pain from such conditions in humans, but that creating animal models of persistent pain "where pain behaviors persist in the absence of a known source of abnormal or nociceptive input" might be ethically and scientifically unjustifiable.

3.2. Animal Models of Human Neuropsychiatric Diseases

Probably one of the areas of neuropsychiatric research that concerns people outside the field the most is the use of, or creation of animal models of neuropsychiatric diseases, particularly those that are known to cause suffering, and sometimes severe psychological distress in humans. The social isolation and maternal separation studies done in the 1960s using infant monkeys and puppies are examples of the kind of psychological research that has generated public criticism. A more recent example is the limb denervation work using nonhuman primates conducted at the Institute for Behavioral Research in Silver Springs, MD, under the direction of Edward Taub, which was brought to public attention in 1981. Dr. Taub was subsequently convicted for failing to provide adequate veterinary care for the monkeys with the experimentally produced deficits (*see Science,* **214,** 1218–1220, December 11, 1981). Although this conviction was later overturned, important questions about the needs for special care for these animals remained unanswered (Rowan, 1984).

3.2.1. Special Needs of Animal Models with Neuropsychiatric Diseases

There is no question that animal models of neurological or psychiatric diseases have special needs beyond those of normal, healthy laboratory animals. These special needs must be recognized and accommodated when such animal models are going to be used in research. Rowan (1984), in his discussion of appropriate care for the deafferented monkeys at the center of the court case against Edward Taub in 1981, concluded that deafferented animals "ideally require intensive nursing care and *future proposals to study deafferented primates (and other animals) should include provisions for extra laboratory animal personnel to care properly for these animals*" (the emphasis is mine). It should be the responsibility of the principal investigator to take into consideration the special needs of the animals before embarking on a research project involving such animals. These special needs will no doubt impact on the research budget in terms of additional animal care time, and in additional materials and equipment. Scientific and ethical reviews of research proposals should include an assessment of these extra considerations for the animals.

Only a couple of examples from the extensive catalog of animal models of human neuropsychiatric diseases are necessary to emphasize the need for attending to the special requirements of these animals. Reserpinized rodents (a model for Parkinson's Disease) become hypothermic and exhibit shivering, and so should be kept warm. This shivering might also be confused with an induced motor dysfunction (tremors) (Heikkila et al., 1989). Jackson (1989) in his discussion of the use of MPTP to produce a Parkinson's Disease-like syndrome in rodents, indicates that very high mortality rates (up to 50% in rats) are common during the administration of MPTP in rats and mice. It is not clear whether this mortality is owing to some unusual effects of the drug in these species or whether it is owing to the severe symptoms induced in the animals—symptoms that could be alleviated with treatment similar to that of Parkinson's Disease in humans. Observations like weight loss, disturbed sleeping patterns, hypomobility, irritability, agitation, fear or aggression, and self-mutilation can be found in the descriptions of several of these animal models, and indicate a real need to be concerned about the mental anguish and suffering the animals are experiencing.

The principle that encompasses this responsibility to attend to the special needs of animal models in neurological or psychiatric disease could be stated as follows: *That any pain, suffering, distress, or deficits in function that negatively affect the animal's well-being, not scientifically "necessary" for the study, should be alleviated or minimized. Cost or convenience should not deter from this. Further, as soon as the study is done, the animal suffering should be terminated.*

The value of "critical anthropomorphism" in assessing the needs of these animals must be re-emphasized here. The experiences of humans with diseases like Parkinson's Disease, anxiety, depression, Huntington's Disease, and so on, can serve as guides to the pain and suffering the animal models may be enduring. The symptomatic treatments used in humans afflicted with these diseases may also have application in the animal models.

3.3. Euthanasia

Acceptable methods for humanely killing laboratory animals are described in the Canadian Council on Animal Care Guide, Volume 1 (1980), and in the 1986 Report of the AVMA Panel on Euthanasia (AVMA, 1986). A controversy arose from the AVMA Report's recommendation on the decapitation of animals. Based on data reported by Mikeska and Klemm (1975) that animals may remain conscious for an average of 13–14 s after decapitation, the Report recommended that the guillotine "should be used only after the animal has been sedated or lightly anesthetized, unless the head will be immediately frozen in liquid nitrogen subsequent to severing." This had direct implications for neuroscientists, and the matter is still being debated. The Canadian Council on Animal Care issued a guideline in 1988 that states that "in all studies where cervical dislocation or guillotining are involved, unless counterindicated by the nature of the study and supported by scientific evidence, light anesthesia or sedation are recommended." The new regulations under the Laboratory Animal Welfare Act of 1985 (USDA, 1989) define euthanasia as "the humane destruction of an animal accomplished by a method that produces rapid unconsciousness and subsequent death without evidence of pain or distress, or a method that utilizes anesthesia produced by an agent that causes painless loss of consciousness and subsequent death."

Other methods of killing animals that are often considered in neuropsychiatric research are the use of microwave radiation and rapid freezing (using liquid nitrogen). Both of these methods have important limitations, and the AVMA Report on Euthanasia (AVMA, 1986) should be consulted by anyone planning the use of these procedures for humanely killing animals.

4. Selecting the Right Animal Model

4.1. General Attributes of Animal Models of Neuropsychiatric Diseases

The assessment of an animal model's usefulness or relevance to human psychology or to neurological research has often been based on criteria proposed by McKinney and Bunney (1969): "The model should resemble the condition it models in its etiology, biochemistry, symptomatology, and treatment." Willner (1984) suggests that although these criteria assess face validity (the similarities to the disease), there is also the predictive validity and construct validity to consider. He defines predictive validity as being concerned with the success of predictions made from that model (and this is very important in the assessment of drug therapies), and construct validity regarding its theoretical rationale. These criteria are also discussed by Greenshaw et al. (1988), and they indicate that predictive validity and face validity criteria are the most useful in discussing animal models of drug action. Jackson (1989) suggests that a valid behavioral model of a central disorder should be similar in clinical signs and in anatomical, pathological, and biochemical lesions produced, to the disease. Further, it should be easy to establish, economical, and ethically acceptable. McGuire et al. (1983) discuss animal models of psychiatric disorders in terms of homologous models (models with phylogenetic closeness to humans) for which the nonhuman primates have many advantages over other species, analogous models (models from different species that share similar functions as a result of parallel evolution), survey, and outcome models. The reader is referred to these publications for detailed discussions of the classification and definition of animal models of neuropsychiatric disorders.

Although no one ideal animal model will be found to be meet all the criteria discussed above, the choice of the appropriate animal model is important from an ethical perspective as well as from a scientific one. "The right animal for the right reasons," a phrase coined by Rowsell (1979), concisely sums up the need for careful selection of the animal model to be used in any field of biomedical research.

4.2. On the Need to Search for Alternative Methods

It is a generally accepted principle that animal use in research is justifiable only if no alternative, nonanimal means of attaining the information are available. The CCAC *Ethics of Animal Experimentation* document states "Animals should be used only if the researcher's best efforts to find an alternative have failed."

This principle is now embraced in the new regulations of the US Laboratory Animal Welfare Act (USDA, APHIS, 1989) and in the British Animals (Scientific Procedures) Act 1986 (HMSO, 1986). Section 5.(5) of the Animal (Scientific Procedures)

Act 1986 states "The Secretary of State shall not grant a project licence unless he is satisfied that the applicant has given adequate consideration to the feasibility of achieving the purpose of the programme to be specified in the licence by means not involving the use of protected animals."

The investigator or licensee must demonstrate that a search for alternatives has been adequately done. What form of assurance will be acceptable is still open to discussion. Some institutional animal care and use committees may accept a statement on their protocol forms signed by the investigator, that such a search has been done, and that the use of live animals in the study is necessary. There is no question that greater emphasis will be placed on this in the future when protocols are reviewed. Whether or not this search for alternatives should include a determination of whether the study could be done in human subjects is debatable.

4.3. Controls over Variables in Animal Experiments

For optimal scientific validity, experiments using animals should be planned and conducted so that all the confounding variables are controlled, thereby ensuring accurate and definitive results using the fewest numbers of animals. The ethical and scientific considerations converge and complement each other here. In order to do this, the investigator must be aware of the many factors that can influence the results of the animal's responses in any given experiment. Franklin and Abbott (1989) touch on this subject, giving several clear examples of this in neuropsychiatric research, and in particular using animal models of pain. For example, the type of restraint (which is by itself stressful) influences the results of pain perception tests. More general discussions on the variables (generally divided into physical, chemical, microbiological, and husbandry factors) that can influence animal experiments are found in Baker et al. (1979) and in Chapter Two of the Foundation for Biomedical Research (1987) publication *The Biomedical Investigator's Handbook,* and the reader is urged to consult these.

4.4. Ethical Responsibilities in Scientific Publications

Submissions of scientific articles of animal-based research for publication in scientific journals should clearly define all the parameters under which the animal experiments were carried out (Ellery, 1985). The accurate definition of the animal model being used, in terms of the conditions under which it was conducted, is important for many reasons, including some ethical ones, like reducing the numbers of animals required in replicating experiments, and reducing the numbers of animals "wasted" because the experiment did not work in the hands of another investigator. A section on editorial responsibilities in publishing animal experiments is included in the book published by Dodds and Orlans (1982).

It is a generally accepted editorial policy that any paper describing animal-based research should indicate that the study was conducted within the existing ethical guidelines of the institution where the work was conducted, and the granting agency funding the research. Journal editorial standards regarding ethical principles must also be observed, and the use of ethically questionable techniques should be justified. Regarding the latter point, for example, if a pain test was used, the investigator should be prepared to justify the need for the pain test used. This would be particularly necessary if a test imposing a greater ethical cost for the animal was used over one that imposed a lower ethical cost.

5. Guidelines and Regulations

5.1. General

This brief section is included to direct the reader to some of the general guidelines and regulations in place that impact on the use of experimental animals. It is not intended to be comprehensive. In most areas there are also regional, local, institutional, and other regulations that apply. The institution and the principal investigator are responsible for compliance with all the pertinent guidelines and regulations on the use of animals in biomedical research, teaching, and testing. Details are contained in the publications cited, and addresses for these publications will be presented where appropriate.

The preceding sections have identified many ethical responsibilities of the principal investigators using animals. In addition the scientist must become familiar with the guidelines and regulations that control the use of animals in biomedical research. In some cases these may relate to conditions under which permission or license is granted to conduct the research.

5.2. Canadian Guidelines and Regulations

5.2.1. Canadian Council on Animal Care

In Canada the guidelines for the care and use of experimental animals are established by the Canadian Council on Animal Care (CCAC). These guidelines are contained in the publications *Guide to the Care and Use of Experimental Animals, Volumes 1 and 2* (1980–1984). These guidelines, as well as other publications of the CCAC are available by writing:

Canadian Council on Animal Care,
1000–151 Slater Street, Ottawa, Ontario K1P 5H3

With the creation of the Canadian Council on Animal Care in 1968, Canada adopted a system of guidelines and recommendations, rather than a national system of legislation and regulations. Ensuring that the guidelines are observed falls on the institution's animal care and use committee. Compliance is monitored by the CCAC's Assessment Program through periodic site visits to all institutions in Canada where experimental animals are used.

5.2.2. Other Canadian Guidelines and Regulations

In the Province of Ontario research institutions must comply with the regulations of the Animals for Research Act 1985. Research institutions in other provinces may also fall under local regulations affecting some aspects of their use of experimental animals.

5.3. American Guidelines and Regulations

5.3.1. The Laboratory Animal Welfare Act 1985

The Laboratory Animal Welfare Act 1985 regulates the use of animals for research, teaching, and testing in the USA. In 1989 new regulations for parts 2 and 3 of the law were enacted (USDA, APHIS, 1989). Some of the regulations that directly impact on the principal investigator are touched on here.

The new regulations under the United States Laboratory Animal Welfare Act (Federal Register, August 31, 1989) clearly spell out some ethical responsibilities of the principal investigator. These regulations state, under Section 2.31 *Institutional Animal Care and Use Committee (IACUC)*, that the principal investigator must demonstrate to the IACUC that he or she "has considered alternatives to procedures that may cause more than momentary or slight pain or distress to the animals, and has provided a written narrative description of the methods and sources,... used to determine that alternatives were not available." Further, the principal investigator must provide to the IACUC, "written assurance that the activities do not unnecessarily duplicate previous experiments."

5.3.2. Guidelines of the National Institutes of Health

The National Institutes of Health (NIH) publishes a guide—*Guide for the Care and Use of Laboratory Animals*—intended for institutional animal facilities and programs. The Guide covers topics like institutional policies, laboratory animal husbandry, veterinary care, facilities, and includes in the appendices a selected bibliography, information on laboratory animal science organizations, and the pertinent legislation in the US. It is recommended reading for all investigators. In the introduction, it states that "it is envisioned that the Guide will encourage scientists to seek improved methods of laboratory animal care and use." The standards in the Guide are used for accreditation of animal care facilities by the American Association for Accreditation of Laboratory Animal Care (AAALAC).

The NIH Guide is available from:

Institute of Laboratory Animal Resources,
Animal Resources Program, Division of Research Resources,
National Institutes of Health, Bethesda, MD 20205

5.4. British Legislation and Regulations

The Animals (Scientific Procedures) Act 1986 and its accompanying regulations, administered by the Home Office, regulate the use of animals in Great Britain. The Home Office guidance notes on the operation of the new Act are most helpful in explaining the intent of the new law as it affects the scientist (the licensee) and procedures done on animals.

6. Miscellaneous Considerations

6.1. Training of Scientists Using Animals for Research and Teaching

In most countries adequate systems are in place for the training and accreditation of laboratory animal care technicians, and of laboratory animal veterinarians. The training of scientists in laboratory animal science has, however, not received the attention it should. The qualifications of persons conducting research on animals have not been strictly scrutinized in the past. Although many institutions provide training for their scientists on an informal basis, only recently have guidelines and regulations addressed this matter.

The Canadian Council on Animal Care (CCAC) addressed their concerns about the need for some level of scientist training in laboratory animal science in the publication of a syllabus outlining a basic course suitable for all scientists entering a research career that might involve animal use (CCAC, 1985). The CCAC recommends that a course, based on the syllabus, should be mandatory for all graduate students in the biomedical sciences. Some Canadian universities (The University of Saskatchewan, for example) now present such mandatory courses for their graduate students.

The training of scientists is also addressed in the new regulations of the US Laboratory Animal Welfare Act (USDA, APHIS, 1989). Section 2.32 states that "It shall be the responsibility of the research facility to ensure that all scientists,... involved in animal care, treatment, and use are qualified to perform their duties. This responsibility shall be fulfilled in part through the provision of training and instruction to those personnel." It goes on to define some of the areas where guidance must be given.

The biomedical scientist must become adequately trained in laboratory animal science. Such training must encompass both the science and ethics of experimental animal use. Society needs assurances that scientists are suitably qualified, through training and experience, to conduct their animal research in a manner consistent with the intent of existing guidelines and regulations.

7. Concluding Remarks

This chapter was written with the underlying belief that ethical principles can be translated into practical guidelines of conduct for scientists using animals in their research. The intent of these guidelines is to promote humane animal experimentation techniques. They are not cast in stone; instead, they must be continuously subject to critical evaluation by the biomedical research community with input from the public.

References

American Veterinary Medical Association (1987) *J. Am. Vet. Med. Assoc.* **191,** 1184–1298.

American Veterinary Medical Association (1986) *J. Am. Vet. Med. Assoc.* **188,** 252–268.

Anon. (1987) *Lab. Anim. Sci.* **37,** 11–13.

Association of Veterinary Teachers and Research Workers (1986) Guidelines for the recognition and assessment of pain in animals. *Vet. Rec.* **118,** 334–338.

Baker H. J., Lindsey J. R., and Weisbroth S. H. (1979) Chapter 8. Housing to Control Research Variables, in *The Laboratory Rat Volume 1—Biology and Disease* (Baker H. J., Lindsey J. R., and Weisbroth S. H., eds.), Academic, New York, pp. 169–192.

Bond N. W. (ed.) (1984) *Animal Models in Psychopathology*. Academic, Sydney, pp. 1–22.

Bond N. W. (1984) Animal Models in Psychopathology: An Introduction, *Animal Models in Pychopathology* (Bond N. W., ed.), Academic, Sydney, pp. 1–21.

Bowd A. D. (1980) Ethical Reservations about Psychological Research with Animals. *Psychol. Rec.* **30,** 201–210.

Burghardt G. M. (1985) Animal awareness—Current perceptions and historical perspective. *Am. Psychol.* **40,** 905–919.

Canadian Council on Animal Care (1980–1984) *Guide to the Care and Use of Experimental Animals, 2 vols.* CCAC, Ottawa, Ontario.

Canadian Council on Animal Care (1985) *Syllabus of The Basic Principles of Laboratory Animal Science.* CCAC, Ottawa, Ontario.

Canadian Council on Animal Care (July 1989) *Categories of Invasiveness in Animal Experiments*. CCAC, Ottawa, 2pp.

Canadian Council on Animal Care (June 1989) *Ethics of Animal Experimentation*. CCAC, Ottawa, 2 pp.

Casey K. L. and Dubner R. (1989) Animal models of chronic pain: Scientific and ethical issues. *Pain* **38,** 249–252.

Dawkins M. S. (1980) *Animal Suffering—The Science of Animal Welfare.* Chapman and Hall, London.

Dodds W. J. and Orlans F. B. (eds.) (1982) *Scientific Perspectives on Animal Welfare*. Academic, New York.

Drewett R., and Kani W. (1981) Chapter 8. Animal Experimentation in the Behavioural Sciences, in *Animals in Research—New Perspectives in Animal Experimentation* (Sperlinger D., ed.), pp. 175–201, Wiley, Chichester, UK.

Ellery A. W. (1985) Guidelines for specification of animals and husbandry methods when reporting the results of animal experiments. *Lab. Animals* **19,** 106–108.

Flecknell P. A. (1984) The relief of pain in laboratory animals. *Lab. Animals* **18,** 147–160.

Foundation for Biomedical Research (1987) *The Biomedical Investigator's Handbook For Researchers Using Animal Models.* Foundation for Biomedical Research, Washington, DC.

Franklin K. B. J. and Abbott F. V. (1989) Techniques for Assessing the Effects of Drugs on Nociceptive Responses, in *Neuromethods 13—Psychopharmacology* (Boulton A. A., Baker G. B., and Greenshaw A. J., eds.), Humana Press, Clifton, New Jersey, pp. 145–216.

Gibson T. E. (ed.) (1985) *The Detection and Relief of Pain in Animals.* Proceedings of the Second Symposium of the BVA Animal Welfare Foundation. British Veterinary Association Animal Welfare Foundation.

Greenshaw A. J., Van Nguyen T., and Sanger D. J. (1988) Animal Models for Assessing Anxiolytic, Neuroleptic, and Antidepressant Drug Action, in *Neuromethods 10—Analysis of Psychiatric Drugs* (Boulton A. A., Baker

G. B., and Coutts R. T., eds.), Humana Press, Clifton, New Jersey, pp. 379–427.

Heikkila R. E., Sonsalla P. K., and Duvoisin R. C. (1989) Biochemical Models of Parkinson's Disease, in *Neuromethods 12—Drugs as Tools in Neurotransmitter Research* (Boulton A. A., Baker G. B., and Juorio A. V., eds.), Humana Press, Clifton, New Jersey, pp. 351–384.

HMSO (1986) Animals (Scientific Procedures) Act 1986. Her Majesty's Stationery Office, Queen's Printers, London.

Iggo A. (1979) Experimental study of pain in animals—Ethical aspects. *Adv. Pain Res. & Therap.* **3,** 773–778.

Jackson D. M. (1989) Drug-Induced Behavioral Models of Central Disorders, in *Neuromethods 12—Drugs as Tools in Neurotransmitter Research* (Boulton A. A., Baker G. B., and Juorio A. V., eds.), Humana Press, Clifton, New Jersey, pp. 385–442.

Keehn J. D. (ed.) (1979) *Psychopathology in Animals.* Academic, New York.

Lineberry C. (1981) Chapter 3. Laboratory Animals in Pain Research, in *Methods of Animal Experimentation Vol. VI.* (Gay W. I., ed.), Academic, New York, pp. 237–311.

Marcuse F. L. and Pear J. J. (1979) Ethics and Animal Experimentation: Personal Views, in *Psychopathology in Animals—Research and Clinical Implications.* (Keehn J. D., ed.), Academic, New York, pp. 305–329.

McGuire M. T., Brammer G. L., and Raleigh M. J. (1983) Animal Models: Are They Useful in the Study of Psychiatric Disorders?, in *Ethopharmacology: Primate Model of Neuropsychiatric Disorders.* Miczek K. A. (ed.), *Progr. Clin. Biol. Res.* **131,** 313–328.

McKinney W. T. and Bunney W. E. (1969) Animal model of depression: Review of evidence and implications for research. *Arch. Gen. Psychiat.* **21,** 490–494.

Mikeska J. A. and Klemm W. R. (1975) EEG evaluation of humaneness of asphyxia and decapitation euthanasia of the laboratory rat. *Lab. Anim. Sci.* **25,** 175–179.

Morton D. B. and Griffiths P. H. M. (1985) Guidelines on the recognition of pain, distress and discomfort in experimental animals and an hypothesis for assessment. *Vet. Rec.* **116,** 431–436.

Orlans F. B., Simmonds R. C., and Dodds W. J. (eds.) (1987) Special Issue—Effective Animal Care and Use Committees. *Lab. Anim. Sci.* **37.**

Regan T. (1983) *The Case for Animal Rights.* University of California Press, Berkeley.

Rollin B. E. (1981) *Animal Rights and Human Morality.* Prometheus Books, Buffalo, New York, pp. 89–148.

Rollin B. E. (1985) The Moral Status of Research Animals in Psychology. *Am. Psychol.* **40,** 920–919.

Rollin B. E. (1989) *The Unheeded Cry—Animal Consciousness, Animal Pain and Science.* Oxford University Press, New York.

Rowan A. N. (1984) Appropriate Care for Deafferented Primates. *J. Med. Primatol.* **13,** 175–181.

Rowan A. N. (1984) *Of Mice, Models & Men—A Critical Evaluation of Animal Research.* State University of New York Press, Albany.

Rowsell H. C. and McWilliam A. A. (1979) The Right Animal for the Right Reasons, in *Proc. Can. Asoc. Lab. Anim. Sci.* 1978–79, 211–220.

Russell W. M. S. and Burch R. L. (1959) *The Principles of Humane Experimental Technique.* Methuen & Co Ltd, London.

Sessle B. J. (1987) Animal Pain Research. *Lab. Anim. Sci.* **37,** 75–77.

Singer P. (1975) *Animal Liberation—A New Ethic for our Treatment of Animals.* Avon Books, New York.

Soma L. R. (1987) Assessment of Animal Pain in Experimental Animals. *Lab. Anim. Sci.* **37,** 71–74.

United States Department of Agriculture, Animal and Plant Health Inspection Service (August 31, 1989) 9CFR Parts 1, 2, and 3: Animal Welfare; Final Rules. *Federal Register* **54(168),** 36112–36163.

United States Department of Health and Human Services, Public Health Services, National Institutes of Health (1985) *Guide for the Care and Use of Laboratory Animals.* **#85-23,** NIH, Maryland.

Willner P. (1984) The validity of animal models of depression. *Psychopharmacology* **83,** 1–16.

Yoxall A. T. (1978) Pain in small animals—its recognition and control. *J. Small Anim. Pract.* **19,** 423–438.

Zimmermann M. (1983) Ethical Guidelines for Investigations of Experimental Pain in Conscious Animals. *Pain* **16,** 109–110.

Animal Models of Alzheimer's Disease

Gary L. Wenk

1. Introduction

Alzheimer's Disease (AD) is a neurodegenerative disorder characterized by a complex array of neuropathological, biochemical, and behavioral sequelae (Folstein and Whitehouse, 1983). AD is a recognized socioeconomic problem that has significant effects on a large percentage of an increasingly more aged population. Numerous experimental studies have been designed to investigate its etiology and possible pharmacotherapies for its treatment. Experimental animal models of AD are designed to reproduce a subset of the neuropathological, biochemical, and behavioral changes that have been identified in the brains of patients with AD. These animal models are, of course, inadequate because they do not completely reproduce all of the pathological and biochemical changes associated with AD; however, each model has been useful for the investigation of specific aspects of the disease. This chapter will outline the many experimental animal models of AD and compare the advantages and disadvantages of specific methodological approaches.

The first section outlines the neuropathological, biochemical, and cognitive changes that are associated with AD. The following sections discuss animal models of selected Alzheimer's-like pathologies produced by intracerebral injections of specific neu-

From: *Neuromethods, Vol. 21: Animal Models of Neurological Disease, I*
Eds: A. Boulton, G. Baker, and R. Butterworth

rotoxins and experimental models of age-related pathologies that develop with normal aging in rodents, nonhuman primates, and humans. Later sections discuss the appropriate choice of animal species to use in these animal models, the methods and protocols for making discrete lesions in specific brain regions, the long-term biochemical, pathological, and behavioral consequences, and finally, the validity of these animal models for the investigations of pharmacotherapies for AD.

2. Alzheimer's Disease

2.1. Neuropathology of Alzheimer's Disease

Numerous clinical criteria have been used to make a diagnosis of AD. However, a definitive diagnosis requires histological confirmation of the presence of specific neuropathological changes. These neuropathological changes affect a variety of neural systems, including brainstem catecholaminergic nuclei, the basal forebrain cholinergic system, amygdala, hippocampus, and specific regions of neocortex. Neurons in these regions exhibit a variety of cytoskeletal abnormalities, including neurofibrillary tangles and unusual cellular constituents, such as Hirano bodies and granulovacuolar degeneration (Terry and Davies, 1980; Price, 1986) and show a decrease in the number of markers for mRNA and protein metabolism (Price, 1986). Neurofibrillary tangles consist of an accumulation of neurofilaments, and occur within degenerating neurites and senile plaques (Kidd, 1963; Gonatas et al., 1967; Perry and Perry, 1985). Senile plaques contain the axons of terminals of affected neurons and are associated with an extracellular deposit of amyloid that is comprised of beta-pleated peptides derived from larger precursor proteins (Perry and Perry, 1985). Senile plaques occur throughout the brain but may concentrate in the temporal and parietal cortex, amygdala, hippocampus, and within certain brainstem nuclei (Meyers et al., 1988). The periphery of plaques may contain reactive cells such as macrophages, astroglia, and microglia

(Perry and Perry, 1985). The abnormal axonal terminals and dendritic processes that are associated with the core of senile plaques involve many different neurotransmitter markers contributed by many different neuronal systems that innervate the affected regions (Struble et al., 1982; Walker et al., 1988a). The specific neurotransmitter substance associated with each senile plaque is probably more closely related to the cytoarchitectural contributions of specific neural systems to a given region than to specific pathological processes associated with each neurotransmitter system (Walker et al., 1988a). Senile plaques also occur in association with normal aging in humans and nonhuman primates (Struble et al., 1985; Walker et al., 1988b).

2.2. Selective Involvement of Specific Neural Systems

Although AD has widespread effects throughout the central nervous system, only a few neural systems are specifically involved and show reliable and consistent pathological changes that are highly correlated with both neuropathological changes and specific cognitive impairments (Arendt et al., 1985; Francis et al., 1985; Etienne et al., 1986). Present evidence suggests that these neural systems are selectively vulnerable to the progressive neurodegenerative processes associated with AD. Therefore, an accurate animal model of AD should reproduce the morphological and biochemical changes in these neural systems. The most consistent and greatest changes have been associated with a population of magnocellular neurons located in the nucleus basalis of Meynert (NBM), medial septal area and diagonal band of Broca (Davies and Maloney, 1976; Whitehouse et al., 1981; Coyle et al., 1983). These basal forebrain magno-cellular neurons provide the primary cholinergic innervation to the neocortex, hippocampus, and other limbic and paralimbic regions (Mesulam et al., 1983). Their degeneration in AD is associated with reduced levels of pre- and postsynaptic cholinergic markers in the cortex, hippocampus, and amygdala (Henke and Lang, 1983; Yates et al., 1983).

Brainstem monoaminergic systems are also affected in AD. Cells in the locus ceruleus and raphe nuclei develop neurofibrillary tangles; degeneration of these cells is associated with a loss of noradrenergic and serotonergic markers, respectively, throughout the brain (Bowen et al., 1979; D'Amato et al., 1987). The terminals of these monoaminergic neurons have been identified in senile plaques throughout the brain (Perry and Perry, 1985; Walker et al., 1988a). In addition, numerous postsynaptic changes in serotonergic receptors have been identified (Cross et al., 1984).

Neuropeptide neurons are also significantly affected in AD. Senile plaques within the amygdala, hippocampus, and cortex often contain markers for a variety of neuropeptides, including somatostatin, substance P, and corticotropin-releasing factor (Struble et al., 1987; Lenders et al., 1989; Nemeroff et al., 1989). The central and medial nuclei of the amygdala, regions with high concentrations of neuropeptide neurons, often show neurofibrillary tangles and the presence of senile plaques as well as significant cell loss (Kemper, 1983).

The degeneration of these neural systems, particularly the basal forebrain cholinergic cells that innervate the neocortex and hippocampus, may underlie some of the cognitive impairments associated with AD (Blessed et al., 1968; Bartus et al., 1982; Arendt et al., 1985). For this reason, many experimental animal models of AD have focused on the behavioral consequences of the lesions in this brain region.

2.3. Cognitive Changes

AD is the most common cause of dementia in patients 45 years and older and is characterized by a progressive loss of higher mental functions, including memory, language, visuospatial perception, and behavior (Folstein and Whitehouse, 1983; McKhann et al., 1984; Roudier et al., 1988). The loss of short-term memory is a prominent feature during the initial stages of the disease (Freedman and Oscar-Berman, 1986) and has been the easiest symptom to reproduce in experimental animal models. Experimental manipulation of many different brain regions has indicated that both cholinergic and noncholinergic neural systems may play a role in the memory impairment associated with AD (Collerton, 1986; Olton and Wenk, 1987; Wenk and Olton, 1987; Wenk et al., 1987).

3. Animal Models of Alzheimer's Disease

3.1. Models of Alzheimer's-Like Neuropathologies

The neuropathological changes associated with AD have been studied using two different experimental animal models; aluminum induced neurofibrillary degeneration and pathologies associated with normal aging.

3.1.1. Aluminum-Induced Pathology

A role for aluminum in the etiology of AD has arisen primarily because of three independent laboratory findings: first, aluminum administration can induce neurofibrillary changes in the neurons of experimental animals; second, aluminum exposure can produce neurological and biochemical changes that lead to impaired memory and cognitive function that is similar to that observed in the early stages of AD; and third, aluminum levels in the brain tissue of AD patients exceeds the levels normally found in age-matched controls.

Aluminum salts injected intrathecally into susceptible species induce neurofibrillary abnormalities in the perikarya and dendrites of neurons in the brain stem and spinal cord (Pendelbury et al., 1988; Terry and Pena, 1965). These neurofibrillary aggregates are composed of normal neurofilament triplet proteins in contrast to the paired helical filaments associated with AD (Bertholf, 1987). The cause of this accumulation may be related to an abnormality in the synthesis, processing, or transport of the neurofilaments within axons and dendrites (Troncoso et al., 1982). Indeed, the aluminum-induced neuropathological preparation may be a better model of impaired neurofilament homeostasis and pathology than a model of the etiology of AD (Troncoso et al., 1982).

Similar neurofibrillary changes can be produced by intracerebroventricular (ICV) or subcutaneous injections of aluminum salts into rabbits, cats, and rats. Neurofibrillary tangles usually develop in frontal and occipital cortex; the concentration of aluminum directly correlates with the number of neurofibrillary tangles in each affected region.

Rabbits are administered aluminum salts by subcutaneous injections (usually into a shaved area of the back) 5 times weekly for 4 wk (25–400 µmoles/kg injection). ICV injections of aluminum salts are usually given as two injections of 5 µmoles into each lateral ventricle or one 50 µL injection of a 1% (w/v) solution (Troncoso et al., 1982). Sterile aluminum lactate is prepared in deionized water, and then filtered (0.22 µm filters) under sterile conditions. The LD 50 for aluminum using this exposure route is about 1600 µmoles/kg/injection.

ICV injections of aluminum produce neurofibrillary tangles in the brainstem, spinal cord, hippocampus, and cortex, whereas systemic injections produce neurofibrillary tangles in many cortical regions, but not in the hippocampus. In addition, there is a significant decrease in the content of Substance P and cholecystokinin and the activity of choline acetyltransferase in the hippocampus and entorhinal cortex, a decline in serotonin (5-hydroxytryptamine, 5-HT) and norepinephrine (NE) levels in the entorhinal cortex, as well as a decline in glutamate, aspartate, and taurine levels in parietal cortex (Beal et al., 1989). This profile of neurochemical changes following aluminum administration is similar, with some exceptions, to that seen in AD.

Subcutaneous injections of aluminum salts significantly impair the classical conditioning of the nictitating membrane response in rabbits. The rabbits either fail to acquire the conditioned response or acquire it less quickly than controls. Aluminum injections also significantly impair the acquisition and performance of a variety of other behavioral tasks. For example, when aluminum is injected directly into the hippocampus of cats, there is an increase in the density of neurofibrillary tangles that correlates with impaired performance in the acquisition of a conditioned avoidance task (Crapper et al., 1980). The results of these behavioral studies are consistent with the hypothesis that aluminum exposure results in learning deficits that are not owing to sensory or motor impairments, but rather to impaired mnemonic processing (Yokel, 1983).

Aluminum intoxication has been implicated in the pathogenesis of the dementia associated with hemodialysis (Dunea et al., 1978; Crapper, McLachlan, and DeBoni, 1980). Similar levels of aluminum were found in tissues from both human dialysis patients with dementia and rabbits treated with aluminum salts (Crapper et al., 1980). A hypothetical causal relationship between elevated levels of aluminum in the brain and the development of neurofibrillary tangles in AD is weakened by evidence that patients with dialysis dementia do not demonstrate neurofibrillary tangles and that AD patients do not have elevated levels of aluminum in their cerebrospinal fluid (CSF).

The role of aluminum in the etiology of AD is unknown and is presently controversial (Trapp et al., 1978; Yokel et al., 1988). The aluminum concentration is elevated in neurons containing neurofibrillary tangles (Crapper et al., 1980) and perhaps within senile plaques. However, aluminum may accumulate in neurons secondarily to intracellular degenerative changes. Further, the neuropathological and behavioral changes following aluminum exposure are similar to those observed in AD; however, the biochemical changes that characterize AD are typically not observed. In addition, although the morphological similarity between neurofibrillary tangles observed in patients with AD and aluminum-induced neurofibrillary degeneration has been well documented, the aluminum-induced changes typically only occur near the sight of the injection, or appear diffusely distributed. In contrast, the neurofibrillary changes observed in AD are found mostly within cortical and hippocampal neurons.

One study of aluminum levels in the cerebrospinal fluid of 180 patients found no correlation between aluminum concentration and age or between any neurological disease. Surprisingly, the concentration of aluminum in one group of AD patients was significantly lower than in a group of age-matched controls (Delaney, 1979). This study found a wide range in the levels of aluminum in cerebrospinal fluid and could not define a specific role for aluminum in any of the diseases investigated.

3.1.2. Age-Related Pathologies as Models

Many of the cognitive impairments observed in patients with AD also occur to a somewhat lesser degree in normal aging. Initially there is a slight amnesia that progresses gradually over a period of many years as general intellectual functions also decline (Craik, 1977). However, the syndrome described for AD clearly differs from normal aging. For example, in nonpathological aging there is no loss of basal forebrain cells and only a moderate development of senile plaques (Chui et al., 1984). Usually the differences in the pathological, biochemical and cognitive changes between AD and nonpathological normal aging are in degree rather in their nature (Ulrich and Stahelin, 1984).

Some of the abnormalities that are associated with AD have also been shown to occur in long-lived nonhuman primates (Walker et al., 1988b). Neuritic plaques have been identified in the cortex of 20-year-old Rhesus monkeys (Kitt et al., 1985; Struble et al., 1985); these plaques are similar to those that occur in AD. Aged monkeys may develop neurofibrillary tangles (L. Cork, personal communication) and often show a decline in the performance of tasks that require learning and memory (Bartus et al., 1978; Walker et al., 1988b). Aged nonhuman primates may therefore be useful models with which to study certain aspects of normal nonpathological aging and the development of specific neuropathological changes, e.g., neuritic plaques. Unfortunately, these animals are difficult to obtain, expensive to purchase and maintain, and only a very small percentage live to be greater than 20 years of age.

Rats also show age-related changes in cholinergic and many noncholinergic biomarkers throughout the brain, including complex alterations in the levels of 5-HT and the catecholamines (Moretti et al., 1987; Sirvio et al., 1988; Godefroy et al., 1989). However, unlike in AD, rats, humans, and nonhuman primates do not show age-related changes in cholinergic markers (Decker, 1987; Wenk et al., 1989b), including choline acetyltransferase, high affinity choline uptake, and many pre- and postsynaptic cholinergic receptors, although the presence or absence of age-related changes in cholinergic markers may be related to the particular strains of rats that have been investigated (Michalek et al., 1989). The lack of changes in cortical cholinergic markers does not correlate with an age-related cell loss in the basal forebrain (Fischer et al., 1989). Further, behaviorally-impaired aged rats often have the most severe pathological changes in the basal forebrain (Fischer et al., 1989) and the greatest decline in normal sleep patterns, brain biochemistry (Markowska et al., 1989) and brain glucose metabolism (Gage et al., 1984b).

Aged rats demonstrate a decrease in the activity of specific energy metabolizing enzyme systems, in particular those related to glycolysis and energy production (Leong et al., 1981; Vanella et al., 1989), similar to that seen in normal human aging and AD. Rats also demonstrate an age-related decline in peripheral sympathetic function and their ability to regulate blood glucose levels that may be related to central mnemonic processes (Martinez et al., 1988; Wenk, 1989). The sympathetic response to cognitive stimuli and glucose regulation and utilization are also impaired in patients with AD (Borson et al., 1989; Hoyer et al., 1988).

Most aging investigations of rodents suffer from the disadvantage that these studies are usually conducted using a genetically homogenous group of subjects, such as the Fisher-344 strain. Studies utilizing only limited inbred strains of laboratory animals may fail to provide a sufficient model for age-related degenerative changes in humans. This concern is supported by evidence that there are considerable strain-related differences in spatial memory abilities (Lindner and Schallert, 1988) in rats. Mice show similar strain-associated differences in the age-related changes in specific central neurotransmitter systems (Ebel et al., 1987).

Although aged animals typically show many of the same behavioral and cognitive impairments that have been identified in the early stages of AD (Gage et al., 1984a; Rapp et al., 1987), the lack of an age-related neuropathology, or alterations in specific neurochemical markers, particularly the forebrain cholinergic system (Decker, 1987; Markowska et al., 1989), may undermine the potential usefulness of the aged rat or nonhuman primate for the study of AD, although they may be excellent models for nonpathological normal aging.

3.2. Lesions as an Interventional Approach

Although young animals do not develop pathological or biochemical changes similar to those seen in AD, it is possible to reproduce a subset of these changes experimentally. One approach is to produce lesions in discrete brain regions.

3.2.1. Choice of Animal Species

The choice of animal species is determined by a number of factors, including: the cost to purchase and house the species; the general level of knowledge on the neuroanatomy (particularly of the basal forebrain cholinergic system) and neurochemistry, and the behavioral capabilities that can be reliably investigated and manipulated in a particular species; and the availability of the species. The last factor is particularly important when using nonhuman primates. Rats are well-suited for these studies because they are relatively inexpensive and are easy to train and test in a variety of behavioral tasks. Analogous neuroanatomical structures with well-defined neurotransmitter systems have been identified in rats for comparison to humans.

3.2.2. Choice of Lesion Locations

Discrete regions of the brain can be selectively destroyed by injection of specific neurotoxins or by the application of electrical current. Recent animal models of the pathology associated with AD have involved lesions of either the basal forebrain cholinergic system, noradrenergic locus ceruleus, serotonergic raphe nuclei, or some combination of these systems. Each of these neural systems may degenerate in AD. The apparent disconnection of the hippocampus by specific pathologies in AD (Hyman et al., 1984) has led to the introduction of experimental models of hippocampal disconnection (Olton, 1986) involving lesions of the fimbria and fornix by application of electrical current (Mitchell et al., 1982), horizontal coronal knife cuts or aspiration of the fimbria or fornix (Owen and Butler, 1981; Dunnett et al., 1982), the injection of specific neurotoxins into the hippocampus (Handlemann and Olton, 1981; Jarrard et al., 1984), transection of the perforant pathways or destruction of entorhinal cortex (Olton et al., 1982).

The remainder of this chapter will focus upon the most extensively investigated animal models of AD, i.e., those involving lesions of the basal forebrain cholinergic system.

3.2.3. Choice of Lesion Methods

The choice of lesion methods depends upon the nature and extent of the proposed lesion. For example, if only cell bodies within the region of the injection are to be destroyed, then injection of selective neurotoxins, such as the excitatory amino acids, can be used (Schwarcz et al., 1978). If fibers of passage as well as cell bodies and afferent terminals within a discrete region are intended to be destroyed, then application of electrical current through a metal electrode can be used to produce an extensive lesion. Most animal models of AD have used either application of electrical currents (i.e., electrolytic lesions) or injections of selective neurotoxins (for review, *see* Wenk and Olton, 1987).

3.2.3.1. Choice of Neurotoxins

The neurotoxins that are typically used are restricted analogs of the endogenous amino acid neurotransmitter glutamate (Olney et al., 1974). These compounds destroy cell bodies, but leave fibers of passage and efferent terminals into the injection region relatively intact (Coyle, 1983). The toxins that have been used include ibotenic (IBO), kainic (KA), quinolinic (QN), and quisqualic (QA) acids and *N*-methyl-D-aspartate. These five excitatory amino acids have been used to produce brain lesions that reproduce specific components of the pathology associated with AD. Each toxin affects a slightly different population of neurons within the injection site (Kohler and Schwarcz, 1983). Cytotoxicity is probably determined by the specific subtype of glutamate receptor that the neuron expresses (Cotman and Iversen, 1987). For example, KA is a potent agonist at the kainate subtype of glutamate receptors and is far less potent at sites that are sensitive to IBO or QA. QA and QN are effective neurotoxins in the basal forebrain and destroy many cholinergic and noncholinergic neurons throughout the ventral pallidum/substantia innominata. However, neither acid effectively destroys neurons in the medial septal area (Perkins and Stone, 1983; Schwarcz and Kohler, 1983).

KA is a potent neurotoxin that destroys cells near the site of its injection. KA also produces a significant amount of cell loss in the hippocampus when it is injected into the basal forebrain (Mason and Fibiger, 1979). This nonspecific injury can be overcome by pretreatment with an anticonvulsant immediately following surgery (Beninger et al., 1986).

3.2.3.2. Preparation and Storage

The neurotoxins are dissolved in phosphate-buffered saline (PBS, pH 7.4). The exact amount of neurotoxin obtained from the supplier (e.g., Sigma or Regis) should be determined prior to use because occasionally the quantity received varies. Typically, 5 mg of IBO is combined with 400 µL of ice-cold PBS. The solution is kept cold on ice and is prepared in a darkened vial because the neurotoxins are sensitive to light and heat and will spontaneously degrade. For example, IBO will be converted spontaneously into the GABA agonist muscimol; this compound does not produce a lesion when injected into the brain. The solution is adjusted to pH 7.7 by the addition of 10 *N* sodium hydroxide (usually about 2–5 µL). These amino acids do not go into solution completely until the pH is greater than 7.0. When the final pH of the solution is adjusted to 7.7, IBO produces a more effective lesion, i.e., significantly more cell loss in the basal forebrain, than when the pH is 7.4. The other neurotoxins, such as KA, QA, or QN, are effective at pH 7.4. The accurate estimation of the final pH requires a pH electrode with a narrow tip; it may be easier to use pH paper during the initial stages of the titration. The final volume is increased to 500 µL with PBS, with a final concentration of 0.025 *M* for IBO, 0.12 *M* for QN, 0.005 *M* for KA, and 0.12 *M* for QA. The solution is then aliquoted into 50 µL volumes that can be stored in polypropylene Eppendorf tubes at –40°C, or below, until used. Unused portions of the toxin can be frozen and thawed repeatedly for at least 6 mo without any significant loss of potency.

3.2.3.3. Optimal Coordinates

In a previous study (Wenk et al., 1984), we investigated the optimal coordinates for the production of extensive basal forebrain lesions in rats. The purpose was to reproduce the extensive cell loss in this brain region that is reported for patients with AD. The entire extent of the basal forebrain, including NBM, medial septal area, and diagonal band of Broca, can be lesioned by multiple injections of the toxin. Multiple small volume injections circumvent many of the problems associated with large volume single injections, such as nonspecific damage to the lateral hypothalamus or ventral pallidum, or the presence of toxin in the third ventricle. Injections into the medial septal area should be as ventral as possible to allow the neurotoxin to reach the cells in the vertical limb of the diagonal band, but the injection should also avoid the risk that some of the toxin may reach the third ventricle. If the injection site is adjusted appropriately, unwanted diffusion of the toxin into the ventricles can be avoided.

Separate injections are placed into the anterior basal forebrain near the anterior commissure. At this site, the toxin destroys cells in the horizontal limb of the diagonal band as well as the anterior NBM. A third injection is placed in the posterior and lateral NBM to destroy those cells that project to occipital cortex and ventral hippocampus (Kitt et al., 1987).

3.2.3.4. Surgical Methods—Rats

Each rat is pretreated with atropine to prevent excess secretions that might impair respiration during surgery and then anesthetized. The rat is placed in the stereotaxic apparatus, the scalp is shaved, incised, and retracted, and holes are drilled in appropriate locations in the skull with a dental drill. The coordinates for multiple injections of a neurotoxin into the NBM are as follows: 0.4 and 0.8 mm posterior to Bregma, 2.6 mm lateral (bilaterally) from the midline, and 6.8 mm below the dorsal surface of the neocortex or 6.9 mm below the dura (Wenk et al., 1989a). A single aliquot of the solution is thawed on ice immediately prior to use and kept cold and protected from light during surgery. The surgical cannula or syringe is filled immediately prior to injection to prevent excess warming of the solution and spontaneous oxidation. Each NBM injection site receives a 0.4–0.5 μL of IBO (the precise volume may vary depending on the concentration of each neurotoxin) injected over a period of 5 min to prevent widespread diffusion. The precise amount of each neurotoxin used to produce the most effective lesion, with the least amount of nonspecific injury, should be determined empirically for each study.

The coordinates for injection of toxin into the medial septal area/vertical limb of the diagonal band region are as follows: 0.8 mm anterior to Bregma, on the midline, and 5.8 mm below the dura. Usually only 0.6 μL of IBO is injected per site.

It is important to avoid destroying neurons within the lateral hypothalamic feeding centers or injecting large volumes of the neurotoxin. If the neurotoxin reaches the lateral hypothalamus, the rat may stop eating, and if the neurotoxin reaches the third ventricle, the rat usually dies during surgery. Although the precise cause has not been investigated rigorously, death may be owing to destruction of the vegetative centers in the floor of the fourth ventricle.

The survival rate for this surgery can be greatly enhanced if the lesions are produced in two stages. In the first stage, the excitotoxin is injected into the basal forebrain in one or two places unilaterally. The second stage of the surgery is performed 1 week later; the neurotoxin is then injected into the contralateral basal forebrain. Complete NBM and medial septal area lesions can be produced using this two-stage lesion procedure. This procedure significantly decreases the mortality rate, but does not comprise the overall effectiveness of the lesions. It is especially valuable in behavioral studies of the retention of an acquired memory, particularly when a great deal of time and effort have been invested into the initial training.

After the injection is made, the needle or cannula should be withdrawn slowly to avoid drawing the neurotoxin to the cortical surface by capillary action. The neurotoxin might destroy intracortical cells and confound the behavioral and biochemical studies.

3.2.3.5. Postoperative Care

The primary phase of risk is during the first few days that the animal is back in the home cage. If there is nonspecific damage to the lateral hypothalamic feeding centers, the animal does not initiate feeding or drinking and must be fed by intubation to prevent starvation. The two-stage surgical procedure outlined above can circumvent many of these difficulties.

During the postoperative period, while the animal is still unconscious, the injection of certain neurotoxins into the basal forebrain region often produces severe tremors, seizures, or excessive chewing of the forepaws. This can be overcome by covering the forepaws with masking tape or cloth. In addition, the surgical procedure will often decrease the body temperature. This can be prevented by performing surgery while the animal lies on a heating pad. In addition, the rat should be placed under a heating lamp during the recovery period.

3.2.3.6. Surgical Methods—Monkeys

The primary problem associated with the use of monkeys as an animal model of AD is the production of lesions in the basal forebrain. Because the magnocellular cholinergic neurons are distributed throughout a large area (Mesulam et al., 1984; Satoh and Fibiger, 1985), it is necessary to make multiple injections of neurotoxin (Aigner et al., 1987; Ridley et al., 1985,1986; Wenk et al., 1986), although a single bilateral injection of IBO has also been used successfully (Irle and Markowitsch, 1987). The concentration and volume of the neurotoxin will influence the degree and anatomical extent of cholinergic cell loss in the basal forebrain. For example, two 1.0 µL injections of IBO (10 µg/µL) decreased choline acetyltransferase (ChAT) activity by 60% in frontal cortex (*see* Section 3.2.4. for explanation of this neurochemical marker), but by only 18–40% in posterior and temporal cortex (Ridley et al., 1985). In contrast, seven 1.3 µL injections of IBO (15 µg/µL) decreased ChAT activity by 63–79% throughout the parietal, sensory, motor, temporal, and frontal cortex (Wenk et al., 1986).

The monkeys are prepared for aseptic surgery using standard procedure. Each monkey is premedicated with dexamethasone and ketamine, anesthetized with pentobarbital, placed in a stereotaxic instrument (David Kopf, Chicago, IL), and the skin, connective tissue, and muscle are incised and retracted to expose the skull. Trephine openings are placed bilaterally in the skull.

The precise location for the injections can be determined in three ways. The first is by comparison to commercially available stereotaxic charts. This method often produces incomplete, asymmetrical areas of cell loss, and the lesioned area often includes nearby structures, including the lateral hypothalamus, amygdala, and thalamus. Injury to these regions can have significant effects on behavior and can undermine the validity of the preparation. Second, the location of various adjacent brain structures can be identified on the basis of their patterns of electrophysiological activity. For example, the presence or absence of unit activity can indicate gray matter and white matter, respectively. The location of the anterior commissure can be determined in this manner. Recently, nuclear magnetic resonance (NMR) imaging has provided very precise identification of the boundaries of the NBM in monkeys. We have recently given 14 injections (1.0 µL each site) of IBO (10 µg/µL) bilaterally into the NBM and vertical and horizontal limb of the diagonal band, using anatomical information obtained from NMR imaging (M. L. Voytko, personal communication). These lesions affect a greater proportion of the NBM and decrease ChAT activity throughout the entire cortex.

3.2.4. Confirmation of Lesion Effectiveness and Specificity

The effectiveness and specificity of the lesions are usually determined by an estimation of the degree of cellular destruction, or the loss of a specific neurotransmitter system that exists within the region of the injections. In experimental animal models of AD, typically some biochemical measure of cholinergic function is determined, either within the basal forebrain or in the frontal cortex and hippocampus. The cholinergic marker that is most often used to estimate the loss of cholinergic terminals in the cortex and hippocampus, or of cholinergic cells in the NBM, is ChAT. Unlike other cholinergic markers, such as acetylcholinesterase, ChAT is specific to cholinergic neurons (Tucek, 1978) and its activity is decreased when the cholinergic cells degenerate or become dysfunctional. The lesions of the basal forebrain are designed to reproduce the extensive loss of cholinergic cells that is associated with AD. The degree to which cholinergic cell loss occurs is a measure of the effectiveness of the lesions.

The specificity of the lesions can be determined by examining markers for noncholinergic cells. The basal forebrain of primates, and possibly of humans and rodents, contains many noncholinergic neural systems, including those that produce and release substance P, neuropeptide Y, leucine enkephalin, somatostatin, neurotensin, and galanin (Walker et al., 1989). Most of these neuropeptides are contained in interneurons that probably project only within the basal forebrain region (Walker et al., 1989) or to the amygdala, e.g., neurotensin (Tay et al., 1989). The neurotoxins discussed previously destroy any neuron within the region of the injection site. However, these toxins are not specific for cholinergic neurons. AF64A was initially introduced as a specific cholinergic toxin (Chrobak et al., 1987,1988; Mantione et al., 1981; Sandberg et al., 1984,1985). However, the cholinergic specificity of this agent has not been supported by many recent investigations (Allen et al., 1988; Jarrard et al., 1984; Levy et al., 1984; McGurk et al., 1987; Villani et al., 1986). An immunocytochemical investigation of the basal forebrain following the injection of a neurotoxin is required in order to determine the specificity of AF64A, or any neurotoxin, for cholinergic cells. Certain basal forebrain cells may not express the appropriate glutamate receptor subtype to confer sensitivity to a specific neurotoxin, therefore some cells may survive exposure to specific neurotoxins, e.g., IBO vs QA (Wenk et al., 1989a).

3.2.5. Long-Term Changes in Biochemistry

The biochemical changes induced by these lesions change over time, particularly in young animals. Therefore, during long-term behavioral testing there is a constantly changing baseline of biochemical function in many neurotransmitter systems. Changes in all relevant neurotransmitter systems should be monitored and correlated with behavior.

3.2.5.1. Long-Term Changes in Cholinergic Markers

The presence of large cholinergic neurons in the NBM provides a way to determine the effectiveness of the injections. The loss of presynaptic markers of cholinergic function, such as ChAT activity, provides an indication of the integrity of the cholinergic cells that remain (Johnston et al., 1981). Changes in the number of presynaptic, high affinity choline uptake sites, determined by [^{3}H]hemicholinium-3 binding or synaptosomal sodium-dependent [^{3}H]choline uptake, provide a dynamic indicator of the activity of these remaining cholinergic cells (Simon and Kuhar, 1975). The number of presynaptic muscarinic (type-2) receptor sites, labeled by [^{3}H]oxotremorine, can be determined as a secondary indicator of the integrity of cholinergic terminals in the amygdala, neocortex, and hippocampus. The loss of presynaptic receptors or high affinity uptake sites also provides a quantitative estimate of the loss of cholinergic terminals. The loss of endogenous levels of acetylcholine can also be determined; however, this requires that the animal be sacrificed by focused microwave radiation in order to inactivate acetylcholinesterase.

Preparation of the brain with microwave radiation precludes the determination of heat-labile markers, such as ChAT and high affinity choline uptake, and is not as effective for larger animals.

Within the first few weeks after the production of these lesions there is a significant decline in the level of all cholinergic markers, including ChAT and acetylcholinesterase activity, sodium-dependent high affinity choline uptake, the density of ChAT-positive immunoreactive cortical fibers, and K^+-stimulated release of [^{3}H]acetylcholine from brain slices. Recovery of these cholinergic markers occurs at different rates, and to varying degrees, depending upon the age of the animal, the extent of the lesion (e.g., unilateral vs bilateral), and the nature of the toxin (QA vs KA vs IBO, and so on). Cortical ChAT activity recovers to control levels after unilateral lesions, usually within 6 mo, but shows little recovery after bilateral lesions (Casamenti et al., 1988; Hepler et al., 1985; Wenk and Olton, 1984). Cortical sodium-dependent high-affinity [^{3}H]choline uptake recovers within 30 d following a unilateral NBM lesion (Casamenti et al., 1988; Pedata et al., 1982), but does not recover after a bilateral NBM lesion (Bartus et al., 1985).

The decreased density of ChAT-positive immunoreactive fibers remains low up to 6 mo after a unilateral NBM lesion. However, the level of ChAT-immunoreactivity per fiber is markedly increased at this time in the deafferented cortical regions (Ojima et al., 1988). This suggests that recovery of cortical ChAT activity is not owing to sprouting (Henderson, 1991; Wenk and Olton, 1984), but is related to an increase in ChAT content in the remaining intact fibers (Ojima et al., 1988). A recovery of high affinity choline uptake suggests that the remaining cholinergic cells have increased their activity. Consistent with this hypothesis is the report that the level of K^+-evoked release of [^{3}H]acetylcholine recovers to control levels in parietal cortex within 4 mo after a unilateral lesion (Gardiner et al., 1987).

There is no recovery of presynaptic muscarinic receptors; postsynaptic receptors undergo little or no change in the number of high and low affinity muscarinic or nicotinic receptor sites (Atack et al., 1989; Wenk and Rokaeus, 1988).

Six mo after unilateral injections of KA into the NBM, cholinergic neurons in the contralateral NBM were hypertrophied (Pearson et al., 1986). This hypertrophy may be subsequent to morphological changes associated with axonal sprouting or with activation of metabolic processes in response to endogenously released growth factors, e.g., NGF (Ojima et al., 1988).

Recovery of these cholinergic markers may not be permanent. Twelve mo after an unilateral NBM lesion produced by IBO, cortical levels of ChAT activity and [^{3}H]hemicholinium-3 binding are significantly reduced, following an initial recovery 9 mo earlier (Hohmann et al., 1987).

3.2.5.2. Long-Term Changes in Noncholinergic Markers

The extent of changes in noncholinergic systems, following the production of NBM lesions, greatly depends upon the manner in which the lesions are produced. Knife cuts and electrocoagulation currents destroy fibers of passage through, and afferent terminals to, the basal forebrain region. These "physical" lesions immediately decrease the levels of many noncholinergic markers throughout the brain (Hohmann and Coyle, 1988). Interpretation of the behavioral changes subsequent to these lesions should not be in terms of loss of cholinergic basal forebrain cells alone. In contrast, studies of neurotoxin-induced lesions have found no changes throughout the brain in many noncholinergic markers immediately after surgery (Johnston et al., 1979,1981; Mufson et al., 1987; Wenk et al., 1987). These studies confirm the specificity and selectivity of the neurotoxins for cell bodies within the injection site.

Long-term noncholinergic changes may involve the serotonergic raphe system (Wenk et al., 1986; Wenk and Engisch, 1986). Four mo after the production of NBM lesions by IBO injections (Wenk and Engisch, 1986; Wenk et al., 1987), and 10 mo after the production of similar lesions in monkeys (Wenk et al., 1986), the number of [^{3}H]ketanserin binding sites (to serotonergic type-2 receptors) was increased throughout the neocortex. In monkeys, the greatest increase in cortical [^{3}H]ketanserin binding sites correlated significantly with the greatest loss of cortical ChAT activity (Wenk et al., 1986).

Three days after a unilateral injection of IBO into the NBM, the cerebral metabolic rate of glucose was significantly decreased (Kiyosawa et al., 1987; London et al., 1984), as determined by 2-deoxy-D-[^{14}C]glucose uptake. However, glucose uptake recovered within 4 wk after surgery in rats and within 13 wk after surgery for baboons (Kiyosawa et al., 1987).

3.2.6. Long-Term Pathological Changes

Three interesting long-term pathological changes have been reported following the production of excitotoxin-induced lesions in the NBM: neuritic plaques, demyelination, and spherical concretions of calcium salts.

3.2.6.1. Neuritic Plaques

A single report has documented the development of neuritic plaque-like objects in the frontal cortex of rats 14 mo after a unilateral injection of IBO into the NBM (Arendash et al., 1987). These neuritic plaque-like structures were observed in the cerebral cortex and hippocampus and consisted of degenerating neuronal terminals and glial cells. No central amyloid deposit was detected by Congo red staining. Silver-stained atrophic neurons in the basolateral nucleus of the amygdala and dorsal hippocampus were also reported (Arendash et al., 1987). An attempt to replicate this finding was unsuccessful (Terry et al., 1988).

3.2.6.2. Demyelination of Fibers of Passage

IBO, KA, and QA are effective neurotoxins that destroy cell bodies within the area of the injection site, but appear to leave fibers of passage intact when examined within the first few weeks after surgery (Coyle and Schwarcz, 1983). However, a recent investigation found that injections of IBO into the medial septal area and lateral geniculate nucleus of the thalamus ultimately resulted in the demyelination of fibers of passage (Coffey et al., 1988). The area of demyelination correlated with extensive gliosis in response to the injury following the IBO injection. This is probably a nonspecific inflammatory response to neuronal loss or injury (Coffey et al., 1988). The results of this study suggested that IBO can secondarily damage fibers of passage within the area of the injection site and may ultimately disrupt axonal transport through the region, or conduction along these fiber systems. These long-term alterations in neural function could significantly influence behavior in tasks that require extensive postoperative testing.

3.2.6.3. Spherical Concretions

The presence of neurotoxins in the basal forebrain may also lead to the deposition of calcium salt crystals as another type of nonspecific response by the brain to cell injury or death (Wenk et al., 1987). This nonspecific change has been reported for other brain regions and is not characteristic of the effects of the excitotoxins on the basal forebrain.

3.3. Pharmacological Models of Alzheimer's Disease

Pharmacological agents that interfere with cognitive processes can be used to impair behavioral performance using the same rationale that was described for experiments using lesions. The action of drugs can be specific for designated neurotransmitter systems or for some other aspect of neural function. For the cholinergic system, scopolamine and atropine have been used to block the action of acetylcholine at its postsynaptic receptor site and impair performance of laboratory animals and humans in a variety of behavioral tasks (Okaichi and Jarrard, 1982; Spangler et al., 1986; Rusted, 1988).

The progressive deterioration in cognitive function associated with AD has been correlated with the loss of cholinergic function and the degeneration of NBM cholinergic neurons (Collerton, 1986; Francis et al., 1985; Smith and Swash, 1978; Whitehouse et al., 1985). Experimental studies on humans, nonhuman primates, and rats found strong similarities between the effects of scopolamine, a muscarinic antagonist, on performance in tasks that require learning and memory and the cognitive deficits seen in AD (Collerton, 1986; Broks et al., 1988; Preston et al., 1988; Dunnett et al., 1989). Alzheimer's patients, as compared to age-matched controls, have a differential sensitivity to scopolamine similar to that seen in rats (Smith, 1988) or monkeys (Aigner et al., 1987; Ridley et al., 1985,1986) with NBM lesions.

Alzheimer's patients may also demonstrate severe attentional deficits (Folstein and Whitehouse, 1983; McKhann et al., 1984). The cholinergic system may play a role in attentional processes (Olton et al., 1988). Scopolamine impairs performance in signal detection and sustained vigilance tasks and may impair selective attention (Wesnes and Warburton, 1983,1984; Dunne and Hartley, 1985).

The transient effects of scopolamine on learning and memory makes this model of AD potentially useful for the investigation of pharmacotherapies designed to reverse cholinergic hypofunction associated with AD (Preston et al., 1988; Thal et al., 1981; Davis and Mohs, 1982; Summers et al., 1987). For example, the potential cognition enhancing drugs Hydergine®, piracetam, and aniracetem can reverse the effects of scopolamine on the 24 h retention of inhibitory avoidance (Piercey et al., 1987; Spignoli and Pepeu, 1987; Saletu et al., 1979).

The effects of scopolamine on performance are transient. In contrast, the amnestic symptoms associated with AD progressively worsen. Furthermore, the degeneration of the basal forebrain cholinergic system only represents a single aspect of the neurodegenerative changes associated with AD. Many other neurotransmitter systems also degenerate; their role in the dementia syndrome remains to be determined. Pharmacological studies similar to those using scopolamine may help define the role that the degeneration of these noncholinergic systems play in AD.

3.4. Choice of Appropriate Behavioral Measures

Aluminum exposure, neurotoxin injections, or scopolamine administration reproduce specific aspects of the pathological and biochemical changes associated with AD. The behavioral tasks chosen to detect the effects of these manipulations should be sufficiently sensitive and specific such that the alteration in performance is owing to the manipulation under investigation. Every psychological task involves many cognitive processes, including perception, sensation, motivation, learning and memory, and motor coordination, to name but a few. An impairment in any of these cognitive processes could impair performance. The behavioral tests that are used should be sensitive so that changes in memory would be reflected by changes in behavior in a specific task. A sensitive task acts, in fact, as an amplifier, so that even small changes in memory will produce a large and easily observable change in performance in a specific behavioral task.

The task should also be selective so that alterations in psychological processes other than the type that is intended to be studied will have little effect on performance. Selective tasks act as filters that minimize the effects of other cognitive processes on performance. Animal models of AD should be tested in a variety of different behavioral tests in order to determine the reason for the abnormal behavior. Studies should compare performance in several different behavioral tasks that require different cognitive, motivational, and attentional processes, that are relatively sensitive and selective for different psychological processes. In addition to the cognitive abnormalities that underlie the impaired performance, it is necessary to determine whether noncognitive processes are involved. Rats with NBM lesions do not have any obvious impairment in their basic sensory-motor skills; however, some studies have reported altered eating and hoarding habits that may represent secondary changes in response to the lesion. Secondary noncognitive measures of function therefore need to be determined.

Behavioral studies of the memory impairments owing to NBM lesions have documented a substantial recovery of performance over time, or following extensive postoperative testing (Bartus et al., 1985). The specific variables that influence the presence or rate of this recovery and the neural processes that function to produce this recovery should be investigated in animal models of AD. Experiments need to provide an extended postoperative testing period, performance needs to be determined immediately at the start of the behavioral testing, i.e., immediately postoperatively, and the rate of change of performance over time needs to be determined, as well as the final asymptotic level of performance at the end of testing. By comparing the changing baseline in behavioral performance with alterations in specific neurotransmitters systems, both cholinergic and noncholinergic, it may be possible to draw relationships between specific biochemical compensatory changes and recovery of performance. An understanding of the mechanisms involved in recovery of performance would provide some indication of an appropriate pharmacotherapy to be used in AD. Such a pharmacotherapy would be designed according to the data obtained from these longitudinal studies.

4. Are Animal Models of Alzheimer's Disease Valid and Useful?

These animal models are valid and useful to the degree that they mimic all components of AD. They are useful if they can make correct predictions about the brain mechanisms or neural systems that are involved in the production of cognitive changes observed in AD and about the effects or consequences of specific pathological changes and how these changes underlie the cognitive impairments and biochemical changes observed. They are also useful if they can aid in the development of therapeutic interventions that might alleviate the symptoms associated with AD.

Experimental models of AD suggest that the loss of basal forebrain cholinergic cells and pharmacological antagonism to cholinergic function may be sufficient to reproduce a component of the amnesia observed in AD. However, AD is a multisystem disorder that affects many different neural systems throughout the brain. The models that have been introduced here account for the loss of the cholinergic function but do not account for the possible interactions between the loss of cholinergic and noncholinergic neural function. Future animal models should investigate the interaction between the different neural systems that degenerate in AD, including both cholinergic and noncholinergic systems.

There are several important differences between these animal models and AD. The behavioral impairments studied in these animal models recover with continued postoperative testing and time, whereas in AD a similar recovery does not occur. The excitatory amino acids and aluminum salts act acutely on the brain and are then removed. In contrast, the neurodegenerative processes that underlie AD are chronic and progressive. Many animal models investigate the effects of specific manipulations in young adult animals. In contrast, the degenerative proc-esses associated with AD occur in aged humans. This is an important distinction since the young nervous system can respond with many compensatory processes following these acute treatments, whereas the aged brain is far less capable of a compensatory response in the presence of a chronic degenerative process. Future models should investigate the effects of lesions or pharmacological challenges in both young and aged animals.

Many animal models of AD have been used to investigate the effectiveness or usefulness of specific pharmacotherapies. These drugs are often effective in the animal model but are ineffective in patients with AD. Most pharmacotherapies are directed only toward the amelioration of the cholinergic deficit associated with AD and tend to ignore the effects of degeneration in noncholinergic neural systems. These animal models may therefore be invalid because they do not mimic all of the pathological and neurochemical components associated with AD. Pharmacotherapies that are effective in animal models that more closely reproduce the full extent of the pathological and biochemical changes are more likely to be clinically effective in AD.

Acknowledgments

The author wishes to thank Anu Durr for typing the initial manuscript and David Olton for many helpful discussions. The preparation of the chapter was supported by Grants NIH AG 05146 and NS 20471 from the US Public Health Service.

References

Aigner T. G., Mitchell S. J., Aggleton J. P., DeLong M. R., Struble R. G., Price D. L., Wenk G. L., and Mishkin M. (1987) Effects of scopolamine and physostigmine on recognition memory in monkeys with ibotenic-acid lesions of the nucleus basalis of Meynert. *Psychopharmacology* **92**, 292–300.

Allen Y. S., Marchbanks R. M., and Sinden J. D. (1988) Non-specific effects of the putative cholinergic neurotoxin ethylcholine mustard aziridinium ion in the rat brain examined by autoradiography, immunocytochemistry and gel electrophoresis. *Neurosci. Lett.* **95**, 69–74.

Arendash G. W., Millard W. J., Dunn A. J., and Meyer E. M. (1987) Long-term neuropathological and neurochemical effects of nucleus basalis lesions in the rat. *Science* **238**, 952–956.

Arendt T., Bigl V., Tennstedt A., and Arendt A. (1985) Neuronal loss in different parts of the nucleus basalis is related to neuritic plaque formation in cortical target areas in Alzheimer's disease. *Neuroscience* **14**, 1–14.

Atack J. R., Wenk G. L., Wagster M. V., Kellar K. J., Whitehouse P. J., and Rapoport S. I. (1989) Bilateral changes in neocortical [^{3}H]pirenzepine

and [^{3}H]oxotremorine-*M* binding following unilateral lesions of the rat nucleus basalis magnocellularis: an autoradiographic study. *Brain Res.* **483**, 367–372.
Bartus R. T., Dean R. L., Beer B., and Lippa A. S. (1982) The cholinergic hypothesis of geriatric memory dysfunction. *Science* **217**, 408–417.
Bartus R. T., Fleming D., and Johnson H. R. (1978) Aging in the rhesus monkey: debilitating effects on short-term memory. *J. Gerontol.* **33**, 858–871.
Bartus R. T., Flicker C., Dean, R. L., Pontecorvo M., Figuerdo J. C., and Fisher S. K. (1985) Selective memory loss following nucleus basalis lesions: long term behavioral recovery despite persistent cholinergic deficiencies. *Pharmacol. Biochem. Behav.* **23**, 125–135.
Beal M. F., Mazurek M. F., Ellison D. W., Kowall N. W., Soloman P. R., and Pendlebury W. W. (1989) Neurochemical characteristics of aluminum-induced neurofibrillary degeneration in rabbits. *Neuroscience* **2**, 329–337.
Beninger R. J., Jhamandas K., Boegman R. J., and El-Defrawy S. R. (1986) Effects of scopolamine and unilateral lesions of the basal forebrain on T-maze spatial discrimination and alternation in rats. *Pharmacol. Biochem. Behav.* **24**, 1353–1360.
Bertholf R. L. (1987) Aluminum and Alzheimer's disease: Perspectives for a cytoskeletal mechanism. *CRC Crit. Rev. Clin. Lab. Sci.* **25**, 195–210.
Blessed G., Tomlinson B. E., and Roth M. (1968) The association between quantitative measures of dementia and of senile change in the grey matter of elderly subjects. *Br. J. Psychiat.* **114**, 797–811.
Borson S., Barnes R. F., Veith R. C., Halter J. B., and Rasking M. A. (1989) Impaired sympathetic nervous system response to cognitive effort in early Alzheimer's disease. *J. Gerontol.*, **44**, M8–12.
Bowen D. M., Spillane J. A., Curzon G., Meier-Ruge W., White P., Goodhardt M. J., Iwangoff P., and Davison A. N. (1979) Accelerated ageing or selective neuronal loss as an important cause of dementia? *Lancet* **1**, 11–14.
Broks P., Preston G. C., Traub M., Poppleton P., Ward C., and Stahl S. M. (1988) Modelling dementia: effects of scopolamine on memory and attention. *Neuropsychology* **5**, 685–700.
Casamenti F., DePatre P. L., Bartolini L., and Pepeu G. (1988) Unilateral and bilateral nucleus basalis lesions: Differences in neurochemical and behavioral recovery. *Neuroscience* **2**, 209–215.
Chrobak J. J., Hanin I., Schmechel D. E., and Walsh T. J. (1988) AF64A-induced working memory impairment: behavioral, neurochemical and histological correlates. *Brain Res.* **463**, 107–117.
Chrobak J. J., Hanin I., and Walsh T. J. (1987) AF64A (ethylcholine aziridinium ion), a cholinergic neurotoxin, selectively impairs working memory in a multiple component T-maze task. *Brain Res.* **414**, 15–21.
Chui H. C., Bondareff W., Zarow C., and Slager U. (1984) Stability of neuronal number in the human nucleus basalis of Meynert with age. *Neurobiol. Aging* **5**, 83–88.
Coffey P. J., Perry V. H., Allen Y., Sinden J., and Rawlins J. N. P. (1988)

Ibotenic acid induced demyelination in the central nervous system: a consequence of a local inflammatory response. *Neurosci. Lett.* **84**, 178–184.
Collerton D. (1986) Cholinergic function and intellectual decline in Alzheimer's disease. *Neuroscience* **19**, 1–28.
Cotman C. W. and Iversen L. L. (1987) Excitatory amino acids in the brain—focus on NMDA receptors. *Trends Neurosci.* **10**, 263–280.
Coyle J. T. (1983) Neurotoxic action of kainic acid. *J. Neurochem.* **41**, 1–11.
Coyle J. T. and Schwarcz R. (1983) The use of excitatory amino acids as selective neurotoxins, in *Handbook of Chemical Neuroanatomy. Vol. 1: Methods in Chemical Neuroanatomy* (Borklund A. and Hokfelt T., eds.), pp. 508–527. Elsevier, New York.
Coyle J. T., Price D. L., and DeLong M. R. (1983) Alzheimer's disease: A disorder of cortical cholinergic innervation. *Science* **219**, 1184–1190.
Craik F. I. M. (1977) Age differences in human memory, in *The Handbook of the Psychology of Aging* (Birren J. E. and Schaire K. W., eds.), pp. 384–420. Van Nostrand Reinhold, New York.
Crapper McLachlan D. R. and DeBoni U. (1980) Aluminum in human brain disease—An overview. *Neurotoxicology* **1**, 3–16.
Crapper D. R., Quittkat S., Krishnan S. S., Dalton A. J., and DeBoni U. (1980) Intranuclear aluminum content in Alzheimer's disease, dialysis encephalopathy, and experimental aluminum encephalopathy. *Acta Neuropathol.* **50**, 19–24.
Cross A. J., Crow T. J., Ferrier I. N., Johnson J. A., Bloom S. R., and Corsellis J. A. N. (1984) Serotonin receptor changes in dementia of the Alzheimer type. *J. Neurochem.* **43**, 1574–1581.
D'Amato R. J., Zweig R. M., Whitehouse P. J., Wenk G. L., Singer H. S., Mayeux R., Price D. L., and Snyder S. H. (1987) Aminergic systems in Alzheimer's disease and Parkinson's disease. *Ann. Neurol.* **22**, 229–236.
Davies P. and Maloney A. J. F. (1976) Selective loss of central cholinergic neurons in Alzheimer's disease. *Lancet* **2**, 1403.
Davis K. L. and Mohs R. C. (1982) Enhancement of memory processes in Alzheimer's disease with multiple-dose intravenous physostigmine. *Am. J. Psychiat.* **139**, 1421–1424.
Decker M. W. (1987) The effects of aging on hippocampal and cortical projections of the forebrain cholinergic system. *Brain Res. Rev.* **12**, 423–438.
Delaney J. F. (1979) Spinal fluid aluminum levels in patients with Alzheimer disease. *Ann. Neurol.* **5**, 580–581.
Dunea G., Mahurkar S. D., Mamdani B., and Smith E. C. (1978) Role of aluminum in dialysis dementia. *Ann. Int. Med.* **88**, 502–504.
Dunne M. P. and Hartley L. R. (1985) The effects of scopolamine upon verbal memory: evidence for an attentional hypothesis. *Acta Psychol.* **58**, 205–217.
Dunnett S. B., Low W. C., Iversen S. D., Stenevi U., and Bjorklund A. (1982) Septal transplants restore maze learning in rats with fornix-fimbria lesions. *Brain Res.* **251**, 335–348.

Dunnett S. B., Rogers D. C., and Jones G. H. (1989) Effects of nucleus basalis magnocellularis lesions in rats on delayed matching and non-matching to position tasks. *Eur. J. Neurosci.* **1**, 395–406.

Ebel A., Strosser M. T., and Kempf E. (1987) Genotypic differences in central neurotransmitter responses to aging mice. *Neurobiol. Aging* **8**, 417–427.

Etienne P., Robitaille Y., Wood P., Gauthier S., Nair N. P. V., and Quirion R. (1986) Nucleus basalis neuronal loss, neuritic plaques and choline acetyltransferase activity in advanced Alzheimer's disease. *Neurosci.* **19**, 1279–1291.

Fischer W., Gage F. H., and Bjorklund A. (1989) Degenerative changes in forebrain cholinergic nuclei correlate with cognitive impairments in aged rats. *Eur. J. Neurosci.* **1**, 34–45.

Folstein M. F. and Whitehouse P. J. (1983) Cognitive impairment of Alzheimer disease. *Neurobehav. Toxicol. Teratol.* **5**, 631–634.

Francis P. T., Palmer A. M., Sims N. R., Bowen D. M., Davison A. N., Esiri M. M., Neary D., Snowden J. S., and Wilcock G. K. (1985) Neurochemical studies of early-onset Alzheimer's disease. *N. Engl. J. Med.* **313**, 7–11.

Freedman M. and Oscar-Berman M. (1986) Selective delayed response deficits in Parkinson's and Alzheimer's disease. *Arch. Neurol.* **43**, 886–890.

Gaal G., Potter P. E., Hanin I., Kakucska I., and Vizi E. S. (1986) Effects of intracerebroventricular AF64A administration on cholinergic, serotonergic and catecholaminergic circuitry in rat dorsal hippocampus. *Neuroscience* **19**, 1197–1205.

Gage F. H., Dunnett S. B., and Bjorklund A. (1984a) Spatial learning and motor deficits in aged rats. *Neurobiol. Aging* **5**, 43–48.

Gage F. H., Kelly P. A. T., and Bjorklund A. (1984b) Regional changes in brain glucose metabolism reflect cognitive impairments in aged rats. *J. Neurosci.* **4**, 2856–2865.

Gardiner I.M., de Belleroche J., Premi B. K., and Hamilton M. H. (1987) Effect of lesion of the nucleus basalis of rat on acetylcholine release in cerebral cortex: time course of compensatory events. *Brain Res.* **407**, 263–271.

Godefroy F., Bassant M. H., Weil-Fugazza J., and Lamour Y. (1989) Age-related changes in dopaminergic and serotonergic indices in the rat forebrain. *Neurobiol. Aging* **10**, 187–190.

Gonatas N. K., Anderson W., and Evangelista I. (1967) The contribution of altered synapses in the senile plaque: an electron microscopic study in Alzheimer's dementia. *J. Neuropathol. Exp. Neurol.* **26**, 25–39.

Handelmann G. E. and Olton D. S. (1981) Recovery of function after neurotoxic damage to the hippocampal CA3 region: Importance of postoperative recovery interval and task experience. *Behav. Neural Biol.* **33**, 453–464.

Henderson Z. (1991) Sprouting of cholinergic axons does not occur in the cerebral cortex after nucleus basalis lesions. *Neuroscience* **1**, 149–156.

Henke H. and Lang W. (1983) Cholinergic enzymes in neocortex, hippocampus and basal forebrain of non-neurological and senile dementia of Alzheimer-type patients. *Brain Res.* **267**, 281–291.

Hepler D. J., Olton D. S., Wenk G. L., and Coyle J. T. (1985) Lesions in nucleus basalis magnocellularis and medial septal area of rats produce qualitatively similar memory impairments. *J. Neurosci.* **5**, 866–873.

Hohmann C. F. and Coyle J. T. (1988) Long-term effects of basal forebrain lesions on cholinergic, noradrenergic and serotonergic markers in mouse neocortex. *Brain Res. Bull.* **21**, 13–20.

Hohmann C. F., Wenk G. L., Lowenstein P., Brown M. E., and Coyle J. T. (1987) Age-related recurrence of basal forebrain lesion-induced cholinergic deficits. *Neurosci. Lett.* **82**, 253–259.

Hoyer S., Oesterreich K., and Wager O. (1988) Glucose metabolism as the site of the primary abnormality in early onset dementia of Alzheimer type? *J. Neurol.* **235**, 143–148.

Hyman B. T., Van Hoesen G. W., Damasio A. R., and Barnes C. L. (1984) Alzheimer's disease: Cell-specific pathology isolates the hippocampal formation. *Science* **225**, 1168–1170.

Irle E. and Markowitsch H. J. (1987) Basal forebrain-lesioned monkeys are severely impaired in tasks of association and recognition memory. *Ann. Neurol.* **22**, 735–743.

Jarrárd L. E., Kant G. J., Meyerhoff J. L., and Levy A. (1984) Behavioral and neurochemical effects of intraventricular AF64A administration in rats. *Pharmacol. Biochem. Behav.* **21**, 273–280.

Johnston M. V., McKinney M., and Coyle J. T. (1979) Evidence for a cholinergic projection to neocortex from neurons in the basal forebrain. *Proc. Natl. Acad. Sci. USA* **76**, 5392–5396.

Johnston M. V., McKinney M., and Coyle J. T. (1981) Neocortical cholinergic innervation in the rat. *Exp. Brain Res.* **43**, 159–172.

Kemper T. L. (1983) Organization of the neuropathology of the amygdala in Alzheimer's disease, in *Banbury Report. 15: Biological Aspects of Alzheimer's Disease*, pp. 31–35, Cold Spring Harbor Lab., Cold Spring Harbor, NY.

Kidd M. (1963) Paired helical filaments in electron microscopy of Alzheimer's disease. *Nature* **197**, 192,193.

Kitt C. A., Mitchell S. J., DeLong M. R., Wainer B. H., and Price D. L. (1987) Fiber pathways of basal forebrain cholinergic neurons in monkeys. *Brain Res.* **406**, 192–206.

Kitt C. A., Struble R. G., Cork L. C., Mobley W. C., Walker L. C., Joh T. H., and Price D. L. (1985) Catecholaminergic neurites in senile plaques in prefrontal cortex of aged nonhuman primates. *Neuroscience* **16**, 691–699.

Kiyosawa M., Pappata S., Duverger D., Riche D., Cambon H., Mazoyer B., Samson Y., Crouzel C., Naquet R., MacKenzie E.T., and Baron J-C. (1987) Cortical hypometabolism and its recovery following nucleus basalis lesions in baboons: A PET study. *J. Cerebral Blod Flow Metabol.* **7**, 812–817.

Kohler C. and Schwarcz R. (1983) Comparison of ibotenate and kainate neurotoxicity in rat brain: A histological study. *Neuroscience* **8**, 819–835.

Lamarca M. V. and Fibiger H. C. (1984) Deoxyglucose uptake and choline acetyltransferase activity in cerebral cortex following lesions of the nucleus basalis magnocellularis. *Brain Res.* **307**, 366–369.

Lenders M.-B., Peers M.-C., Tramu G., Delacourte A., Defossez A., Petit H., and Mazzuca M. (1989) Dystrophic peptidergic neurites in senile plaques of Alzheimer's disease hippocampus precede formation of paired helical filaments. *Brain Res.* **481**, 344–349.

Leong S. F., Lai J. C. K., Lim L., and Clark J. B. (1981) Energy-metabolizing enzymes in brain regions of adult and aging rats. *J. Neurochem.* **37**, 1548–1556.

Levy A., Kant G. J., Meyerhoff J. L., and Jarrard L. E. (1984) Noncholinergic neurotoxic effects of AF64A in substantia nigra. *Brain Res.* **305**, 169–172.

Lindner M. D. and Schallert T. (1988) Aging and atropine effects on spatial navigation in the Morris water task. *Behav. Neurosci.* **102**, 621–634.

London E. D., McKinney M., Dam M., Ellis A., and Coyle J. T. (1984) Decreased cortical glucose utilization after ibotenate lesion of the rat ventromedial globus pallidus. *J. Cerebr. Blood Flow Metab.* **4**, 381–390.

Mantione C. R., Fisher A., and Hanin I. (1981) The AF64A-treated mouse: possible model for central cholinergic hypofunction. *Science* **13**, 579–580.

Markowska A. L., Stone W. S., Ingram D. K., Reynolds J., Gold P. E., Conti L. H., Pontecorvo M. J., Wenk G. L., and Olton D. S. (1989) Individual differences in aging: Behavioral and neurobiological correlates. *Neurobiol. Aging* **10**, 31–43.

Martinez J. L., Schulteis G., Janak P. H., and Weinberger S. B. (1988) Behavioral assessment of forgetting in aged rodents and its relationship to peripheral sympathetic function. *Neurobiol. Aging* **9**, 697–708.

Mason S. T. and Fibiger H. C. (1979) On the specificity of kainic acid. *Science* **20**, 1339–1341.

McKhann G., Drachman D., Folstein M., Katzman R., Price D., and Stadlan E. M. (1984) Clinical diagnosis of Alzheimer's disease. *Neurology* **34**, 939–944.

McGurk S. R., Hartgraves S. L., Kelly P. H., Gordon M. N., and Butcher L. L. (1987) Is ethylcholine mustard aziridinium ion a specific cholinergic neurotoxin? *Neuroscience* **22**, 215–224.

Mesulam M. M., Mufson E. J., Levey A. I., and Wainer B. H. (1983) Cholinergic innervation of cortex by the basal forebrain: cytochemistry and cortical connections of the septal area, diagonal band nuclei, nuclei basalis (substantia innominata) and hypothalamus in rhesus monkey. *J. Comp. Neurol.* **214**, 170–197.

Mesulam M. M., Mufson E. J., Levey A. I., and Wainer B. H. (1984) Atlas of cholinergic neurons in the forebrain and upper brainstem of the macaque based on monoclonal choline acetyltransferase immunohistochemistry and acetylcholinesterase histochemistry. *Neuroscience* **12**, 669–686.

Meyers D., Armstrong R. A., Smith C. U. M., and Carter R. A. (1988) The spatial arrangement pattern of senile plaques in senile dementia of the Alzheimer type (SDAT). *Neurosci. Res. Comm.* **2**, 99–106.

Michalek H., Fortuna S., and Pintor A. (1989) Age-related differences in brain choline acetyltransferase, cholinesterase and muscarinic receptor sites in two strains of rats. *Neurobiol. Aging* **10**, 143–148.

Mitchell S. J., Rawlins J. N. P., Steward O., and Olton D. S. (1982) Medial septal area lesions disrupt theta rhythm and cholinergic staining in medial entorhinal cortex and produce impaired radial arm maze behavior in rats. *J. Neurosci.* **2**, 292–302.

Moretti A., Carfagna N., and Trunzo F. (1987) Effects of monoamines and their metabolites in the rat brain. *Neurosci. Res.* **12**, 1035–1039.

Mufson E. J., Kehr A. D., Wainer B. H., and Mesulam M. -M. (1987) Cortical effects of neurotoxic damage to the nucleus basalis in rats: persistent loss of extrinsic cholinergic input and lack of transsynaptic effect upon the number of somatostatin-containing, cholinesterase-positive, and cholinergic cortical neurons. *Brain Res.* **417**, 385–388.

Nemeroff C. B., Kizer J. S., Reynolds G. P., and Bissette G. (1989) Neuropeptides in Alzheimer's disease: a postmortem study. *Reg. Peptides* **25**, 123–130.

Ojima H., Sakurai T., and Yamasaki T. (1988) Changes in choline acetyltransferase immunoreactivity and the number of immunoreactive fibers remaining after lesions of the magnocellular basal nucleus of rats. *Neurosci. Lett.* **95**, 31–36.

Okaichi H. and Jarrard L. E. (1982) Scopolamine impairs performance of a place and cue task in rats. *Behav. Neural Biol.* **35**, 1982.

Olney J. W., Rhee V., and Ho O. L. (1974) Kainic acid: a powerful neurotoxic analogue of glutamate. *Brain Res.* **77**, 507–512.

Olton D. S. (1986) Interventional approaches to memory: Lesions, in *Learning and Memory: A Biological View* (Martinez J. L. and Kesner R., eds.), pp. 379–397, Academic, New York.

Olton D. S., Walker J. A., and Wolf W. A. (1982) A disconnection analysis of hippocampal function. *Brain Res.* **233**, 241–253.

Olton D. S. and Wenk G. L. (1987) Dementia: Animal models of the cognitive impairments produced by degeneration of the basal forebrain cholinergic system, in *Psychopharmacology: The Third Generation of Progress* (Meltzer H.Y., ed.), pp. 941–953, Raven, New York.

Olton D. S., Wenk G. L., Church R. M., and Meck W. H. (1988) Attention and the frontal cortex as examined by simultaneous temporal processing. *Neuropsychology* **26**, 307–318.

Owen M. J. and Butler S. R. (1981) Amnesia after transection of the fornix in monkeys: long-term memory impaired, short-term memory intact. *Behav. Brain Res.* **3**, 115–123.

Pearson R. C. A., Neal J. W., and Powell T. P. S. (1986) Hypertrophy of cholinergic neurones of the basal nucleus in the rat following damage of the contralateral nucleus. *Brain Res.* **382**, 149–152.

Pedata F., LoConte G., Sorbi S., Marconcini Pepeu I., and Pepeu G. (1982)

Changes in high affinity choline uptake in rat cortex following lesions of the magnocellular forebrain nuclei. *Brain Res.* **233**, 359–367.

Pendlebury W. W., Beal M. F., Kowall N. W., and Soloman P. R. (1988) Neuropathologic, neurochemical and immunocytochemical characteristics of aluminum-induced neurofilamentous degeneration. *Neurotoxicology* **9**, 503–510.

Perkins M. N. and Stone T. W. (1983) Quinolinic acid: regional variations in neuronal sensitivity. *Brain Res.* **259**, 172–176.

Perry E. K. and Perry R. H. (1985) New insights into the nature of senile (Alzheimer-type) plaques. *Trends Neurosci.* **10**, 301–303.

Piercey M. F., Vogelsang G. D., Franklin S. R., and Tang A. H. (1987) Reversal of scopolamine-induced amnesia and alterations in energy metabolism by the nootropic piracetam: Implications regarding identification of brain structures involved in consolidation of memory traces. *Brain Res.* **42**, 1–9.

Preston G. C., Brazell C., Ward C., Broks P., Traub M., and Stahl S. M. (1988) The scopolamine model of dementia: determination of central cholinomimetic effects of physostigmine on cognition and biochemical markers in man. *J. Psychopharmacol.* **2**, 67–79.

Price D. L. (1986) New perspectives on Alzheimer's disease. *Ann. Rev. Neurosci.* **9**, 489–512.

Rapp P. R., Rosenberg R. A., and Gallagher M. (1987) An evaluation of spatial information processing in aged rats. *Behav. Neurosci.* **101**, 3–12.

Ridley R. M., Baker H. F., Drewett B., and Johnson J. A. (1985) Effects of ibotenic acid lesions of the basal forebrain serial reversal learning in marmosets. *Psychopharmacology* **86**, 438–443.

Ridley R. M., Murray T. K., Johnson J. A., and Baker H. F. (1986) Learning impairment following lesion of the basal nucleus of Meynert in the marmoset: Modification by cholinergic drugs. *Brain Res.* **376**, 108–116.

Roudier M., Marcie P., Podrabinek N., Lamour Y., and Davous P. (1988) Correlations between memory, language, agnosia, and apraxia in 80 patients with senile dementia of the Alzheimer type. *Drug Develop. Res.* **14**, 231–234.

Rusted J. M. (1988) Dissociative effects of scopolamine on working memory in healthy young volunteers. *Psychopharmacology* **96**, 487–492.

Saletu B., Grunberger J., Linzmayer L., and Anderer P. (1979) Proof of CNS efficacy and pharmacodynamics of nicergoline in the elderly by acute and chronic quantitative pharmaco-EEG and psychometric studies, in *Drug Treatment and Prevention in Cerebrovascular Disorders* (Tognoni G. and Garattini S., eds.), pp. 245–272, Elsevier/North Holland, Amsterdam.

Sandberg K., Hanin I., Fisher A., and Coyle J. T. (1984) Selective cholinergic neurotoxin: AF64A's effects in rat striatum. *Brain Res.* **293**, 49–55.

Sandberg K., Schnaar R. L., McKinney M., Hanin I., Fisher A., and Coyle J. T. (1985) AF64A: An active site directed irreversible inhibitor of choline acetyltransferase. *J. Neurochem.* **44**, 439–445.

Satoh K. and Fibiger H. C. (1985) Distribution of central cholinergic neurons in the baboon (Papio): I. General morphology. *J. Comp. Neurol.* **236**, 197–214.

Schwarcz R. and Kohler C. (1983) Differential vulnerability of central neurons of the rat to quinolinic acid. *Neurosci. Lett.* **38**, 85–90.

Schwarcz R., Scholz D., and Coyle J. T. (1978) Structure-activity relations for the neurotoxicity of kainic acid derivatives and glutamate analogues. *Neuropharmacology* **17**, 145–151.

Simon J. R. and Kuhar M. J. (1975) Impulse-flow regulation of high affinity choline uptake in brain cholinergic nerve terminal. *Nature* **255**, 162–163.

Sirvio J., Valjakka A., Jolkkonen J., Jervonen A., and Riekkinen P. J. (1988) Cholinergic enzyme activities and muscarinic binding in the cerebral cortex of rats of different age and sex. *Comp. Biochem. Physiol.* **90C**, 245–248.

Smith C. M. and Swash M. (1978) Possible biochemical basis of memory disorder in Alzheimer Disease. *Ann. Neurol.* **3**, 471–473.

Smith G. (1988) Animal models of Alzheimer's disease: experimental cholinergic denervation. *Brain Res. Rev.* **13**, 103–118.

Spangler E. L., Rigby P., and Ingram D. K. (1986) Scopolamine impairs learning performance of rats in a 14-unit T-maze. *Pharmacol. Biochem. Behav.* **25**, 673–679.

Spignoli G. and Pepeu G. (1987) Interactions between oxiracetam, aniracetam and scopolamine on behavior and brain acetylcholine. *Pharmacol. Biochem. Behav.* **27**, 491–495.

Struble R. G., Cork L. C., Whitehouse P. J., and Price D. L. (1982) Cholinergic innervation of neuritic plaques. *Science* **216**, 413–415.

Struble R. G., Powers R. E., Casanova M. F., Kitt C. A., Brown E. C., and Price D. L. (1987) Neuropeptidergic systems in plaques of Alzheimer's disease. *J. Neuropathol. Exp. Neurol.* **46**, 567–584.

Struble R. G., Price Jr., D. L., Cork L. C., and Price D. L. (1985) Senile plaques in cortex of aged normal monkeys. *Brain Res.* **361**, 267–275.

Summers W. K., Majovski L. V., Marsh G. M., Tachiki K., and Kling A. (1987) Oral tetrahydroaminoacridine in long-term treatment of Alzheimer's disease. *New Engl. J. Med.* **315**, 1241–1245.

Tay S. S. W., Williams T. H., and Jew J.Y. (1989) Neurotensin immunoreactivity in the central nucleus of the rat amygdala: An ultrastructural approach. *Peptides* **10**, 113–120.

Terry R. D. and Davies P. (1980) Dementia of the Alzheimer type. *Ann. Rev. Neurosci.* **3**, 77–95.

Terry R. D. and Pena C. (1965) Experimental production of neurofibrillary degeneration. *Neuropathol. Exp. Neurol.* **24**, 200–210.

Terry R. D., Mandel R. J., Buzsaki G., Gage F. H., and Thal L. J. (1988) Characterization of the effects of nucleus basalis lesions in rats 14 months post-lesion. *Soc. Neurosci. Abstr.* **1**, 1007.

Thal L. J., Rosen W., and Sharpless N. S. (1981) Choline chloride fails to improve cognition in Alzheimer's disease. *Neurobiol. Aging* **2**, 205–208.

Trapp G. A., Miner G. D., Zimmerman R. L., Mastri A. R., and Heston L. L. (1978) Aluminum levels in brain in Alzheimer's disease. *Biol. Psychiat.* **13**, 709–718.

Troncoso J. C., Price D. L., Griffin J. W., and Parhad I. M. (1982) Neurofibrillary axonal pathology in aluminum intoxication. *Ann. Neurol.* **12**, 278–283.

Tucek S. (1978) *Acetylcholine Synthesis in Neurons*. Chapman and Hall, London.

Ulrich J. and Stahelin H. B. (1984) The variable topography of Alzheimer type changes in senile dementia and normal old age. *Gerontology* **30**, 210–214.

Vanella A., Villa R. F., Gorini A., Campisi A., and Giuffrida-Stella A. M. (1989) Superoxide dismutase and cytochrome oxidase activities in light and heavy synaptic mitochondria from rat cerebral cortex during aging. *J. Neurosci. Res.* **22**, 351–355.

Villani L., Contestabile A., Migani P., Poli A., and Fonnum F. (1986) Ultrastructural and neurochemical effects of the presumed cholinergic toxin AF64A in the rat interpeduncular nucleus. *Brain Res.* **379**, 223–231.

Walker L. C., Kitt C. A., Cork L. C., Struble R. G., Dellovade T. L., and Price D. L. (1988a) Multiple transmitter systems contribute neurites to individual senile plaques. *J. Neuropathol. Exo. Neurol.* **47**, 138–144.

Walker L. C., Kitt C. A., Struble R. G., Wagster M. V., Price D. L., and Cork, L. C. (1988b) The neural basis of memory decline in aged monkeys. *Neurobiol. Aging* **9**, 657–666.

Walker L. C., Koliatsos V. E., Kitt C. A., Richardson R. T., Rokaeus A., and Price D. L. (1989) Peptidergic neurons in the basal forebrain magnocellular complex of the rhesus monkey. *J. Comp. Neurol.* **280**, 272–282.

Wenk G. L. (1989) An hypothesis on the role of glucose in the mechanism of action of cognitive enhancers. *Psychopharmacology* **99**, 431–438.

Wenk G. L., Cribbs B., and McCall L. (1984) Nucleus basalis magnocellularis: Optimal coordinates for selective reduction of choline acetyltransferase in frontal neocortex by ibotenic acid injections. *Ex. Brain Res.* **56**, 335–340.

Wenk G. L. and Engisch K. L. (1986) [^{3}H] Ketanserin (serotonin type 2) binding increases in rat cortex following basal forebrain lesions with ibotenic acid. *J. Neurochem.* **47**, 845–850.

Wenk G. L., Engisch K. L., McCall L. D., Mitchell S. J., Aigner T. G., Struble R. L., Price D. L., and Olton D. S. (1986) [^{3}H]Ketanserin binding increases in monkey cortex following basal forebrain lesions with ibotenic acid. *Neurochem. Int.* **9**, 557–562.

Wenk G. L., Hughey D., Boundy V., Kim A., Walker L., and Olton D. (1987) Neurotransmitters and memory: Role of cholinergic, serotonergic, and noradrenergic systems. *Behav. Neurosci.* **101**, 325–332.

Wenk G. L., Markowska A. L., and Olton D. S. (1989a) Basal forebrain lesions and memory: Alterations in neurotensin, not acetylcholine, may cause amnesia. *Behav. Neurosci.* **103**, 765–769.

Wenk G. L. and Olton D. S. (1984) Recover of neocortical choline acetyltransferase activity following ibotenic acid injection in the nucleus basalis of Meynert in rats. *Brain Res.* **293**, 184–186.

Wenk G. L. and Olton D. S. (1987) Basal forebrain cholinergic neurons andAlzheimer's disease, in *Animal Models of Dementia* (Coyle J.T., ed.), pp. 81–101, Liss, New York.

Wenk G. L., Pierce D. J., Struble R. G., Price D. L., and Cork L. C. (1989b) Age-related changes in multiple neurotransmitter systems in the monkey brain. *Neurobiol. Aging* **10**, 11–19.

Wenk G. L. and Rokaeus A. (1988) Basal forebrain lesions differentially alter galanin levels and acetylcholinergic receptors in the hippocampus and neocortex. *Brain Res.* **460**, 17–21.

Wesnes K. and Warburton D. M. (1983) Effects of scopolamine on stimulus sensitivity and response in a visual vigilance task. *Neuropsychobiology* **9**, 154–157.

Wesnes K. and Warburton D. M. (1984) Effects of scopolamine and nicotine on human rapid information processing performance. *Psychopharmacology* **82**, 147–150.

Whitehouse P. J., Price D. L., Clark A. W., Coyle J. T., and DeLong M. R. (1981) Alzheimer Disease: Evidence for selective loss of cholinergic neurons in the nucleus basalis. *Ann. Neurol.* **10**, 122–126.

Whitehouse P. J., Struble R. G., Hedreen J. C., Clark A. W., and Price D. L. (1985) Alzheimer's disease and related dementias: selective involvement of specific neuronal systems. *CRC Critical Rev. Clin. Neurobiol.* **1**, 319–339.

Yates C. M., Simpson J., Gordon A., Maloney A. F. J., Allison Y., Ritchie I. M., and Urquhart A. (1983) Catecholamines and cholinergic enzymes in pre-senile and senile Alzheimer-type dementia and Down's syndrome. *Brain Res.* **280**, 119–126.

Yokel R. A. (1983) Repeated systemic aluminum exposure effects on classical conditioning of the rabbit. *Neurobehav. Toxicol. Teratol.* **5**, 41-46.

Yokel R. A., Provan S. D., Meyer J. J., and Campbell S. R. (1988) Aluminum intoxication and the victim of Alzheimer's disease: similarities and differences. *Neurotoxicology* **9**, 429–442.

Animal Models of Huntington's Disease

Dwaine F. Emerich and Paul R. Sanberg

1. Introduction

Huntington's disease (HD) is an inherited, progressive neurodegenerative disorder transmitted by a single autosomal dominant gene. The symptomology of the disease was first described by George Huntington in 1872 as consisting of a progressive dementia coupled with bizarre uncontrollable movements and abnormal postures. HD is found in nearly all ethnic and racial groups with slight variations in prevalence rates. Overall, the prevalence rate of HD in the US is approx 50/1,000,000 (Reed and Chandler, 1958; Sanberg and Coyle, 1984). Although HD may occur during the juvenile years (Korenyi and Whittier, 1973); the manifestation of the disorder typically occurs in middle life, about 35–45 years of age. From the time of onset, an intractable course of mental deterioration and progressive motor abnormalities begins with death usually occurring within 15 years.

Because of its dramatic symptomology, relentless course, and hereditary nature, HD has gained a substantial amount of attention. Despite the interest in HD, little is known about the underlying etiology of the disorder. Although research is beginning to unravel the pathophysiology and molecular biology involved, there is still a wide gap between our understanding of the neural substrates of HD and our ability to prevent or alleviate them. Our increasing knowledge of the neural pathology in HD has revealed a complex mosaic of related and interdependent neurochemical and histopathological alterations. In addition, genetic research has recently isolated the chromosome

From: *Neuromethods, Vol. 21: Animal Models of Neurological Disease, I*
Eds: A. Boulton, G. Baker, and R. Butterworth

responsible for the expression of HD. This knowledge may soon lead to the exact identification of the gene or genes responsible for the disorder. The elucidation of the genetic abnormality in HD together with a more complete understanding of the pathology associated with it provides new avenues of treatment for the disorder. The precise characterization of the genetic abnormality in HD will likely lead to treatments based on molecular genetic studies, and could involve the deletion or replacement of specific genetic components. Work with animal models of HD is also providing new treatment strategies for HD. For instance, putative pharmacological treatments have been developed from the suggestion that endogenous excitotoxic compounds underlie the progressive neural degeneration in HD. In addition, the transplantation of fetal neural tissue has been shown to reverse the neurochemical and behavioral alterations observed in a variety of models of HD.

An important approach for understanding and treating neurodegenerative diseases, such as HD, is the development of appropriate animal models that closely mimic the behavioral and neurobiological sequelae of HD. Such a model would aid in the further elucidation of the biological and behavioral expression of HD, and also suggest unique therapeutic strategies for its treatment. This chapter evaluates the current status of the development of animal models of HD and possible treatments for the disorder. First, the behavioral pathology of HD and the current status of the genetic abnormality underlying HD are discussed. Next, the organization of the basal ganglia is described with reference to its neurochemical and cellular pathology in HD. Animal models of HD are then described with emphasis on pharmacologically induced dyskinesia and excitotoxin-mediated damage to the striatum. Because of the promise that excitotoxin models hold for understanding the etiology of HD, considerable attention is paid to the kainic and quinolinic acid models and what aspects of HD they may be used to mimic. Finally, the suitability of pharmacological treatments and neural transplantation as possible therapies in excitoxic models is addressed with discussion of the advantages and disadvantages of each.

2. Behavior and Genetics of Huntington's Disease

The onset of the motor and mental features of HD is variable, and the symptoms typically manifest themselves over a protracted time-course. Although the onset of these symptoms typically occurs at approx 35–40 years of age, it may range from childhood to the eighth decade. Clinically, the physical features of HD consist of constant choreiform movements of the entire body, which are irregular and involuntary (Sanberg and Coyle, 1984; Hefter et al., 1987). These movements may be unilateral for a time, but later recruit all the limbs (Pinel, 1976). The face often looks quite grotesque because of the constant writhing contortions of facial muscles. Dysarthria, dysphagia, and disturbances in ocular motility may also develop, and the individual's speech becomes quite unintelligible (Starr, 1967; Dix, 1970; Davis, 1976; Young et al., 1986; Podoll et al., 1988) .

The expression of the mental symptoms in HD is more variable than the motor abnormalities. The primary mental features of HD are generally considered similar to dementia (Dewhurst et al., 1979; Mann et al., 1980; Sanberg and Coyle, 1984; Cummings and Benson, 1988; Brandt et al., 1988; Heindel et al., 1988). The initial signs may consist of lack of grooming, eccentric traits, irritability, impairments in memory, emotional instability, and delusions of paranoid grandeur. Personality disorders, such as outbursts of rage or violent temper, may occur. Conversely, some patients may be slow and apathetic. The early mental symptoms of HD can mimic those of schizophrenia and bipolar affective illness, leading to many HD patients being misdiagnosed as having schizophrenia (Klawans et al., 1972; Bowman and Lewis, 1980; Van Putten and Menkes, 1973). It is generally agreed that the mental symptoms occur prior to the motor symptoms of HD, which probably also contributes to misdiagnosis (Bruyn, 1968; James et al., 1969; Girotti et al., 1988).

Often, the HD patient does not appear demented because of a general preservation of intellect and language functions. This general intellectual preservation is likely the result of the sparing of cortically mediated cognitive functions. In other neurodegenerative disorders, such as Alzheimer's disease, the myriad of cognitive alterations occur against a backdrop of substantial cortical degeneration (Coyle et al., 1983; Brandt et al., 1988; Butters et al., 1988). Accordingly, the mnemonic impairments observed in HD are referred to as a subcortical dementia. Taken together, the HD patient exhibits a selective disruption of learning and memory processes that are associated with a relative sparing of other cognitive functions (Scholz and Berlemann, 1987; Brandt et al., 1988; Heindel et al., 1988; Saint-Cyr et al., 1988; Collewijn et al., 1988).

From the time of onset, the motor and mental symptoms of HD progress at an increasingly disabling rate. Initially, the individual may exhibit subtle shifts in personality or cognition that occur together with a mild chorea. As the neuronal degeneration of the disease progresses, the afflicted individual undergoes more intense and abrupt personality changes combined with substantial cognitive impairments. As the mental features of HD change over time, the motor effects change from a choreic dyskinesia to a more disabling dystonic and parkinsonian-like syndrome (Shoulson, 1986).

The behavioral and neural pathology of HD is ultimately the result of the inheritance of an autosomal dominant gene with full penetrance. Accordingly, a complete understanding of the disorder can only be acheived by unraveling the nature of the genetic defect in HD. In the last 20 years, the chromosomal location of the genes responsible for a variety of neurological disorders has been discovered. For instance, the defects associated with retinoblastoma, mitochondrial myopathy, and familial amyloidotic polyneuropathy have been largely characterized. Furthermore, we are gaining substantial insight into the genetic nature of such disorders as Alzheimer's disease and Duchenne muscular dystrophy (Martin, 1989).

Efforts to map the chromosomal location of HD intensified in the early 1980s with the discovery of a large concentration of HD patients in Venezuela. A large pedigree containing approx 5000 members with several hundred members afflicted with HD and an additional 1500 at risk was collected near Lake Maracaibo. Initial studies using restriction fragment length polymorphisms (RFLP) and linkage analysis in blood samples from the

Venezuelian family and families located in the United States were successful in isolating a polymorphic marker that cosegregated with the HD allele (Gusella et al., 1983). The probe, referred to as G8, was assigned to location D4S10 of the human genome and lies on the short arm of human chromosome 4. Although the determination of the chromsome of the HD gene limits the location of the gene itself, the precise location of the HD gene remains unknown. New markers, D4S43, D5462, and D4S95, have recently been demonstrated to be closer, but still proximal, to the HD gene (Hayden et al., 1988; Smith et al., 1988; Pohl et al., 1988; Robbins et al., 1989; Skraastad et al., 1989). Currently, the HD gene is estimated to be approx 1–1.5 $\times 10^6$ base pairs from the teleomere. In addition to the difficulties encountered in determining the exact location of the HD gene, there has been little success in finding any cytogenetic abnormality in the D4S10 region (Martin, 1989).

In part, the difficulties encountered in locating the HD gene 10 are reflections of the limitations of gene mapping techniques. By having a large number of polymorphic markers near the HD gene, the search becomes narrowed and the resolution of genetic map greater. However, the precision of the map is necessarily limited because the frequency of recombination decreases as the markers used become closer to the actual teleomere (Gilliam et al., 1988). Therefore, as investigators close in on the location of the gene itself, the search becomes increasingly more difficult. A second problem concerns the recombination through which the HD gene travels. A determination of the location of a particular marker may reveal that a recombination has occurred, but the appearance of actual clinical symptoms would be required to determine how the HD allele has traveled. Accordingly, the failure of an individual to express the HD phenotype could reflect either the absence of the HD allele or a late onset of the disease. Given these considerations, investigators are currently turning to more precise techniques, including pulsed field gel analysis, chromosome jumping, and cloning techniques to gain efficiency in the search for the HD gene.

Presymptomatic testing of HD is now possible and has met with some success (Martin, 1989). However, the location of the HD gene will allow presymptomatic or prenatal testing without the need of detailed family pedigrees. This would not only have the practical considerations of reduced cost and effort to obtain family information, but would increase the accuracy of the test by bypassing the issue of recombination. With the availability of presymptomatic testing come serious moral and ethical considerations. Individuals at risk for HD would have to weigh the possibility of knowing that they are risk free with the possibility of accelerating the suffering that would normally begin with the expression of the disease.

The availability of accurate presymptomatic testing may be especially important when designing possible interventive treatment strategies. Certainly HD is a genetic disorder, and treatment that fails to correct this fundamental defect will not cure the disease. However, it is crucial to develop primary or adjunctive treatment stategies that may minimize or reverse the pathological and behavioral consequences of the disorder. This is true not only for those already affected with HD, but all of those at risk prior to the development of genetic techniques that may prevent the disease from occurring. Vital to the development and evaluation of possible treatments for HD are appropriate animal models of the disease. With cooperation between clinicians and basic researchers, animal models may facilitate the development and suitability of different prophylactic treatments, such as the blockade of NMDA receptors or use of fetal neuronal tissue in HD.

3. Anatomy and Neurochemistry of the Basal Ganglia

3.1. Organization of the Basal Ganglia

The basal ganglia consists primarily of the caudate, putamen, and globus pallidus. The subthalamic nucleus, substantia nigra, nucleus accumbens, and amygdala have also been considered part of the basal ganglia. Although these structures are not universally considered components of the basal ganglia, they provide important anatomical links, and likely play a vital role in its functional output under normal conditions and in disease states, such as HD.

The anatomical circuitry of the basal ganglia is complex and only briefly outlined here. The caudate and putamen (striatum) are fused at the anterior segment, and contain cells that project to both the internal and external components of the globus pallidus as well as the pars reticulata of the substantia nigra. The globus pallidus, which lies medial to the putamen and lateral to the internal capsule, and substantia nigra have diverse connections with the subthalamic nucleus. The subthalamic nucleus receives afferent connections from the external segment of the globus pallidus. In turn, the subthalamic nucleus projects to both segments of the globus pallidus and to the pars reticulata of the substantia nigra. Because the basal ganglia does not have direct afferent or efferent connections with the spinal cord, its influence on motor processes is indirect. Instead, the primary input to the basal ganglia arises in disperse neocortical regions, and the primary output is directed to the prefrontal and premotor area of the frontal cortex. Accordingly, the frontal cortex plays a substantial role in mediating the motor functions of the basal ganglia.

The striatum consists of at least six different cell types that are distinguishable on the basis of their size, dendritic spines and arborizations, and axonal trajectories (Table 1). The four types of projection neurons are similar in that they all use GABA as their neurotransmitter, but they differ markedly in their peptide content as well as the region to which they project. The projection neurons make up approx 90% of all striatal cells. In contrast to the projection neurons, the smaller striatal interneurons are less prevalent. The larger aspiny interneurons contain acetylcholine (ACh), whereas the small aspiny neurons are identifiable by the presence of somatostatin and neuropeptide Y. Table 1 summarizes the various types of striatal neurons.

Recent anatomical work also suggests that the striatum is inhomogeneous and consists of islands of cells called striosomes, which stain weakly for acetylcholinesterase and are interspersed by a stronger staining matrix (Graybiel, 1983; Graybiel et al., 1981, 1989; Joyce et al., 1986; Rhodes et al., 1987; Gerfen et al., 1987). Striosome and matrix neurons may also be differentiated on the basis of their innervation. The striosomes receive afferent input

Table 1
Classification of Striatal Neurons

Morphology	Transmitter/ peptide*	Projection/ interneuron	Effect in Huntington's disease
Medium spiny	GABA, sub. P, dyn.	Globus pallidus interna	Decreased
Medium spiny	GABA, sub. P, dyn.	Sub. nigra pars reticulata	Decreased
Medium spiny	GABA, dyn., enk.	Globus pallidus externa	Decreased
Medium spiny	GABA	Sub. nigra pars compacta	Decreased
Large aspiny	Acetylcholine	Interneuron	Spared
Small aspiny	Somat., neuropep. Y	Interneuron	Spared

*Sub. P = substance P; dyn. = dynorphin; enk. = enkephalin; somat. = somatostatin; neuropep. Y = neuropeptide Y.

from the medial substantia nigra and medial frontal cortex, whereas the matrix neurons receive input from the substantia nigra and ventral tegmental area as well as the motor, sensory, and association cortices. In addition to their differential staining for acetylcholinesterase, the striosome and matrix neurons contain different neurotransmitters. Both cell types utilize GABA, but the matrix neurons also contain somatostatin and neuropeptide Y, whereas the striosome neurons contain dynorphin and neurotensin (Graybiel, 1983).

3.2. Afferent Connections of the Striatum

All of the afferent inputs to the basal ganglia terminate in the striatum. The largest of these connections, the corticostriate projection, appears to use glutamate as its major neurotransmitter (Divac et al., 1977; Reubi and Cuenod, 1979; Walker, 1983). This system arises in all cortical areas and innervates the majority of the striatum, exerting an excitatory influence over the numerous small spiny interneurons. Alterations in these glutamate afferents have been proposed as an etiological factor underlying the histopathological and neurochemical alterations observed in HD. This hypothesis is based in large part on recent demonstrations of intrastriatal injections of excitotoxins, such as kainic or quinolinic acid, producing morphological, biochemical, and behavioral alterations reminiscent of HD. This will be discussed below.

Other inputs to the striatum include:

1. A dense dopaminergic projection from the substantia nigra (Fibiger et al., 1972; McGeer et al., 1987);
2. The thalamostriatal tract, which has been suggested to use either ACh, glutamate, or aspartate as its major neurotransmitter (McGeer et al., 1987);
3. A cholecystokinin (CCK) projection from the pyriform cortex and amygdala (Meyer et al., 1982);
4. Small projections from both the raphe nucleus (serotonin) (Chan-Palay, 1977; Steinbusch, 1984) and the locus ceruleus (norepinephrine) (Moore and Card, 1984); and
5. The GABAergic pallidostriatal tract (McGeer et al., 1987).

3.3. Efferent Connections of the Striatum

The efferent connections of the striatum are relatively simple compared to its afferent input. There is substantial evidence for an enkephalin projection from the striatum to the globus pallidus (Brann and Emson, 1980; Del Fiacco et al., 1982; Yang et al., 1983), as well as a substance P projection to the globus pallidus and substantia nigra (Jessel et al., 1978; Staines et al., 1980). Striatal efferents also terminate in the substantia nigra, using GABA, dynorphin, and substance P as their neurotransmitters (Vincent et al., 1982; Araki et al., 1985). Striatal inputs to the globus pallidus have also been reported to use GABA and dynorphin (McGeer et al., 1987).

The globus pallidus (pars interna) gives rise to the major efferents from the basal ganglia, and projects to the thalamus via the ansa lenticularis and lenticular fasciculus. The thalamic nuclei that receive these inputs also receive convergent efferents from the cerebellum that, in turn, project back to the prefrontal cortex. Therefore, the basal ganglia receives an enormous cortical and thalamic input and, in turn, projects back to these same regions. Because there are no nuclei in the basal ganglia in a position to exert a direct motor influence on the spinal cord, the basal ganglia modulates the activity of the cortex and thalamus. In this way, the various structures of the basal ganglia act as integrators and modulators of information to and from the cortex as well as from the cerebellum.

4. Pathology of Huntington's Disease

4.1. Anatomical Pathology

Classically, HD is associated with a gross generalized atrophy of the cortex and basal ganglia affecting both the gray and white matter (Dulap, 1927). Histologically, this is accompanied by an extensive gliotic reaction, and loss of small neurons in both the striatum and in layers 3, 5, and 6 of the frontal and parietal cortices. The severe damage to the neostriatum results in a compensatory, secondary hydrocephalus with a gross dilation of the ventricular system (Dulap, 1927; Barr et al., 1978; Bruyn et al., 1979; Lange, 1981; Vonsattel et al., 1985).

The severe atrophy of the striatum is uniformly exhibited among Huntington's patients and averages approximately a 60–90% decrease in mass occurring in cases of juvenile onset. Histopathologically, these decreases in striatal volume are related to a severe loss of the medium-sized spiny projection neurons, the major output neurons of the striatum (Reiner et al., 1988). Neurochemically, the loss of these neurons is associated with decreases in GABA, substance P, dynorphin, and enkephalin (Bird, 1980; Buck et al., 1981; Ferranti et al., 1987a). On the other hand, local circuit aspiny neurons reactive for NADPH-diaphorase and somatostatin are relatively spared (Beal et al., 1988; 1989). The large aspiny neurons, which are likely cholinergic, although spared in the early stages of the disorder, may exhibit degenerative changes as the disease progresses (Ferranti et al., 1987a; Roberts and DiFiglia, 1989) (*see* Table 1).

The globus pallidus also exhibits a marked atrophy (typically 50%), which is owing to the loss of strio-pallidal fibers as well as a substantial decrease in the number of pallidal neuronal perikarya (Lange et al., 1976). Given that the globus pallidus is a major output from the caudate and putamen, the atrophy of this nucleus is not unexpected and likely reflects a transneuronal degeneration subsequent to striatal degeneration.

The sulcal widening and loss of neuronal weight in the cerebral cortex indicate that the cortex also undergoes a substantial deterioration (Forno and Jose, 1973). However, unlike the relatively ubiquitous cell loss observed in the striatum, the degenerative process appears to be more restricted in the cortex with cell loss occurring in the third, fifth, and sixth cortical layers. Interestingly, studies using retrograde tracing techniques indicate that the small pyramidal cells of these deep cortical layers are the source of cortical innervation to the striatum (Oka, 1980). Together with the observation by Cross et al. (1986) that high-affinity glutamate uptake is reduced in the striatum of HD patients, it appears that HD is associated with severe striatal atrophy concomitant with degeneration of both its major inputs and outputs.

Although HD has been generally considered to be a neurodegenerative disorder characterized by the restricted loss of certain populations of neurons, it is becoming clear that the neural degeneration associated with HD is quite widespread (*see* Table 2). The pathological changes in neural regions removed from the basal ganglia are, however, typically not as severe. For example, the pars reticulata of the substantia nigra (Gebbink, 1968; Oyanagi and Ikuta, 1987; Oyanagi et al., 1989), thalamic nuclei (Forno and Jose, 1973; Dom et al., 1976), subthalamic nucleus (Lange et al., 1976), cerebellum (Jervis, 1963; Bruyn, 1968; Castaigne et al., 1976; Rodda, 1981), hypothalamus (Bruyn, 1968, 1973), hippocampus (Forno and Jose, 1973; Mattson et al., 1974; McIntosh et al., 1978; Averback, 1980), and a variety of brainstem regions, including the superior olive and red nucleus (Roizin et al., 1976), have all been shown to exhibit reactive gliosis and evidence of neural degeneration. On the other hand, the nucleus accumbens and pars compacta of the substantia nigra appear relatively normal (Bots and Bruyn, 1981). It appears then that the neuropathology of HD is not restricted to the striatum, but rather affects a variety of neural regions. A more detailed understanding of the functional interactions between the basal ganglia and these regions may provide knowledge regard-

Table 2
Regional Pathology in Huntington's Disease Brain*

Region	Pathology
Gross brain	Weight and size reduced
Ventricular system	Dilation
Caudate and putamen	Volume reduced, loss of neurons, gliosis
Globus pallidus and claustrum	Volume reduced, loss of neurons, gliosis
Nucleus accumbens	Slight loss of neurons and gliosis, identation of neuronal elements
Thalamic nuclei	
Centromedian	Loss of neurons and gliosis
Dorsomedial	Slight gliosis
Ventral anterior	Loss of neurons
Ventral lateral	Loss of neurons
Hypothalmic nuclei	
Lateral	Loss of neurons
Ventromedial	Shrinkage and loss of neurons
Paraventricular	Loss of neurons
Supraoptic	Loss of neurons
Cortex	Atrophy, loss of neurons, gliosis
Hippocampus	Atrophy, gliosis, neurofibrillary tangles
Cerebellum	
Dentate nucleus	Loss of neurons and gliosis
Purkinje layer	Loss of neurons and gliosis
Inferior olive	Atrophy, loss of neurons and gliosis
Subthalamic nucleus	Loss of neurons and gliosis
Substantia nigra pars reticulata	Atrophy, loss of neurons and gliosis
Brainstem nuclei	
Superior olive	Loss of neurons
Occulomotor	Gliosis
Midbrain tegmentum	Gliosis
Paramedian reticular	Loss of neurons
Dorsal vagus	Gliosis
Aquaduct area	Gliosis
Red nucleus	Loss of neurons
Spinal cord	Gliosis

*Modified from Sanberg and Coyle, 1984.

ing the relationship of the pathological changes and the myriad of motor and cognitive abnormalities associated with HD.

4.2. Neurochemical Pathology

The discovery of dopamine in the striatum and the depletion of this transmitter in Parkinson's disease coupled with the suggestion that the symptomologies of Parkinson's and Huntington's disease are at opposites ends of a behavioral continuum suggested that increased levels of dopamine may underlie the motor abnormalities in HD. Unfortunately, attempts to determine whether dopaminergic parameters are altered in HD have met with difficulty. Initial studies suggested either little change or a slight elevation of dopamine levels in striatum, substantia nigra, and nucleus accumbens (Bernheimer et al., 1973; Chase, 1973; Bird and Iversen, 1974; McGeer and McGeer, 1976a; Mann et al., 1980; Bird, 1980; Spokes, 1980). It was suggested that this was the result of the generalized atrophy of the striatum, which resulted in an apparent preservation of dopaminergic neurons. However, it was quickly pointed out that this explanation was probably insufficient and that other factors, such as the loss of GABA feedback fibers from the striatum to substantia nigra, may contribute to the apparent preservation of dopamine. Interestingly, Kish et al. (1987) recently examined dopamine levels in the regionally subdivided striatum of Huntington's patients. Both dopamine and HVA were found to be significantly reduced in the caudal caudate. It is therefore conceivable that subtle alterations in striatal dopaminergic parameters are masked by the tendency of investigators to examine neurochemical changes in the striatum as a whole. Taken together, it appears that striatal dopaminergic indices may not be dramatically altered, but compared to the marked loss of other neurochemical parameters, the net effect may be a relative overactivity of striatal dopamine.

Because dopaminergic antagonists have been used clinically to ameliorate chorea, it has been suggested that dopamine receptors are supersensitive. However, radioligand studies indicated that the number of striatal dopaminergic receptors are actually reduced in HD (Reisine et al., 1977). The reduction in the number of dopamine binding sites could reflect either a compensatory action in response to elevated levels of dopamine in the striatum or the loss of receptors located on dying striatal neurons. Alternatively, a reduced inhibition of dopaminergic transmission as a result of degenerating GABA systems may partially account for the altered pharmacological sensitivity. Melamed et al. (1982), however, reported that the nigrostriatal pathway adequately adapts to the loss of striatal neurons, which normally inhibit dopamine activity, both in HD and kainate-lesioned rats. It therefore remains unclear why dopaminergic compounds exert such powerful effects.

The GABAergic neurons of the basal ganglia undergo a severe deterioration as indexed by measures of transmitter levels and enzymatic activity. Perry and colleagues (1973) reported that levels of GABA are reduced in the substantia nigra, caudate, putamen, and globus pallidus in HD. Since these studies, decreases in glutamate decarboxylase have also been found in the choreic striatum, further suggesting decreases in striatal GABA concentrations. There are also decreases in glutamate decarboxylase in the substantia nigra and globus pallidus, likely owing to degeneration of striatonigral and striatopallidal neurons (McGeer et al., 1973; Bird and Iversen, 1974; Schwarcz et al., 1977). Paralleling the decreases in striatal GABA and glutamate decarboxylase are decreases in GABAergic receptors.

The striatal content of ACh, as measured by the activity of the synthetic enzyme choline acetyltransferase (ChAT), is also greatly reduced in HD (McGeer et al. 1973; Bird and Iversen, 1974; Stahl and Swanson, 1974; Aquidonius et al., 1975; McGeer and McGeer, 1976a; Spokes, 1980), as are the number of muscarinic receptors in the striatum (Hiley and Bird, 1974; Enna et al., 1976a,b; Wastek et al., 1976; Wastek and Yamamura, 1978). It is unclear, however, whether the decreases in ChAT represent the actual degeneration and death of striatal cholinergic neurons in HD. Ferranti et al. (1987a,b) recently reported that, although ChAT activity was decreased in striatum of Huntington's patients, there was a relative preservation of AChE-reactive cells. Because ChAT and AChE colocalize within cholinergic neurons, it was suggested that AChE containing cell bodies are preserved, but ChAT-rich cholinergic terminals degenerate in the disease. The decreases in cholinergic parameters may also be the consequence of alterations in nearby dopaminergic neurons. Since the action of dopamine on striatal cholinergic neurons is likely inhibitory, it may be that a general overactivity of dopaminergic neurons decreases cholinergic indices. Sanberg and Creese (1981) reported that a single injection of the dopaminergic antagonist bromocriptine produced a persistent decrease in striatal ChAT activity. If a similar situation occurred in the Huntington's brain, i.e., dopamine overactivity and decreased ChAT activity, the decreases in cholinergic indices may not have reflected the actual death of those neurons, but rather a dopamine-mediated decrease in activity.

The cortical innervation of the striatum uses glutamate as its neurotransmitter. It has recently been proposed that an endogenous excitotoxic agent, such as glutamate, may play a role in the selective neuronal degeneration observed in HD. If true, then one might expect to see changes in glutamate parameters in individuals afflicted with the disease. Initial studies by Gray et al. (1980) indicated that cultured fibroblasts from HD patients were significantly more susceptible to glutamate than fibroblasts from age-matched controls. Carter (1982) reported that the activity of glutamine synthetase (the enzyme that converts glutamate to glutamine) was reduced 20–30% in the frontal cortex and putamen of HD patients. Similarly, Wong et al. (1982) reported that ornithine aminotransferase, which converts ornithine to glutamate, is decreased in the cortex and striatum of HD patients. More recently, Cross et al. (1986) reported marked reductions in high-affinity glutamate uptake sites in the striatum of HD patients. It is conceivable that an impaired uptake system would locally increase glutamate until it reached toxic levels. Interestingly, McBean and Roberts (1985) demonstrated that disruption of glutamate uptake following damage to the corticostriatal tract rendered striatal cells more vulnerable to a subsequent injection of glutamate. Finally, alterations in glutamate receptors have been observed in HD. Greenamyre et al. (1985) reported decreases in striatal glutamate receptors in the HD striatum. Likewise, Young et al. (1988) reported dramatic decreases of 93% in the number of NMDA receptors in the putamen of relatively young patients. These observations are important in that they suggest that alterations in receptors for an endogenous excitotoxic agent, perhaps glutamate, occur early in the disease state and may play a role in the progression of neural degeneration.

A variety of other neurochemical alterations are prevalent in HD (*see* Bird, 1980 for a review). For instance, the undecapeptide substance P was one of the first peptides examined in HD. The basal ganglia contains one of the highest levels of substance P of any brain region where it appears to exert an excitatory influence over striatal afferents projecting to the globus pallidus and substantia nigra (Mroz et al., 1977). Substance P is decreased by as much as 90% in the pallidum and substantia nigra (Kanazawa et al., 1979; Buck et al., 1981). On the other hand, striatal levels of substance P have been reported to be either normal or slightly decreased (Gayle et al., 1978; Beal et al., 1989). Likewise, substance P appears to be unaffected in the cortex, thalamus, and hypothalamus.

Cholecystokinin (CCK) levels have also been evaluated in HD. CCK was once thought to be predominantly a gastrointestinal peptide, but was subsequently found to be present in high concentrations in the central nervous system, especially the cortex and striatum (Innis et al., 1979; Innis and Snyder, 1980). An initial study by Emson et al. (1980) revealed that CCK immunoreactivity was decreased by approx 50% in the globus pallidus and substantia nigra. A subsequent study by Hays et al. (1981) demonstrated that the number of CCK receptors is decreased in the basal ganglia and cerebral cortex of Huntington's patients. Moreover, it was suggested that decreases in CCK profiles may be unique to HD.

Enkephalin is an endogenous neurotransmitter that is found in particularly high concentrations in the globus pallidus. The cell bodies of origin for these enkephalinergic processes are localized within the striatum. Arregui et al. (1979) conducted a detailed analysis of enkephalin levels in human brain, and observed that the lateral pallidum and nucleus accumbens contained the highest levels of any region examined. Although the levels in the medial pallidum and substantia nigra were approx 50% lower, they still were higher than other brain regions examined. Arregui et al. (1979) further demonstrated that enkephalin levels were decreased by as much as 50% in the medial and lateral pallidum as well as the substantia nigra. Enkephalin levels were not altered in either the caudate or putamen.

Angiotensin II is an octapeptide that has been reported to be decreased in HD (Arregui et al., 1977,1978). The activity of angiotensin II converting enzyme is decreased by as much as 60–90% in the caudate, putamen, globus pallidus, and substantia nigra of HD brains. In contrast, several neurotransmitters and neuropeptides, including norepinephrine, somatostatin, neurotensin, neuropeptide Y, and thyrotropin-releasing hormone, are present at normal or slightly elevated levels in HD.

Together, these data indicate that every neuronal system known to originate in the striatum exhibits decrements in the concentration of its presynaptic neuronal markers in HD. Furthermore, when the degree of striatal atrophy is taken into account, the total deficits of the markers for these neuronal systems support a profound loss consistent with the known cellular pathology. The specificity of these synaptic neurochemical changes, however, has been confirmed by the demonstration that the spared dopaminergic afferent system does not show comparable decrements in either the substantia nigra or the striatum. Additionally, markers for other neuronal systems, such as serotonin and somatostatin, that project to the striatum from other brain areas are not decreased (but *see* Kish et al., 1987). These findings clearly demonstrate that the neuronal vulnerability of HD is not specified by any one striatal neurotransmitter, but affects virtually all of the identified intrinsic and efferent systems or the corpus striatum.

5. Animal Models of Huntington's Disease

An important approach for understanding neurological diseases is to develop appropriate animal models. Such models help to elucidate the biological and behavioral manifestations of clinical disorders, and suggest novel therapeutic strategies for their prevention or treatment. Although these models may.not reproduce the complex etiologies, pathophysiology, or behavioral abnormalities associated with disease of the nervous system, they do provide a practical way to explore specific questions concerning structure and function. Animal models of HD based on studies using rodents have substantially escalated our understanding of HD and its underlying pathology. Because the anatomy and neurochemistry of the rodent and human striatum share considerable homology, the rodent brain is well suited for studies aimed at determining the behavioral consequences of select alterations in striatal function. Most revelant to the present discussion, pharmacological manipulations or lesions of the striatum produce marked and easily quantifiable alterations in motor and cognitive function. Although these behavioral abnormalities may not exactly reproduce those seen in HD, they do provide a functional measure of the integrity of the striatum and a model for the evaluation of treatments to remediate those behavioral effects. Some of the models of HD currently used (i.e., excitotoxic models) appear to mimic closely both the neural pathology and behavioral alterations that are characteristic of HD. With these models, there has been considerable progress made in developing treatments that alleviate or prevent the pathological and behavioral consequences of striatal damage. Pharmacological treatments have proven useful in preventing the excitotoxicity produced following injections of compounds, such as quinolinic acid, into the striatum and hippocampus. In addition, the transplantation of fetal tissue into excitoxically lesioned animals has been shown to promote neurochemical and functional recovery. Rodent models may also help to determine some of the etiological factors in HD. Although animal models may not permit the determination of the genetic lesion in HD, they can help unravel the cascade of events subsequent to the genetic aberration that contributes to the ongoing neural degeneration in HD. For example, rodent models have been especially useful for examining the potential contribution of excitotoxins to the select neural degeneration in HD. It appears then that animal models of HD and other neurological disorders may be useful for examining the anatomical, neurochemical, and behavioral alterations of the disease. In addition, these models provide new avenues for developing and testing various treatments for HD.

5.1. Dyskinesia Models Based on Striatal Neurotransmitter Imbalances

A variety of avenues have been explored to develop an animal model of HD. Although models focusing on hyperkinesias in animals resulting from genetic conditions (Koestner, 1973) and various types of brain lesions (Klawans and Rubovits, 1974; Neill et al., 1974) have been used, the majority of models have centered around motor abnormalities produced by short-term stimulation of some of the neurochemical changes found in HD. These models are briefly outlined below.

5.1.1. Dopamine

Dopamine has a well-established role as a mediator of motor function in the striatum. Moreover, one of the neurochemical hallmarks of HD is normal or increased levels of dopamine relative to other transmitter systems in the striatum (Bernheimer et al., 1973; Bird and Iversen, 1974; McGeer and McGeer, 1976a; Mann et al., 1980). If this relative overactivity contributes to the behavioral pathology observed in the disease, then pharmacological alterations of striatal dopamine systems might be one way of evaluating the contribution of dopamine to these behavioral alterations. Costall and Naylor (1975) injected dopamine directly into the striatum of rats, and noted a pronounced increase in gnawing behavior and biting together with an abnormal protrusion of the tongue, hyperactivity, acute twisting of the neck and head, whole body rocking, and abrupt head and neck jerks. The onset of these behaviors was approx 1–2 h and subsided within 6 h post-injection. They further reported that the hyperactivity was reversed by the administration of the dopaminergic antagonists haloperidol or fluphenazine, whereas the remaining abnormalities were reversed by pimozide and oxiperomide administration. Cools and Van Rossum (1976) reported similar behavioral alterations following intrastriatal injections of amphetamine and apomorphine. In addition, Dill et al. (1976)

reported that injections of the dopamine metabolite 3-methoxy-tyramine (3-MT) into the striatum of both rats and squirrel monkeys produced a "chorea-like" syndrome consisting of forelimb and hindlimb movements, chewing, grimacing, neck tremor and torsions, and in some cases, convulsions. The specificity of these dopaminergic effects was further demonstrated by a neuroleptic-induced reversal of these movement abnormalities.

5.1.2. Acetylcholine

There are marked decreases in ChAT activity and the number of muscarinic receptors in the striatum of HD patients. Although it is unclear whether the intrinsic cholinergic neurons of the striatum actually die in the disease, cholinergic parameters appear reduced relative to other neurotransmitters, such as dopamine and serotonin. Accordingly, striatal injections of cholinergic antagonists may provide a useful model for the behavioral alterations that result from this transmitter imbalance in HD. McKenzie and Viik (1975) reported that intrastriatal injections of alcuronium elicited a pronounced dyskinetic syndrome consisting of bursts of tremor in the contralateral forelimb, tremors of facial muscles, rearing, and chewing. Similarly, D-tubocurarine, another cholinergic antagonist, produced a contralateral rotation of the head, facial grimacing, and teeth chattering, which was blocked by the concurrent administration of neostigmine.

Unfortunately, injections of cholinergic agonists appear to produce essentially the same behavioral disturbances. Intrastriatal injections of carbachol or acetylcholine produced tremor of the limbs and facial muscles, as well as chewing and salivation (Dill et al., 1968; McKenzie et al., 1972; Murphy and Dill, 1972; McKenzie and Viik, 1975). Given these considerations, the utility of dyskinesia models based on alterations of cholinergic function that mimic those observed in HD is unclear.

5.1.3. GABA

GABA levels have been consistently shown to be decreased in HD, indicating a marked degeneration of GABAergic neurons. To examine the contribution of the GABAergic system to the chorea associated with HD, Standefer and Dill (1978) evaluated the dyskinesia produced by a series of GABAergic antagonists, including picrotoxin and bicuculline. Although both elicited similar responses picrotoxin produced tremor, twitching, chewing, and convulsions at much lower doses than bicuculline. All indices of dyskinesia were blocked by the administration of GABA. McKenzie and Viik (1975) reported that the dyskinesia produced by D-tubocurarine is blocked by administration of glutamic acid. Furthermore, the effects of glutamic acid were potentiated by the cojoint administration of the glutamic acid decarboxylase inhibitor allyglycine, suggesting that glutamate may play a role in the dyskinesia produced by GABA manipulations.

5.1.4. Other

Rezek and colleagues (1977) reported that striatal injections of somatostatin produced a generalized behavioral excitation, stereotyped movements, tremors, reductions of rapid eye movements, and decreased slow-wave sleep when doses ranging from .01–.1 µg were injected. Higher doses (1–10 µg) produced lack of coordination and severe contralateral motor abnormalities, including hemiplegia and rigidity so severe that the animals often would lie motionless.

Intrastriatal injections of mescaline result in motor disturbances characterized by choreiform movements, tremor, and generalized convulsions (Dill, 1972). Because isoproterenol, dopamine, propranolol, and haloperidol blocked the appearance of these effects, it was suggested that mescaline produces striatally mediated locomotor abnormalities by disrupting the balance between cholinergic and catecholaminergic transmission.

Electrical stimulation of the striatum has been reported to produce a contralateral stereotypy similar to that following injection of dopamine. Since this effect was potentiated by administration of serotonin, it was suggested that alterations in serotonergic transmission may play a modulatory role in striatally mediated dyskinesia (Cools and Van Rossum, 1976).

Recently Toth and Lajtha (1989) reported a detailed account of the motor effects of intrastriatal injections of excitatory amino acids. Injections of L-glutamate, L-aspartate, NMDA, quisqualate, and kainate all produced varying degrees of choreiform movements of the head, forelimb, and trunk, as well as barrel rolling, masticatory movements, wet dog shakes, and salivation. Interestingly, the motor effects of L-glutamate were blocked by the administration of L-glutamic acid diethyl ester, but not by haloperidol, GABA, or glycine. These results indicated that the motor effects of L-glutamate were mediated by activation of glutaminergic receptors, and were not related to alterations in the dopaminergic or GABAergic systems. These results are of considerable interest given that glutamate analogs produce alterations in striatal neurochemical and morphological indices that resemble those observed in HD *(see below)*.

There are several inherent problems with pharmacological models of HD or chorea-like syndromes. Many of the compounds used in these studies have limited CNS specificity, and affect more than one transmitter or receptor type. In addition, certain compounds exhibit a high degree of behavioral toxicity. For instance, although manipulations of the GABAergic system produce locomotor abnormalities, they are often accompanied by seizures. Moreover, pharmacological manipulations of striatal transmission have a short duration of action and cannot reproduce the chronicity of the pathology in HD. Finally, the dyskinesia produced in these models often bears little homology to the motor impairments observed in HD. This is not to say that the behavioral alterations observed following these treatments are of no relationship to those in HD. In fact, it is unlikely that any rodent model of HD would be able to reproduce with precision the chorea found in HD. Considering that rats are quadrapeds and humans are bipeds, it is not surprising that chorea as such does not develop in these models. However, although pharmacological manipulations of striatal function do not reproduce the chorea of HD, the resulting locomotor abnormalities may be generally considered to resemble those in the disease.

Given the above considerations, it is difficult to establish causal or even suggestive relationships between drug-induced neural changes and specific behavioral alterations. Therefore, investigators have begun to focus on models of HD based on the relatively selective toxicity of excitotoxic compounds, such as kainic and quinolinic acid.

6. Excitotoxin Models of Huntington's Disease

Some investigators have attempted to mimic the symptomology of HD by mechanically or electrolytically lesioning various brain areas, such as the striatum (Neill et al., 1974). However, this remains an unsatisfactory model in that lesion techniques almost invariably damage supportive structures as well as fibers that pass through and terminate in the damaged area. A better strategy for investigating the relationship between striatal damage and locomotor abnormalities, such as those observed in HD, might involve the use of selective cytotoxic compounds.

Selective toxic compounds have been widely used in neurobiology to examine the functional properties of the nervous system. Toxins have been used to examine the function and molecular biology of ion channels, axoplasmic transport processes, neurotransmitter systems, and the principles of synaptic transmission (McGeer et al., 1987). In a related context, neurotoxins have been successfully used to examine the covariation between altered neurotransmitter dynamics and behavior, and to develop animal models of neurological disorders.

Glutamate is one of the major excitatory neurotransmitters found in the CNS. It can act, however, as a potent neurotoxin. This finding has prompted an interest in the possible endogenous role of excitotoxicity in the neuronal death found in human neurodegenerative disorders (Olney and de Gubareff, 1978). Given that the corticostriate projections utilize glutamate (Fonnum et al., 1981), a number of attempts have been made to develop animal models of HD based on the relatively specific cytotoxic effects of excitotoxic compounds. These compounds include structural analogs of glutamate, such as kainic acid (KA) and ibotenic acid (IA), and the endogenous tryptophan metabolite quinolinic acid (QA). When injected into the brains of rats, in extremely small doses, these compounds produce a marked and locally restricted toxic effect while sparing axons of passage and afferent nerve terminals. The behavioral, neurochemical, and anatomical consequences of excitotoxicity resemble those observed in HD and have led to the speculation that an aberrant overproduction or breakdown of endogenous excitotoxic compounds is an etiological factor in HD.

6.1. The Kainic Acid Model

6.1.1. Neuropathology

Intrastriatal injections of nanomolar quantities of KA result in the degeneration of local, intrinsic neurons while sparing axons of passage and terminals of extrinsic origin. The consequent neurochemical and pathological alterations bear a considerable analogy to that described in the striatum of HD patients upon postmortem examination (Coyle and Schwarcz, 1976; McGeer and McGeer, 1976b; Coyle et al., 1977,1978; Zaczek et al., 1978; Fields et al., 1978; Coyle, 1979; McGeer et al., 1979; McGeer and McGeer, 1982). The neurochemical alterations induced by KA injections include decreases in GABA, ACh, and angiotensin levels, as well as decreases in receptors for GABA, serotonin, and dopamine. In contrast, striatal levels of dopamine and serotonin have been reported to be unaffected. These biochemical changes are accompanied by reductions in the number of intrinsic striatal neurons, ventricular dilatation, and decreases in striatal volume (*see* Coyle, 1979 for a review). Secondary to these alterations is a transneuronal degeneration of the globus pallidus and the pars reticulata of the substantia nigra.

Importantly, KA does not selectively spare neurons that are reactive for somatostatin or neuropeptide Y (Table 3). Because this population of neurons seem to be spared in HD, the KA model does not appear to reproduce faithfully the entire constellation of histological changes observed in the disease. The specificity of KA has also been criticized, because it can produce hippocampal damage (Pisa et al., 1980,1981; Sanberg et al., 1979, 1981b). However, this occurs as a result of the epileptic sensitivity of hippocampal neurons and may be prevents by prior treatment with the anticonvulsant diazepam (Sanberg, 1980; Ben-Ari et al., 1980). Importantly, other excitotoxins, such as IA, produce a profile of striatal damage that is similar to that produced by KA without damage to the hippocampus or neighboring struc-

Table 3
Biochemical and Morphological Changes in the Striatum Reported in Huntington's Disease and Rats Intrastriatally Injected with Kainic or Quinolinic Acid

	Huntington's disease	Kainic acid	Quinolinic acid
Neurochemical indices			
GABA	Decreased	Decreased	Decreased
Dopamine	Normal	Normal	Normal
Serotonin	Normal	Normal	Normal
Acetylcholine	Normal/decreased	Decreased	Normal/ decreased
Somatostatin	Normal/increased	Decreased	Normal
NeuropeptideY	Normal/increased	Decreased	Normal
Substance P	Decreased	Decreased	Decreased
Angiotensin	Decreased	Decreased	?
Enkephalin	Decreased	Decreased	?
Receptor binding			
GABA	Decreased	Decreased	?
Muscarinic	Decreased	Decreased	?
Serotonin	Decreased	Decreased	?
Dopamine	Decreased	Decreased	Decreased
Morphology			
Ventricles	Dilated	Dilated	Dilated
Mass	Decreased	Decreased	Decreased
Gliosis	Increased	Increased	Increased

tures. Since the behavioral consequences of IA are also similar to those following KA, it is likely that the effects of KA are the result of striatal and not hippocampal damage. Given these considerations, it appears that intrastriatal injections of KA produce a profile of histological and neurochemical alterations that closely mimic some, but not all, of those observed in patients afflicted with HD.

6.1.2. Behavioral Pathology

The behavioral hallmarks of HD are abnormal, uncontrollable, and constant choreiform movements. Unfortunately, KA-induced chorea is not as recognizable as that observed in

HD. However, the KA-lesioned rat does exhibit an abnormal profile of locomotor activity, including increases in swing time and a decrease in stance time compared to control rats. Thus, the lesioned rats put their paws on the ground for shorter periods of time and swing them longer than normal rats. This has been suggested to be analagous to the locomotor pattern observed in patients with HD (Hruska and Silbergeld, 1979). Another similarity to HD that exists in the motor activity of KA-lesioned rats is that, compared to controls, their locomotor activity is markedly potentiated during arousal (Mason and Fibiger, 1979). HD patients also show greater choreiform movements during their awake period and when aroused (Sanberg and Johnston, 1981; Sanberg and Coyle, 1984). In addition, rats with striatal KA lesions tested in photoactometers have shown an exaggerated motor response to amphetamine, whereas the response to apomorphine did not differ significantly from controls. These animals also exhibit a greater locomotor response to the cholinergic antagonist scopolamine, but a decreased cataleptic response to the dopaminergic antagonist haloperidol, as compared to controls (Sanberg, 1980; Sanberg et al., 1981b). Clinical studies have suggested similar valences in response to comparable drug challenges in HD patients (*see* Sanberg and Johnston, 1981 for a review).

A variety of psychological and psychiatric symptoms occur against the neurobiological backdrop of HD. Although these symptoms have been attributed to cortical atrophy, recent evidence indicates that HD dementia is different from dementias associated solely with cortical abnormalities. Consequently, the term "subcortical dementia" has become synonymous with HD dementia (Coyle et al., 1983; Cummings and Benson, 1988). Accordingly, studies with KA-lesioned rats indicate that marked impairments in learning, memory, and affect occur without concomitant cortical damage. For example, intrastriatal injections of KA result in: impaired acquisition and retention of a passive avoidance and spatial alternation task (Sanberg et al., 1978,1979; Pisa et al., 1980; Dunnett and Iversen, 1981), impaired retention of delayed alternation (Pisa et al., 1980; Dunnett and Iversen, 1981) and short-term memory of reward alternation (Sanberg et al., 1979; Pisa et al., 1981), resistance to extinction of operant learning (Sanberg et al., 1979), and failure to initiate activity spontaneously in a spontaneous alternation task (Pisa et al., 1980). These alterations occur in conjunction with decreases in body weight and regulatory deficits that are symptomatic in HD patients (Sanberg et al., 1981a,1986; Sanberg and Fibiger, 1979).

6.2. The Quinolinic Acid Model

6.2.1. Neuropathology

Quinolinic acid (2,3-pyridine dicarboxylic acid), a metabolite of tryptophan, has attracted a great deal of attention recently because of its powerful excitotoxic properties and wide distribution in both rat and human brain (Schwarcz and Kohler, 1983; Foster et al., 1983; Schwarcz et al., 1983,1984; Schwarcz and Shoulson, 1987). In addition, high concentrations of its catabolic enzyme, quinolinic acid phosphoribosyltransferase (QPRT), and immediate anabolic enzyme, 3-hydroxyanthranilic acid (3HAO), have been detected within the caudate suggesting that it normally serves a role in striatal functioning (Foster et al., 1985; Okuno et al., 1987; Reynolds et al., 1988; Schwarcz et al., 1989).

QA has been reported to exert a more selective degenerative effect in the striatum than KA or IA, which more closely resembles the pathology of HD (Ellison et al., 1987; Beal et al., 1986,1988,1989). Like KA, QA injections cause depletions of GABAergic neurons while relatively sparing cholinergic neurons and axons of extrinsic origin. Beal and colleagues (1986, 1988c,1989) have reported that intrastriatal injections of QA also spare somatostatin- and neuropeptide Y-containing neurons; whereas KA or IA does not spare these neuronal populations. Because this pattern of cell loss closely mimics that observed in HD, it has been suggested that this model most closely reproduces the neuropathology observed in the disease (Table 3). However, there is controversy over whether these cell populations are actually spared following OA. Davies and Roberts (1987, 1988) injected QA into the striatum of rats and found no evidence for a sparing of somatostatin-containing neurons. Likewise, Boegman et al. (1987) reported that QA did not spare neuropeptide

Y-immunoreactive neurons following intrastriatal QA. Finally, Sanberg and colleagues (1990) were not able to find any evidence of a sparing of NADPH-d-containing neurons following QA. In an attempt to reconcile these discrepancies, Beal and colleagues (1989) conducted a detailed analysis of QA's toxicity, and compared its profile of neurotoxicity to that produced by other toxins, including KA, quisqualic acid, NMDA, AMPA, and homocysteic acid. They reported that doses of QA, NMDA, and homocysteic acid that decreased striatal GABA levels by 50% or more produced a sparing of somatostatin-neuropeptide Y levels. On the other hand, such compounds as KA, quisqualic acid, and AMPA, which are not NMDA agonists, produced a more generalized toxicity and did not spare somatostatin-neuropeptide Y concentrations. Although all neuronal profiles were decreased within the primary lesion region, somatostatin and neuropeptide Y neurons were relatively spared within the transition zone. Unfortunately, the region in which the cell counts were conducted was very small and yielded a minimal number of neurons reactive for NADPH-d in both lesioned and control striata (approx six). The absence of a robust population of neurons in the region of analysis coupled with a small number of striata analyzed (4/group) limits the reliability of statistical analysis. This may be especially problematic when evaluating an effect that is relative to accompanying alterations within the same tissue. Since it is generally conceded that all cell populations are lost in the lesion core, the presence or absence of specific populations of neurons might be better evaluated by analyzing the entire transition region. If there is a relative sparing of these neurons in regions removed from the lesion zone, then this region may better represent the Huntington's brain in that, if an endogenous excitotoxin, such as QA, plays a role in the cell death in HD, it might be expected that the levels of QA would be most similar to those observed in the transition zone in the QA animal model.

Experiments using cultured striatal neurons support the selective nature of QA. Koh and Choi (1988) incubated cultured striatal neurons with various doses of either NMDA or QA, and evaluated the possibility that certain populations of striatal neurons would be more resistant than others. Although high doses of NMDA or QA produced a general neurotoxic effect, low to moderate doses exerted a more selective pattern of toxicity. At these doses, only striatal neurons containing NADPH-d or ACh were resistant to NMDA and QA. Similar effects have been reported in cortical as well as striatal neurons (Koh et al., 1986; Whetsell and Schwarcz, 1989).

Roberts and DiFiglia (1989) recently reported that there are progressive pathological changes following QA that may bear a homology to the ongoing deterioration found in HD. Rats received a single injection of QA, and were sacrificed 2, 7, or 30 wk later. Histological analysis revealed that there was a progressive shrinkage of the striatum over time with cell loss maximal in the region of injection and declining within the proximal transition zone. Although large, presumably cholinergic, neurons were relatively spared, there was a dramatic reduction in the number of these neurons between 2–30 wk postlesion. Despite this progressive neuronal decline, those neurons that did survive retained their normal ultrastructural appearance and maintained some synaptic inputs. These time-related changes suggest that there is a prolonged neuronal death following excitotoxic damage and that this pattern of progressive neuronal loss may be similar to that observed in HD.

If QA or a similar endogenous neurotoxin is involved in the pathogenesis of HD, then it might be expected that the neostriatum would be among the structures that are most vulnerable to the excitotoxic effects of QA injections. On the other hand, Schwarcz and Kohler (1983) have demonstrated that the cerebellum, amygdala, substantia nigra, septum, and hypothalamus are less susceptible to QA. Although these structures are affected in HD, they are to a lesser degree than the striatum. Moreover, it appears that neonatal, but not mature, animals are resistant to the toxic effects of QA corresponding to the typical onset of HD occurring in middle age. Together, these features make QA, or a similar endogenous agent, a promising candidate for the underlying etiological factor in HD.

If an abnormal production of QA or similar excitotoxic agents underlies the pathological consequences of HD, then it would be important to demonstrate altered levels of these compounds in the HD brain. Foster et al. (1985) reported a trend toward elevated QPRT in the striatum and substantia nigra of HD brains. Moreover, Schwarcz et al. (1989) reported similar increases in rat striatum following QA, suggesting a possible altered metabolism in response to abnormally high levels of QA in the CNS. In the studies by Schwarcz et al. (1989), IA and ablations of the cortico-striatal pathway produced significant elevations of striatal QPRT and 3-HAO, whereas the local administration of 6-OHDA did not affect QA metabolism. These data suggest that the source of endogenous QA is nonneuronal, and the authors speculated that glial cells may be the striatal source of QA. In fact, several lines of evidence support this notion. For instance, 3-HAO has been detected in astroglial cells (Okuno et al., 1987). In addition, the increases in QPRT and 3-HAO reported by Schwarcz and colleagues (1989) parallel the onset of reactive gliosis following striatal damage. Finally, HD is associated with a marked gliotic reaction that is not dissimilar to that observed in the excitotoxin-lesioned striatum. Together, these data suggest that the initial insult in HD may trigger a reactive gliosis that perpetuates and maintains the ongoing neural degeneration observed in the disease. Although these findings suggest a role for QA as a possible etiologic agent in HD, a more detailed understanding of the synthesis, catabolism, and excretion of endogenous QA is required to comprehend its normal role and pathogenesis in HD. These considerations notwithstanding, the QA model appears to reproduce many of the pathological features of HD faithfully.

6.2.2. Behavioral Pathology

Although histological and neurochemical experiments support the use of QA as a model for HD, there have been few behavioral experiments to examine the suitability of the model. Sanberg and colleagues (1989a) recently examined the locomotor effects of intrastriatal injections of QA. Although injections of 75 nmol of QA did not alter locomotor activity, 150 and 225 nmol produced significant increases in activity at both 2 and 4 wk postsurgery. In addition, these same animals exhibited a transient weight loss following both 150 and 225 nmol of QA. The alterations in locomotor activity and weight were qualitatively similar to those following 3 nmol of KA, but were smaller in overall magnitude.

Calderon et al. (1988) reported that QA lesions abolish dopamine-mediated catalepsy. Bilateral QA (150 and 225 nmol) decreased the cataleptic response to the D_1 receptor antagonist SCH23390 and the D_2 receptor antagonist haloperidol at 14–16 wk following surgery. These data are also similar to those observed following KA, and indicate that the D_1 and D_2 dopaminergic receptors, which mediate the catalepsy response, are sensitive to the neurotoxicity of QA.

Although these data are similar to those obtained following KA and resemble the behavioral pathology observed in HD, they do not adequately permit the evaluation of the behavioral specificity of QA. The existence of no qualitative differences between KA and QA in these experiments may (1) support the lack of neural specificity of QA observed by some investigators (Davies and Roberts, 1987,1988; Boegman et al., 1987) or (2) indicate that the behavioral analysis employed to date is insufficient, and that additional, more precise testing paradigms are required to validate the utility of the QA model. Regardless, the validity of QA as a model of HD must be based in part on the behavioral consequences of the compound and the resulting homology to HD itself.

6.3. Other Excitotoxic Models

The KA and QA models led investigators to search for other endogenous compounds with similar actions. Along these lines, Reike and colleagues (1984a,b,1989) have examined the neurotoxic effects of L-pyroglutamate (L-PGA) and the similarity of its effects on striatal morphology to those in HD. L-PGA is the cyclized internal amide of L-glutamic acid, and is involved in the intracellular transport of various amino acids and the synthesis and breakdown of glutathione. Infusions of L-PGA into the striatum produce a necrotic lesion core consisting of macrophages and neutrophils. Surrounding the primary lesion is a pynknotic and spongiose region of vacuolated neuropil and spiny neurons reactive for NADPH and somatostatin. Behaviorally, these animals exhibit circling and postural asymmetries of the head and neck. Although limited, these studies suggest that L-PGA may produce a profile of neural and behavioral toxicity with some similarity to that observed in HD.

Although there is little evidence to suggest a causative role of L-PGA in the neuropathology of HD, some observations deserve mention. Reike et al. (1983) reported that a small sample of HD patients had increased plasma levels of L-PGA compared to matched controls. Interestingly, Uhlhous and Lange (1988) reported similar increases in plasma L-PGA. However, these same investigators reported that striatal levels of L-PGA were decreased in HD patients. The significance of this discrepancy is unclear, since the cellular origin of striatal L-PGA is unknown. It is conceivable that striatal levels of L-PGA appear to decrease because of the progressive loss of L-PGA-containing neurons. Aside from these observations, little is known concerning the role of L-PGA in HD. These data do suggest, however, that a variety of endogenous compounds may contribute to the pathology of the HD brain.

6.4. What Excitotoxin Models Tell Us About Huntington's Disease

6.4.1. Glutamate and Huntington's Disease

The similarities between the KA and QA models and HD itself suggest the existence of an endogenous molecule with excitotoxic properties in the disease. The strongest candidate at present is glutamate or an analog. Glutamate is a potent agonist for all excitatory amino acid receptors and is present in virtually all brain regions. In support of the excitotoxic hypothesis, the striatum receives a substantial innervation of glutaminergic fibers (Fonnum et al., 1981,1984) and contains one of the highest densities of excitatory amino acid receptors of any major brain region (Cotman et al., 1987,1988). Postmortem studies have also demonstrated significant reductions in the numbers of NMDA, KA, and quisqualate receptors in HD patients. Notably, the binding sites for KA in the striatum have been reported to decrease by 50–60%, whereas the number of NMDA receptors in the putamen may be decreased by as much as 90% in HD (Henke, 1979; Beaumont et al., 1979; London et al., 1981).

In an attempt to determine if any glutamatergic abnormalities occur in HD, Carter (1982) measured the activity of glutamine synthetase, which converts glutamate to glutamine, in HD patients and age-matched controls. He demonstrated a 20–30% decrement in the activity of the enzyme in the frontal cortex, putamen, temporal cortex, and cerebellum. Since this enzyme plays an important part in the process of glial uptake and conversion of glutamate to glutamine, he has argued that these decreases may relate to an excitotoxic-induced neuronal degeneration in HD.

Because ornithine aminotransferase may be a synthesizing enzyme for glutamate from ornithine, and this enzyme may be partially concentrated in glutamatergic neurons, Wong et al. (1982) investigated ornithine aminotransferase activity in postmortem cortical and striatal tissues from HD patients. They found that the specific activity was reduced by 34–49% in the frontal cortex, parietal cortex and striatum, possibly reflecting deterioration of the corticostriatal glutamatergic neurons in HD. In this regard, it is noteworthy that Kim et al. (1980) showed a decrease in the amount of glutamate in the CSF of long-term HD patients.

Gray et al. (1980) demonstrated that cultured HD fibroblasts degenerated at concentrations of glutamate that do not significantly affect fibroblasts from age-matched controls. This finding, together with those of Wong et al. (1982), are important because they support an increased sensitivity to the toxic effects of glutamate in HD patients. Although there is comparatively little direct evidence supporting a role for glutamate in the pathology of HD, this theory remains one of the most attractive etiological considerations in the disorder. In large part, this is owing to the discovery and elucidation of specific populations of receptors *(see below)* for glutamate and related excitotoxic compounds.

6.4.2. NMDA Receptors

Because of the similarities among the neurochemical, histopathological, and behavioral sequelae of HD and the QA lesion, there is interest in characterizing the mechanism of toxicity of QA, since this may elucidate the fundamental process involved in the neuronal degeneration of the disorder. Recently a great deal of attention has been focused on the *N*-methyl-D-aspartate (NMDA) receptor as a mediator of a variety of neuroplastic processes, as well as the cell loss observed in a variety of neurodegenerative disorders. NMDA receptors are one of at least three receptor subtypes that mediate the responsivity to excitatory amino acids, such as glutamate and aspartate, and are found throughout the brain: The highest concentration occurs in limbic regions and the basal ganglia (Cotman et al., 1987; Cotman and Iversen, 1987). QA has a high affinity for the NMDA receptor and, consequently, produces its toxic effects via an interaction with these receptors. The exact mechanism of this neurotoxicity is speculative, but likely involves the prolonged opening of calcium channels and the ensuing influx of calcium in levels high enough to activate lipases and proteases that irreversibly damage mitochondria.

The similarity of QA lesions to the neuropathology of HD together with the characterization of an endogenous receptor for QA strongly suggests either overproduction of the endogenous neurotoxic molecule or an abnormality in the postsynaptic membrane that renders the neurons vulnerable. Although direct evidence for QA or another endogenous NMDA agonist contributing to the pathology of HD is scarce, Young and colleagues (1988) reported a massive (93%) reduction in NMDA receptors in the putamen of HD patients. Moreover, these changes were observed in a relatively young population of patients. If alterations in NMDA receptors indeed underlie the neuropathology of HD, then it might be expected that changes in these receptors would be evident early in the disease. These results are also consistent with the observation that QA, injected into the striatum, produces a similar (92%) decrease in striatal NMDA receptors (Greenamyre and Young, 1989). In addition, cultured cortical and striatal cells have been reported to die rapidly when exposed to QA or NMDA itself, whereas cells containing NADPH-d remain unaffected (Koh and Choi, 1986,1988). Because a selective sparing of NADPH-d-containing cells occurs in HD, these data support the theory that excess exposure to quinolate or other NMDA agonists underlies the pathology and subsequent behavioral abnormalities in HD.

6.5. Therapeutic Strategies with Excitotoxin Models

6.5.1. Pharmacological Treatments

The possible involvement of NMDA receptors in the neuropathology of HD provides new avenues of therapeutic intervention. It may be possible to negate the toxic effects of overactivated NMDA so receptors by at least two general methods: (1) by preventing the presynaptic release of glutamate or (2) by administering select NMDA receptor antagonists. These possible treatments are outlined below.

The excitotoxicity of such compounds as QA may be partially dependent on the integrity of the glutamatergic input to the striatum. Supporting this notion are the findings that lesioning the cortico-striatal pathway prevents the KA- and QA-induced alterations in striatal morphology (Coyle et al., 1978; Schwarcz et al., 1984). Since this pathway utilizes glutamate as its neurotransmitter, these data imply that removal of the endogenous source of glutamate underlies the neural protection afforded by prior decortication. It may also be possible to modify striatal glutamate levels pharmacologically and prevent excitotoxicity. For instance, it is known that adenosine or adenosine analogs decrease presynaptic glutamate release. Arvin et al. (1988) recently reported that prior treatment with 2-chloroadenosine prevents KA-induced toxicity in rat striatum. If excitotoxicity contributes to the neuropathology in HD, then pharmacologically modifying striatal glutamate levels may represent a useful strategy for preventing or slowing the neurodegeneration associated with the disease.

Blockade of the NMDA receptor is another promising therapeutic approach for excitotoxicity. The NMDA receptor may be completely blocked by the administration of 2-amino-7-phosphonohepanate (APH) or 2-amino-5-phosphovalerate (APV) (Schwarcz et al., 1984). The neuropathological and seizure-inducing effects of QA following intrahippocampal injections are prevented by prior administration of APH (Schwarcz et al., 1984). In addition, both APH and APV prevent QA-induced striatal damage, whereas other compounds, such as allopurinol, ketamine, nimodipine, and baclofen, are without effect (Beal et al., 1989). Unfortunately, the clinical value of these compounds may be limited in that they are highly charged and cross the blood–brain barrier poorly.

MK-801 is an anticonvulsant agent that is a potent NMDA receptor antagonist (Wong et al., 1986; Foster and Wong, 1987; Kemp et al., 1987; Cotman and Iversen, 1987; Kloog et al., 1988; Huettner and Bean, 1988). Following systemic or oral administration, MK-801 is readily absorbed and crosses the blood–brain barrier easily. Recent evidence indicates that MK-801, in a dose-dependent manner, effectively prevents excitotoxic neuronal damage. Following systemic administration, MK-801 has been reported to prevent QA-induced striatal and hippocampal cell loss (Gill et al., 1987; Foster et al., 1987; Beal et al., 1988a,1989; Kochhar et al., 1988). Moreover, MK-801 prevents the neuronal damage and decreases in ChAT following injections of NMDA into either of these structures. MK-801 also prevented the increased locomotor activity, decreased catalepsy response to haloperidol, and caused alterations in feeding behavior and convulsions following intrastriatal QA (Giordano et al., 1990). However, because this compound acts at the phencyclidine site, it may produce undesirable psychotomimetic effects that severely question its clinical efficacy. In addition, Norman et al. (1990) recently reported that chronic treatment with MK-801 increased the QA-induced loss of D_1 dopamine receptors in the striatum. If the blockade of NMDA receptors with MK-801 were a useful treatment for HD, then long-term use of the compound would be required. However, if chronic treatment with MK-801 results in a supersensitivity of NMDA receptors, the neurotoxic actions of endogenous ligands for these receptors might be enhanced.

Glycine has also been reported to potentiate the response of NMDA receptors in cultured mouse cortical neurons and in oocytes (Reynolds et al., 1987; Wong et al., 1987). Johnson and Ascher (1987) reported that glycine increased NMDA-mediated responses through a strychnine-insensitive glycine recognition site that is part of the NMDA receptor/ion channel complex. Klechner and Dingledine (1988) further demonstrated that glycine is actually required for the activation of NMDA receptors. Together these data suggest that the use of glycine antagonists may represent a useful strategy for blocking the neuropathology subsequent to NMDA activation. The endogenous tryptophan metabolite kynurenic acid blocks the activation of NMDA receptors by acting as a selective antagonist at the glycine modulatory site. Kemp and colleagues (1988) reported that 7-chlorokynurenic acid antagonizes NMDA-induced depolarizations in rat cortical slices, and this antagonism is reversed by glycine administration. Furthermore, kynurenic acid blocks both the neurotoxic and seizure inducing effects of intrahippocampal injections of QA (Foster et al., 1984). These findings are particularly intriguing, because they suggest that kynurenic acid modulates the delicate balance between glutamate transmission and excitotoxicity.

6.5.2. Neural Transplantation

The transplantation of fetal neural tissue into the adult brain was initially seen as a way to study the development and regenerative capacity of the nervous system. However, it was discovered that neural grafting could attenuate the functional deficits induced by a variety of central lesions. Based on these encouraging observations, it was proposed that neural grafting might be an appropriate treatment for various neurodegenerative disorders. The logic of replacing lost populations of neurons with fetal neurons was inherently appealing, and animal studies clearly indicated that grafted fetal tissue could integrate with the host brain and promote functional recovery following acute injury or in animal models of Alzheimer's, Parkinson's, or Huntington's diseases (Low et al., 1982; Gage and Bjorklund, 1986; Sanberg et al., 1986; Nilsson et al., 1987; Bjorklund et al., 1988; Hoffer et al., 1988; Sladek et al., 1988; Sladek and Gash, 1988; Norman et al., 1988a,b; Sanberg et al., 1989a,b).

A variety of studies have indicated that transplanted striatal tissue readily survives and grows within the lesioned striatum (Deckel et al., 1983,1986a,b; 1988a,b; Deckel and Robinson, 1986,1987; McGeer et al., 1984; Isacson et al., 1984,1985,1986, 1987a,b; McAllister et al., 1984,1985; McAllister, 1987; Sanberg et al., 1987a,c; Norman et al., 1988a,b; Difiglia et al., 1988; Wictorin et al., 1989a,b,c). Within approx 6–10 wk, striatal grafts appear as well-delineated, mature-looking tissue (McAllister et al., 1984,1985; Difiglia et al., 1988). Morphological analysis has also indicated that the grafted tissue displays a patchy distribution of AChE, met-enkephalin, and substance P positive regions, which resemble that normally observed in the intact striatum (Isacson et al., 1987a). The host-transplant anatomical similarity is paralleled by a significant recovery of both ACh and GABA within the previously depleted striatum (Schmidt et al., 1981; Isacson et al., 1986; Segovia et al., 1989). GABA release is also partially restored in the globus pallidus and substantia nigra by striatal grafts following ibotenic acid lesions (Sirinathsinghji et al., 1988). Moreover, these grafts may be innervated by efferents originating in the host neuropil (McGeer et al., 1984; Pritzel et al., 1986; Clarke et al., 1988a,b; Wictorin et al., 1988, 1989). Accordingly, the transplantation of fetal striatal tissue has generated substantial interest as a possible treatment for the behavioral sequelae associated with HD. Tables 4 and 5 (*see* pp. 104–107) summarize the experiments examing the anatomical, biochemical, and behavioral consequences of striatal transplants into excitotoxin-lesioned striatum.

Although behavioral recovery has been reported to occur following integration of transplants into the striatum, there are several difficulties inherent in these studies. For example, survival of the transplanted tissue does not necessarily correlate with behavioral recovery. In addition, a number of factors, such as graft location and the time of behavioral testing following transplantation, are critical factors in determining the extent of behavioral recovery. Some of these issues are discussed in the following section.

The reversal of behavioral abnormalities resulting from excitotoxin lesions has been demonstrated in several studies. In particular, investigators have focused on the normalization of the increases in locomotor behavior observed in lesioned rats. Unfortunately, the extent of behavioral recovery varies across studies and likely reflects inconsistencies in testing paradigms. For instance, Deckel and colleagues (1983,1986a,b) examined changes in spontaneous locomotor activity during the daytime, whereas Sanberg et al. (1986,1987a,1988) and Isacson et al. (1984,1986,1987a,b) examined changes in nocturnal activity. Although all of these studies reported that striatal grafts reverse the increases in activity following excitotoxic lesions, the improvements observed during daytime testing are often smaller in magnitude and sometimes fail to reach statistical significance. It appears then that the testing paradigm used yields somewhat variable results and complicates the comparisons between the extent of recovery in these studies.

The dependent measures used to assess locomotor activity may differentially influence the interpretation of transplant-induced recovery. The normalization of locomotor behavior is usually reflected in the number of photobeams interrupted. However, locomotor may be reflected in a multitude of behavioral indices, such as the velocity of movement and the total distance traveled. With this in mind, the Digiscan Animal Activity Monitoring System has been used to provide a detailed description of the topography of the animal's locomotor profile (*see* Sanberg et al., 1985,1986 for details). This system provides a detailed account of a variety of locomotor indices, including the velocity of movement, distance traveled, number of movements, stereotypy, and so forth. The importance of such a system is reflected in the report that KA differentially affects these variables and that there is a similar differential recovery of these alterations following fetal striatal transplants (Giordano et al., 1988). Together, these studies clearly indicate that the locomotor abnormalities in excitotoxic-lesioned animals are at least partially recovered, but also that additional investigations are required to determine the extent and nature of this recovery. Moreover, it is apparent that, when evaluating behavioral recovery, the behavior under investigation must be clearly defined.

Following excitotoxic damage to the striatum, animals exhibit not only locomotor abnormalities, but also impairments in learning and memory (Sanberg et al., 1978; Sanberg and Johnston, 1981). It is, therefore, important to determine whether the behavioral recovery observed in locomotor testing extends to cognitively oriented tasks. To address this issue, animals have

Table 4
Behavioral and Pharmacological Studies
in Striatal Transplants in Excitotoxin-Lesioned Striatum

Study	Host sex	Lesion type	Days post-lesion	Donor specifications	Delayed/reward alteration	Locomotor testing	Pharmacological testing
Deckel et al. (1983)	F	BI KA	7	E18 solid 1.5 mm^3/side		Daytime hyperactivity reduced	
Isacson et al. (1984)	M	UNI IA	7	E13-15 cell suspension 4 µL		Nocturnal hyperactivity reduced to control level	
Deckel et al. (1986b)	F	BI KA	7	E18 solid 1.5 mm^3	Partial improvement	Daytime hyperkinesis reduced	
Deckel et al. (1986a)	F	BI KA	7	E18 solid 1.5 mm^3		Reduced daytime hyperkinesis	No reversal of responsiveness to apomorphine and amphetamine
Deckel and Robinson (1986)	M	BI KA	7	E16 solid 1.5, 3 or 6 µL/side		Induced spontaneous hypoactivity both noct./day. Both varied directly with graft volume	Reversed hyperactivity to apomorphine and amphetamine
Isacson et al. (1986,1987)	M	BI IA	7	E14-15 cell suspension 3 µL/side	Recovered in NS implants; GP implants no effect	Nocturnal hyperactivity control level; GP implant hyperactive with food dep.	

Giordano et al. (1986,1988)	M	BI KA	28	E17 solid 4 μl intraparenchymal placement	Nocturnal hyperactivity reduced to control level	
				Intraventricular placement	No reduction in nocturnal hyperactivity	
Sanberg et al. (1986)	M	BI KA	28	E17 solid 4 × 1 μL	Nocturnal hyperactivity reduced to control level	Reversal of amphetamine hyper-respons-iveness
Norman et al. (1987,1988)	M	UNI KA	28	E17-19 solid 4 × 1 μL		Change in topo-graphy; reduction of apomorphine rotation
Dunnett et al. (1988)	M	UNI IA	7	E14-15 cell suspension		Reduction of apo-morphine/amphe-tamine rotation behavior
Giordano et al. (1988)	M	BI QA	42	E17-19 solid 4 × 1		Transient recovery of haloperidol and SCH23390-induced catalepsy
Hagenmeyer-Houser and Sanberg (1987)	M	None	0	E17 solid 8 μL	Profound hyperactivity	
Lu et al. (1990)	M	None	0	E15-17 solid 1.5 3 or 6 μL/side	Hyperactivity	
Dunnett et al. (1988)	F	UNI IA	14	E15 cell suspension 2 × 2 μL		Reversal of apomor-phine and amphet-amine hyper-responsiveness

Table 5*
Striatal Transplants, Anatomical and Physiological Aspects

Study	Host sex	Lesion type	Days post-lesion	Donor specifications	Anatomical/ physiological	Biochemical/metabolic findings
Schmidt et al. (1981)	F	UNI KA	5	E14 cell suspensions 5 μL	Transplant survives; medium neurons numerous; host DA fibers in graft	ChAT and GAD increase
Isacson et al. (1984)	F	UNI KA	5–7	E-13-15 cell suspension 5 μL		GAD normal; normalize metabolic hyperactivity in target tissues; increase metabolism in striatum
McGeer et al. (1984)	?	UNI KA	21	≤6 h neonatal solid 1 μL	Graft neurons migrate; (+) revascularization catechola-mine; in graft neurons	Increase glucose utilization
McAllister et al. (1984)	?	UNI KA	5	E14 cell suspension 5 μL	Graft neurons survive; no integration	
McAllister et al. (1984)	?	UNI KA	5	E14 cell suspension 5 μL	Medium spiny I and II survive; rare aspiny III	
McAllister et al. (1985)	?	UNI IA	5	E14 cell suspension 5 μL	Medium spiny I and II survive	
McAllister et al. (1986)	?	UNI/BI KA	5	E14 cell suspension 5 μL	Afferent/efferent connections between host/graft absent	
Isacson et al. (1985)	F	UNI IA	5–7	E14-15 cell suspension 4-5 μL		ChAT/GAD increase
Emson et al. (1985)	?	UNI IA	5–7	E14-15 cell suspension 4-5 μL	CCK, MET-ENK and SUB P present in graft	
McAllister and Norelle (1986)	?	UNI KA	5	E14-15 cell suspension 5 μL	All neurons in transplant from donor	
Deckel et al. (1986)	F	BI KA	7	E17 solid 1.5 mm^3	Relative absence of dopamine receptors in transplant	

Deckel and Robinson (1986)	F	BI KA	7	E18 solid 1.5, 3, or 6 μL		Cortical NE varied directly with graft vol.; DOPAC/DA minimal changes
Pritzel et al. (1986)	F	UNI/IA	4–6	E14-15 cell suspension 5 μL	Afferent/efferent connections between host/graft present	
Zhou et al. (1986)	?	UNI IA	?	E14 solid	No efferents to SN; sparse DA terminals in graft	
Walsh et al. (1986)	?	UNI KA	?	E14 solid	IPSPs + EPSPs in graft neurons with host NS stimulation	
Deckel et al. (1988)	F	BI KA	7	E17 solid 2 × 3 μL	Dopamine and M_1 receptors decreased; M_2 receptors increased in grafts; D_2 receptors distributed in patches	
Clarke et al. (1988)	F	UNI IA	14	E14 cell suspension 2 × 2 μL	Afferent connections between host/graft	
Wictorin et al. (1988)	F	UNI IA	5–7	E15 cell suspension 4 × 1.25 μL	Afferent connections from SN, amygdala, and raphe	
Wictorin et al. (1989)	F	UNI IA	7–10	E14-15 cell suspension 2 μL	Synaptic contacts between host corticostriatal path and graft; connections between GP, SN, and graft	
Wictorin and Bjorklund (1989)	F	UNI IA	7–10	E14-15 cell suspension 2 × 2 μL	Graft innervated by afferents from cortical and subcortical regions	
Zhou et al. (1989)	?	None	0	E14 cell suspensions	Grafts contain GABA and cholinergic neurons, SUB P neurons present; numerous medium-sized neurons	
Norman et al. (1989)	M	UNI KA	?	E17 solid 4 × 1 μL	Sparse dopamine receptors in graft	

*NE=norepinephrine; DA=dopamine; DOPAC=dihydroxyphenyl-acetic acid; IPSP= inhibitory postsynaptic potential; EPSP = excitatory postsynaptic potential; 3,4 E =fetal age (in days); UNI=unilateral; BI=bilateral; KA=kainic acid; IA = ibotenic acid: ChAT = choline acetyltransferase; GAD=glutamic acid decarboxylase; M=male; F=female; CCK=cholecystokinin; MET-ENK = met-enkephalin; SUB =substance P; NS=neostriatal; SN = substantia nigra; GP=globus pallidus.

been tested on a delayed rewarded alternation test in a T-maze. Controls are able to acquire this task with minimal amounts of training, whereas lesioned animals exhibit a marked and persistent impairment (Divac et al., 1978). Injections of fetal striatal tissue partially reverse the impaired alternation behavior following striatal damage, and therefore appear to improve learning and memory abilities (Deckel et al., 1986a,b; Isacson et al., 1987a,b). Interestingly, rats that exhibit a partial reversal of locomotor hyperactivity and T-maze performance may show no recovery of sensorimotor changes, including vibrissae orientation, responsivity to olfactory stimuli, blunt and sharp touch, muscle tone, or limb strength (Deckel et al., 1986). These data suggest that fetal striatal tissue transplants promote functional recovery, but that the extent of this recovery is task dependent.

The excitotoxin-lesioned striatum provides a useful model for evaluating the pharmacological properties of striatal transplants. The utility of this model is founded on the observation that dopamine exerts a tonic influence on striatal neurons and drugs that alter dopaminergic transmission also modulates locomotor activity in rats. Following bilateral lesions of the striatum, rats demonstrate an exaggerated hyperactivity in response to amphetamine (Mason et al., 1978; Sanberg et al., 1986) that is reversed following striatal transplants (Sanberg et al., 1986). Similar results have been obtained following unilateral striatal lesions. In this model, rats will rotate in response to dopaminergic agonists, such as amphetamine and apomorphine. Striatal transplants have been reported to produce a time-dependent decrease in rotational behavior following both amphetamine and apomorphine (Norman et al., 1987,1988a,b; Dunnett et al., 1988). It is important to point out, however, that although the basic findings in these studies are similar, there are methodological differences that may produce somewhat different results. The studies by Norman et al. (1987,1988a,b) assessed rotational behavior in an open field, which is in contrast to the rotometer used by Dunnett et al. (1988). The use of these two different procedures provides different information (i.e., the open field may be used to examine both the number and topography of rotations, whereas the rotometer restricts movement and assesses pivotal-type rotations). Interestingly, the animals tested in the open field exhibited changes from pivotal-type rotations to a wide circle ambulation as early as 3–5 wk posttransplantation (Norman et al., 1987,1988). On the other hand, changes in the rotometer were not found for 8 mo posttransplantation (Dunnett et al., 1988). Because changes in the topography of rotation are not detectable in the rotometer, it is possible that topographical changes in locomotion occur earlier than actual reductions in rotation following striatal transplants.

Because transplanted tissue reverses the enhanced sensitivity to dopaminergic agonists following excitotoxin lesions of the striatum, it might be expected that the transplanted tissue would express dopamine receptors. However, the studies conducted to date have produced dissimilar results. Isaacson and colleagues (1987a,b) and Deckel et al. (1986a,1987) have demonstrated that transplanted tissue contains patches of D_2 dopamine receptors resembling those observed in the intact striatum. Deckel et al. (1986,1987) also reported a correlation between the lack of behavioral recovery and sparse presence of dopamine receptors revealed by [^{3}H] spiperone autoradiography in transplanted tissue. Conversely, Norman et al. (1988a,b,1989a,b) reported a transplant-related recovery of apomorphine-induced rotational behavior in transplant animals without associated D_1 or D_2 receptors. These studies also revealed a deficit in [^{3}H] forskolin binding to the stimulatory guanine nucleotide regulatory subunit/adenylate cyclase complex within the transplanted tissue. These results led Norman et al. (1988a,b,1989a,b) to conclude that tissue that is neurochemically dissimilar to the host is capable of reversing the functional deficits associated with striatal damage.

The above discussion illustrates the difficulty in correlating behavioral and neurobiological recovery processes. However, a number of possible mechanisms for functional recovery following fetal striatal transplants might be postulated. One of the most frequently cited mechanisms is an anatomical connectivity between the host and transplant. Although intuitively appealing, attempts to demonstrate these anatomical connections have been controversial. Initial studies revealed that transplanted tissue grows and differentiates into mature-looking tissue, but does not make afferent or efferent connections with the surrounding tissue (Schmidt et al., 1981; McAllister et al., 1985; McAllister, 1987; Walker et al., 1987; Zemanick et al., 1987). Similarly, HRP injected into the host striatum or surrounding tissue at 5 wk posttransplant is not transported into either the transplant or the host tissue, respectively (Walker and McAllister, 1987). On the other hand, tyrosine hydroxylase (TH) positive neurons have been demonstrated to innervate grafted striatal tissue at 8–10 mo posttransplantation (Clarke et al., 1988a,b). In addition, fluorescent beads are retrogradely transported from the transplant to the surrounding neuropil (Wictorin et al., 1988a). Similarly, connections between the host cortex and transplant have been identified following injections of the retrograde tracer Phaseolus vulgaris leukoagglutinin (Wictorin et al., 1988b). These studies, therefore, indicate that anatomical connections may indeed form between the host and transplanted tissue. However, it remains difficult to reconcile these data with the available behavioral data. For example, behavioral recovery has been reported to occur at time-points much earlier than those at which reorganization occurs (Norman et al., 1988a,b). Even though TH-positive neurons have been reported to penetrate the grafted tissue as early as 6–7 wk posttransplant (Zhou et al., 1986), behavioral recovery may still occur substantially earlier.

Electrophysiological studies may aid in determining the time-course of anatomical integration and its relation to functional recovery. Several convergent lines of data indicate physiological connections between host and transplanted striatal neurons. These include:

1. Recordings of inhibitory and excitatory postsynaptic potentials in grafts following stimulation of the host striatum (Walsh et al., 1988);
2. Stimulation of the corpus callosum, producing electrophysiological responses in transplanted tissue in a slice preparation (Rutherford et al., 1987); and
3. Frontal cortex stimulation, producing similar responses in both the normal striatum and in striatal transplants following both KA and IA (Wilson et al., 1987).

The majority of data indicates that anatomical integration between host and transplanted tissue occurs. However, this is not a universal phenomenon, and not all studies find evidence of integration at the time when behavioral recovery occurs. These findings have led to the suggestion that a diffusible neurotrophic factor released by or within the vicinity of the transplant partially underlies functional recovery (Deckel et al., 1986b; Tulipan et al., 1986).

Supporting this contention is the observation that adrenal medulla grafts promote behavioral recovery in animal models of Parkinson's disease in a manner not dependent on the reconstitution of anatomical circuitry or the release of catecholamines (Stromberg et al., 1985; Freed et al., 1986; Becker and Freed, 1988; Norman et al., 1989a,b). Interestingly, these grafts may be equally effective at promoting recovery whether placed into the lesioned striatum or the adjacent lateral ventricle. Similar degrees of recovery have been accounted for by postulating that the adrenal tissue releases a trophic factor (such as nerve growth factor) and that placement into the lateral ventricle allows the factor to diffuse readily into the nearby striatum. However, studies by Giordano et al. (1988) and Sanberg et al. (1987a) recently demonstrated that striatal tissue placed into the lesioned striatum, but not the lateral ventricles, was required for behavioral recovery. It was suggested that the ventricular placements provided a limited behavioral recovery, because the levels of neurotrophic activity may have been insufficient to promote higher levels of recovery. The transplants themselves may not be the source of trophic activity. Rather, reactive macrophages or glial cells may be providing the necessary trophic support. Glial cells have been shown to promote neurite outgrowth, restore ionic balance, and decrease levels of excitotoxins, such as glutamate (Muller and Seifert, 1982; Bridges et al., 1985; Kromer and Cornbrooks, 1985; Neito-Sampedro and Cotman, 1985). Additionally, Kesslak et al. (1986) recently reported that the transplantation of cultured astrocytes is as effective as cortical tissue in promoting behavioral recovery following cortical aspirations.

Although the mechanism by which transplanted tissue exerts its beneficial effects is largely unknown, it remains a viable approach for promoting behavioral and neurobiological recovery. As encouraging as initial studies have been, there is evidence to suggest that transplanted tissue may exert a deleterious as well as beneficial effect. For instance, Hagenmeyer-Houser and Sanberg (1987) injected striatal tissue into intact host striatum and observed a profound hyperactivity. Similarly, Deckel and Robinson (1987) reported that, although striatal transplants ameliorate the behavioral deficits resulting from excitotoxic striatal lesions, transplants into normal striatum produced a "lesion-like effect" and disrupted locomotor activity. Moreover, Lu et al. (1990) have recently confirmed these results and further suggest transplanted tissue from the striatum disrupts locomotor behavior in a manner dependent on the initial volume of tissue injected. These studies also indicated that transplanted tissue from the striatum as well as other neural regions is capable of producing locomotor abnormalities following injection into intact striatum. Whether these transplant-induced abnormalities are the result of the physical or neurochemical characteristics of the transplants is unknown. Regardless of the underlying mechanism of these unwanted consequences, the possible contribution of the region from which the transplant is derived as well as the amount of tissue used should be given serious consideration. This is especially true given the current debates centering around the continuation of tissue transplantation in human neurodegenerative disorders. The detrimental effects of transplants may be especially difficult to detect in individuals with an already severe disruption of function. It is conceivable that the negative effects of transplantation would not manifest themselves directly, but might slow the progression of or limit the extent of behavioral recovery in laboratory animals or humans. It seems then that further research is necessary in order to decipher the negative and positive effects of neural transplantation. This research should also seriously consider the refinement of existing behavioral paradigms in order to evaluate the subtle topographic changes in behavior that result from transplants of fetal tissue.

7. Conclusions

Since the pioneering studies of Bird and Iversen (1974) and Perry et al. (1973) demonstrating a selective alteration in GABAergic parameters in HD, there has been a remarkable escalation in our understanding of the pathophysiology of the disease. Our concepts of the neurochemical and morphological sequelae of HD have grown from these early observations to the present-day complicated mosaic of neurobiological changes in the striatum and related structures. Although the cooperation of clinicians, basic neuroscientists, and families of HD patients has contributed to this progress, an equally important contribution has been made from the development of animal models of HD. Investigators have taken advantage of new excitotoxic models that appear to mimic the neurobiological and behavioral characteristics of HD with remarkable homology. Paramount to the development and validity of these models were several observations, including:

1. The characterization of an excitatory corticostriatal pathway that uses glutamate as its neurotransmitter;
2. The availability of excitotoxic compounds, which produce a pattern of neural degeneration resembling that observed in HD;
3. The discovery of receptors for excitatory amino acids in human and rat brain; and
4. Measurable concentrations of endogenous excitotoxins, such as QA in normal and HD patients.

These findings have substantially bolstered the hypothesis that a fundamental deficit in HD is a dysfunction of glutamatergic transmission, which results in the slow progressive neural degeneration characteristic of HD.

The development of these models holds great promise both for understanding the etiology of HD and for the development of therapeutic strategies (i.e., pharmacological alterations of NMDA receptors, glutamate transmission, or neural transplantation).

However, since HD is a genetically transmitted disorder, these animal models will not provide information concerning the genetic defect in HD. The uncovering of the genetic abnormality in HD will occur through molecular genetic research (Gilliam et al., 1988). In this way, we are most likely to reach the ultimate goal in the treatment of HD, which is to prevent its expression. Recombinant DNA techniques have identified a restriction-fragment-length polymorphism, which is mapped to chromosome 4 and is linked to the HD gene (Gusella et al., 1983). This fragment identifies a neighboring piece of DNA with a 5% recombination rate with the gene deficit, and may soon help determine the exact identity of the normal counterpart of the HD gene and with it the ensuing knowledge of its pathology in the disease. Based on these findings, tests have been developed that can identify a presymptomatic individual. Accordingly, this knowledge will suggest therapies specifically designed to compensate for the changes induced by the aberrant HD gene. For example, if the genetic abnormality in HD is the result of a loss of function, then treatment could revolve around the replacement of a missing constituent. Recent advances in the development of "designer genes," which produce specific gene products, such as growth factors or enzymes, hold great promise as replacement strategies. These designer neurons could be implanted into the appropriate neural site, perhaps reverse or slow the progression of the disease state, and afford the individual a higher level of functioning. Alternatively, if the HD allele produces an increased or inappropriate expression of a specific gene product, then an appropriate therapeutic strategy would be to reduce the expression of the HD gene.

Acknowledgments

The authors appreciate the support of the PHS Grant R01 NS 25647, the Huntington's Disease Society of America, the Tourette Syndrome Association, the Pratt family and friends, the Smokeless Tobacco Research Foundation, and Omnitech Electronics, Inc. Thanks are also expressed to Andrew B. Norman and Magda Giordano for helpful discussions.

References

Aquilonius S. M., Eckernas S. A., and Sundwall A. (1975) Regional distribution of choline acetyltransferase in the human brain: changes in Huntington's chorea. *J. Neurol. Neurosurg. Psychiat.* **38**, 669–677.

Araki M., McGeer P. L., and McGeer E. G. (1985) Striatonigral and pallidonigral pathways studied by a combination of retrograde horseradish peroxidase tracing and a pharmacohistochemical method for gamma aminobutyric acid transaminase. *Brain Res.* **331**, 17–24 .

Arregui A., Emson P. C., and Spokes E. G. (1978) Angiotensin-converting enzyme in substantia nigra: reduction of activity in Huntington's disease and after intrastriatal kainic acid in rats. *Eur. J. Pharmacol.* **52**, 121–124.

Arregui A., Iversen L. L., Spokes E. G. S., and Emson P. C. (1979) Alterations in post-mortem brain angiotensin-converting enzyme activity and some neuropeptides in Huntington's disease. *Adv. Neurol.* **23**, 517–525.

Arregui A., Bennett J. P., Jr., Bird E. D., Yamamura H. I., Iversen L. L., and Snyder S. H. (1977) Huntington's chorea: selective depletion of activity of angiotensin converting enzyme in the corpus striatum. *Ann. Neurol.* **2**, 294–298.

Arvin B., Neville L. F., and Roberts P. J. (1988) 2-chlorodenosine prevents kainic acid-induced toxicity in rat striatum. *Neurosci. Lett.* **93**, 336–340.

Averback P. (1980) Histopathology of acute cell loss in Huntington's chorea brain. *J. Pathol.* **132**, 55–61.

Barr A. N., Heinze W. J., Dobben G. D., Valvassori G., and Sugar O. (1978) Bicaudate index in computerized tomography of Huntington's disease and cerebral atrophy. *Neurology* **28**, 1196–1200.

Beal M. F., Kowall N. W., Ellison D. W., Mazurek M. F., Swartz K. J., and Martin J. B. (1986) Replication of the neurochemical characteristics of Huntington's disease by quinolinic acid. *Nature* **321**, 168–171.

Beal M. F., Kowall N. W., Swartz K. J., Ferranti R. J., and Martin J. B. (1988a) Systemic approaches to modifying quinolinic acid striatal lesions in rats. *J. Neurosci.* **8**, 3901–3908.

Beal M. F., Ellison D. W., Mazurek M. F., Swartz K. J., Malloy J. R., Bird E., and Martin J. B. (1988b) A detailed examination of substance P in pathologically graded cases of Huntington's disease. *J. Neurol. Sci.* **84**, 51–61.

Beal M. F., Mazurek M. F., Ellison D. W., Swartz K. J., McGarvey U., Bird E. D., and Martin J. B. (1988c) Somatostatin and neuropeptide Y concentrations in pathologically graded cases of Huntington's disease. *Ann. Neurol.* **23**, 562–569.

Beal M. F., Kowall N. W., Swartz K. J., Ferranti R. J., and Martin J. B. (1989) differential sparing of somatostatin-neuropeptide Y and cholinergic neurons following striatal excitotoxic lesions. *Synapse* **3**, 38–47.

Beaumont K., Maurin Y., Reisine T. D., Fields J. Z., Spokes E., Bird E. D., and Yamamura H. I. (1979) Huntington's disease and its animal model alterations in kainic acid binding. *Life Sci.* **24**, 809–816.

Becker J. B. and Freed W. J. (1988) Adrenal medulla grafts enhance functional recovery of the striatal dopamine system following substantia nigra lesions. *Brain Res.* **462,** 401–406.

Ben-Ari Y., Tremblay E., Ottersen O. P., and Meldrum B. S. (1980) The role of epileptic activity in hippocampal and "remote" cerebral lesions induced by kainic acid. *Brain Res.* **191,** 79–97.

Bernheimer H., Birkmayer W., Hornykiewicz O., Jellinger K., and Seitelberger F. (1973) Brain dopamine and syndromes of Parkinson and Huntington-clinical morphological and neurochemical correlation. *J. Neurol. Sci.* **20,** 415–455.

Bird E. D. (1980) Chemical pathology of Huntington's disease. *Ann. Rev. Pharmacol. Toxicol.* **20,** 533–551.

Bird E. D. and Iversen L. L. (1974) Huntington's chorea: postmortem measurement of glutamic acid decarboxylase, choline acetyltransferase, and dopamine in basal ganglia. *Brain* **97,** 457–461.

Bjorklund A., Lindvall O., Isacson O., Brundin P., Wictorin K., Strecker R. E., Clarke D. J., and Dunnett S. B. (1988) Mechanisms of action of intracerebral neural transplants: studies on nigral and striatal grafts to the lesioned striatum. *Trends Neurosci.* **10,** 509–516.

Boegman R. J., Smith Y., and Parent A. (1987) Quinolinic acid does not spare striatal neuropeptide Y-immunoreactive neurons. *Brain Res.* **415,** 178–182.

Bots G., Th. A. M., and Bruyn G. W. (1981) Neuropathological changes of the nucleus accumbens in Huntington's chorea. *Acta Neuropathol.* **55,** 21,22.

Bowman M. and Lewis M. S. (1980) Site of subcortical damage in diseases which resemble schizophrenia. *Neuropsychology* **18,** 597–601.

Brandt J., Folstein S. E., and Folstein M. F. (1988) Differential cognitive impairment in Alzheimer's disease and Huntington's disease. *Ann. Neurol.* **23,** 555–561.

Brann M. R. and Emson P. C. (1980) Microiontophoretic injection of fluorescent tracer combined with simultaneous immunofluorescent histochemistry for the demonstration of efferents from the caudate putamen projecting to the globus pallidus. *Neurosci. Lett.* **16,** 61–66 .

Bridges R. J., Neito-Sampedro M., and Cotman C. W. (1985) Stereospecific binding of L-glutamate to astrocyte membranes. *Soc. Neurosci. Abstr.* **11,** 110.

Bruyn G. W. (1968) Huntington'g chorea, historical, clinical and laboratory synopsis. *Handbook Clin. Neurol.* **6,** 298.

Bruyn G. W. (1973) Neuropathological changes in Huntington's chorea. *Adv. Neurol.* **1,** 399–403.

Bruyn G. W., Bots G., Th. A. M., and Dom R. (1979) Huntington's chorea: current neuropathological status. *Adv.Neurol.* **23,** 83–93.

Buck S. H., Burks T. F., Brown M. R., and Yamamura H. I. (1981) Reduction in basal ganglia and substantia nigra substance P levels in Huntington's disease. *Brain Res.* **209,** 464–469.

Calderon S. F., Sanberg P. R., and Norman A. B. (1988) Quinolinic acid lesions of rat striatum abolish D_1- and D_2-dopamine receptor mediated catalepsy. *Brain Res.* **450,** 403–407.

Carter C. J. (1982) Glutamine synthetase activity in Huntington's disease. *Life Sci.* **31,** 1151–1159.

Castaigne P., Escourelle R., and Gray F. (1976) Huntington's chorea and cerebellar atrophy (a case report with clinical and pathological data). *Rev. Neurol. (Paris)* **132,** 233.

Chan-Palay V. (1977) Indolamine neurons and their processes in the normal rat brain and in chronic diet-induced thiamine deficiency demonstrated by uptake of ^{3}H-serotonin. *J. Comp. Neurol.* **176,** 467–494.

Chase T. N. (1973) Biochemical and pharmacological studies of monoamines in Huntington's chorea. *Adv. Neurol.* **1,** 533–542 .

Clarke D. J., Dunnett S. B., Isacson O., and Bjorklund A. (1988a) Striatal grafts in the ibotenic acid lesioned neostriatum: ultrastructural and immunocytochemical studies. *Prog. Brain Res.* **78,** 47–53.

Clarke D. J., Dunnett S. B., Isacson O., Sirinathsinghji D. J. S., and Bjorklund A. (1988b) Striatal grafts in rats with unilateral neostriatal lesions. I. Ultrastructural evidence of afferent synaptic inputs from the host nigrostriatal pathway. *Neuroscience.* **24,** 791–801.

Collewijn H., Went L. N., Tamminga E. P., and Verget Van der Vlis M. (1988) Oculomotor defects in patients with Huntington's disease and their offspring. *J. Neurol. Sci.* **86,** 307–320.

Cools A. R. and Van Rossum J. M. (1976) Intrastriatal administration of monoamines: behavioral effects. *Pharmacol. Therap. B.* **2,** 129–136.

Costall B. and Naylor R. J. (1975) Neuroleptic antagonism of dyskinetic phenomena. *Eur. J. Pharmacol.* **33,** 301–312.

Cotman C. W. and Iversen L. L. (1987) Excitatory amino acids in the brain—focus on NMDA receptors. *Trends Neurosci.* **10,** 261–265.

Cotman C. W. and Monaghan D. T. (1988) Excitatory amino acid neurotransmission: NMDA receptors and Hebb-type synaptic plasticity. *Ann. Rev. Neurosci.* **11,** 61–80.

Cotman C. W., Monaghan D. T., Ottersen O. P., and Storm-Mathisen J. (1987) Anatomical organization of excitatory amino acid receptors and their pathways. *Trends Neurosci.* **10,** 273–279.

Coyle J. T. (1979) An animal model for Huntington's disease. *Biol. Psychiatry* **14,** 251–276.

Coyle J. T. and Schwarcz R. (1976) Lesion of striatal neurones with kainic acid provides a model for Huntington's chorea. *Nature* **263,** 244–246.

Coyle J. T., Molliver M. E., and Kuhar M. J. (1978) In situ injection of kainic acid: A new model for selectively lesioning neural cell bodies and sparing axons of passage. *J. Comp. Neurol.* **180,** 301–323.

Coyle J. T., Price D. L., and Delong M. R. (1983) Alzheimer's disease: a disorder of cholinergic innervation of cortex. *Science* **219,** 1184–1190.

Coyle J. T., Schwarcz R., Bennett J. P., and Campochiarc P. (1977) Clinical neuropathologic and pharmacologic aspects of Huntington's disease: correlates with a new animal model. *Prog. Neuropsychopharmacol.* **1,** 13–30.
Cross A. J., Slater P., and Reynolds G. P. (1986) Reduced high-affinity glutamate uptake sites in the brains of patients with Huntington's disease. *Neuroscience Lett.* **67,** 198–202.
Cummings J. L. and Benson D. F. (1988) Psychological dysfunction accompanying subcortical dementias. *Ann. Rev. Med.* **39,** 53–61.
Davies S. W. and Roberts P. J. (1987) No evidence for preservation of somatostatin-containing neurons after intrastriatal injections of uinolinic acid. *Nature* **327,** 326–329.
Davies S. W. and Roberts P. J. (1988) Model of Huntington's disease. *Science* **241,** 474,475.
Davis A. (1976) Emily—a victim of Huntington's chorea. *Nursing Times* **72,** 449.
Deckel A. W. and Robinson R. G. (1986) Transplantation of different volumes of fetal striatum: effects on locomotion and monamine biochemistry. *Soc. Neurosci. Abstr.* **16,** 1479.
Deckel A. W. and Robinson R. G. (1987) Receptor characteristics and behavioral consequences of kainic acid lesions and fetal transplants of the striatum, in *Cell and Tissue Transplantation into Adult Brain,* vol. 495 (Azmitia E. C. and Bjorklund A., eds.), Annals of the New York Academy of Sciences, New York, pp. 556–580.
Deckel A. W., Moran T. H., and Robinson R. G. (1986a) Behavioral recovery following kainic acid lesions and fetal implants of the striatum occurs independent of dopaminergic mechanisms. *Brain Res.* **363,** 383–385.
Deckel A. W., Moran T. H., Coyle J. T., Sanberg P. R., and Robinson R. G. (1986b) Anatomical predictors of behavioral recovery following fetal striatal transplants. *Brain Res.* **365,** 249–258.
Deckel A. W., Moran T. H., and Robinson R. G. (1988a) Receptor characteristic and recovery of function following kainic acid lesions and fetal transplants of the striatum. I. Cholinergic systems. *Brain Res.* **474,** 27–38.
Deckel A. W., Moran T. H., and Robinson R. G. (1988b) Receptor characteristic and recovery of function following kainic acid lesions and fetal transplants of the striatum. II. Dopaminergic systems. *Brain Res.* **474,** 39–47.
Deckel A. W., Robinson R. G., Coyle J. T., and Sanberg P. R. (1983) Reversal of long term locomotor abnormalities in the kainic acid model of Huntington's disease by day 18 fetal striatal implants. *Eur. J. Pharmacol.* **93,** 287,288.
Del Fiacco M., Paxinos G., and Cuello A. C. (1982) Neostriatal enkephalin-immunoreactive neurons project to the globus pallidus. *Brain Res.* **231,** 1–17.
Dewhurst K., Oliver J., Trick K. L. K., and McKnight A. L. (1979) Neuropsychiatric aspects of Huntington's disease. *Confinia Neurol.* **31,** 258–268.
Difiglia M., Schiff L., and Deckel A. W. (1988) Neuronal organization of fetal striatal grafts in kainate and sham lesioned rat caudate nucleus: light and electron microscopic observations. *J. Neurosci.* **8,** 1112–1130.

Dill R. E. (1972) Mescaline; receptor interaction in the rat striatum. *Arch. Int. Pharmacodynam. Therap.* **195,** 320–329.

Dill R. E., Dorris R. L., and Phillips-Thonnard I. (1976) A pharmacologic model of Huntington's chorea. *J. Pharm. Pharmaol.* **28,** 646 –648.

Dill R. E., Nickey W. M., Jr., and Little M. D. (1968) Dyskinesia in rats following chemical stimulation of the neostriatum. *Tex. Rep. Biol. Med.* **26,** 101–106.

Divac J., Fonnum F., and Storm-Mathisen J. (1977) High-affinity uptake of glutamate in terminals of corticostriatal axons. *Nature* **266,** 377,378.

Divac L., Markowitsch H. J., and Pritzel M. (1978) Behavioral and anatomical consequences of small intrastriatal lesions in the rat. *Brain Res.* **151,** 523–532.

Dix M. R. (1970) Clinical observations upon the vestibular responses in certain disorders of the central nervous system. *Adv. Oto-Rhino-Laryngol.* **17,** 118.

Dom R., Malfroid M., and Baro F. (1976) Neuropathology of Huntington's chorea. *Neurology* **26,** 64–68.

Dulap C. B. (1927) Pathological changes in Huntington's chorea with special reference to the corpus striatum. *Arch. Neurol. Psychiary* **18,** 867–943.

Dunnett S. B. and Iversen S. D. (1981) Learning impairments following selective kainic acid induced lesions within the neostriatum of rats. *Behav. Brain Res.* **2,** 189–209.

Dunnett S. B., Isacson O., Sirinathsinghji D. J. S., Clarke D. J., and Bjorklund A. (1988) Striatal grafts in rats with unilateral neostriatal lesions. III. Recovery from dopamine-dependent motor asymmetry and deficits in skilled paw reaching. *Neuroscience* **24,** 813–820.

Ellison D. W., Beal M. F., Mazurek M. F., Malloy J. R., Bird E. D., and Martin J. B. (1987) Amino acid neurotransmitter abnormalities in Huntington's disease and the quinolinic acid model of Huntington's disease. *Brain* **110,** 1657–1673.

Emson P. C., Rehfeld J. F., Langevin H., and Rossor M. (1980) Reduction in cholecystokinin-like immunoreactivity in the basal ganglia in Huntington's disease. *Brain Res.* **198,** 497–500.

Emson P. C., Dawbarn D., Rosser M. N., Rehfeld J. F., Brundin P., Isacson O., and Bjorklund A. (1985) Cholecystokinin content in the basal ganglia in Huntington's disease. The expression of cholecystokinin immunoreactivity in striatal grafts to ibotenic acid-lesioned rat striatum, in *Neuronal Cholecystokinin,* Vanderhaegen J. and Crawley J. N. (eds.) The New York Academy of Sciences, New York, pp. 488–494.

Enna S. J., Bennett J. P., Bylund D. B., Snyder S. H., Bird E. D., and Iversen L. L. (1976a) Alterations of brain neurotransmitter receptor binding in Huntington's chorea. *Brain Res.* **116,** 531–537.

Enna S. J., Bird E. D., Bennett J. P., Bylund D. B., Snyder S. H., and Iversen L. L. (1976b) Huntington's chorea changes in neurotransmitter receptors in the brain. *New Engl. J. Med.* **29,** 1305–1309.

Ferranti R. J., Beal M. F., Kowall N. W., Richardson E. P., and Martin J. B. (1987a) Sparing of acetylcholinesterase-containing striatal neurons in Huntington's disease. *Brain Res.* **415,** 178–182.

Ferranti R. J., Kowall N. W., Beal M. F., Martin J. B., Bird E. D., and Richardson E. P. (1987b) Morphologic and histochemical characteristics of a spared subset of striatal neurons in Huntington's disease. *J. Neuropath. Exp. Neurol.* **46,** 12–27.

Fibiger H. C., Prudritz R. E., McGeer P. L., and McGeer E. G. (1972) Axonal transport in nigro-striatal and nigro-thalamic neurons: effects of medial forebrain bundle lesions and 6-hydroxydopamine. *J. Neurochem.* **19,** 1697–1708.

Fields J. Z., Reizine T. D., and Yamamura H. I. (1978) Loss of striatal dopaminergic receptors after intrastriatal kainic acid injection. *Life Sci.* **23,** 569–574.

Fonnum F., Storm-Mathisen J., and Divac I. (1984) Biochemical evidence for glutamate as neurotransmitter in corticostriatal and corticothalamic fibers in rat brain. *Neuroscience* **6,** 863–873.

Forno L. S. and Jose C. (1973) Huntington's chorea:a pathological study. *Adv. Neurol.* **1,** 453–470.

Foster A. C. and Wong E. H. F. (1987) The novel anticonvulsant MK-801 binds to the activated site of the *N*-methyl-D-aspartate receptor in rat brain. *Brit. J. Pharmacol.* **91,** 403–409.

Foster A. C., Collins J. F., and Schwarcz R. (1983) On the excitotoxic properties of quinolinic acid 2,3-piperidine dicarboxylic acids and structurally related compounds. *Neuropharmacology* **22,** 1331–1342.

Foster A. C., Gill R., Kemp J. A., and Woodruff G. N. (1987) Systemic administration of MK-801 prevents *N*-methyl-D-aspartate-induced neuronal degeneration in rat brain. *Neurosci. Lett.* **76,** 307–311.

Foster A. C., Vezzani A., French E. D., and Schwarcz R. (1984) Kynurenic acid blocks neurotoxicity and seizures induced in rats by the related brain metabolite quinolinic acid. *Neurosci. Lett.* **48,** 273–278.

Foster A. C., Whetsell W. O., Bird E. D., and Schwarcz R. (1985) Quinolinic acid phosphoribosyltransferase in human and rat brain: activity in Huntington's disease and in quinolinate-lesioned rat striatum. *Brain Res.* **336,** 207–214.

Freed W. J., Cannon-Spoor H. E., and Krauthamer E. (1986) Intrastriatal adrenal medulla grafts in rats. Long term survival and behavioral effects. *J. Neurosurg. 65,* 664–670.

Gage F. H. and Bjorklund A. (1986) Cholinergic septal grafts into the hippocampal formation improve spatial learning and memory in aged rats by an atropine sensitive mechanism. *J. Neurosci.* **6,** 2837–2847.

Gayle J. S., Bird E. D., Spokes E. G., Iversen L. L., and Jessell T. (1978) Human brain substance P distribution in controls in Huntington's chorea. *J. Neurochem.* **30,** 633,634.

Gebbink T. B. (1968) Huntington's chorea. Fibre changes in the basal ganglia. *Handbook Clin. Neurol.* **6,** 399.

Gerfen C. R., Herkenham M., and Thibault J. (1987) The neostriatal mosaic. II. Patch– and matrix-directed mesostriatal dopaminergic system. *J. Neurosci.* **7,** 3915–3934.

Gill R., Foster A. C., and Woodruff G. N. (1987) Systemic administration of MK-801 protects against ischemia-induced hippocampal neurodegeneration in the gerbil. *J. Neurosci.* **7,** 3343–3349.

Gilliam T. C., Gusella J. F., and Lehrach H. (1988) Molecular genetic strategies to investigate Huntington's disease. *Adv. Neurol.* **48,** 465–469.

Giordano M., Houser S. H., and Sanberg P. R. (1988) Intraparenchymal fetal striatal transplants and recovery in kainic acid lesioned rats. *Brain Res.* **446,** 183–188.

Giordano M., Ford L., Norman A. B., and Sanberg P. R. (1990) MK-801 prevents guinolinica acid-induced behavioral deficits and neurotoxicity in the striatum. *Brain Res. Bull.* **24,** 433–465.

Giordano M., Houser S. H., Russell K. H., and Sanberg P. R. (1986) Location of fetal striatal transplants as a determinant for behavioral recovery of striatal lesioned rats. *Soc. Neurosci. Abstr.* **12,** 1479.

Girotti F., Marano R., Soliver P., Geminiani G., and Scagliano G. (1988) Relationship between motor and cognitive disorders in Huntington's disease. *J. Neurol.* **235,** 454–457.

Gray P. N., May P. C., and Elkin J. (1980) L-glutamate toxicity in Huntington's disease. *Biochem. Biophys. Res. Commun.* **95,** 707–714.

Graybiel A. M. (1983) Biochemical anatomy of the striatum, in *Chemical Neuroanatomy* (Emson, P. C., ed.), Raven, New York, pp. 427–503.

Graybiel A. M., Lui F-C., and Dunnett S. B. (1989) Intrastriatal gafts derived from fetal striatal primordia. I. Phenotypy and modular organization. *J. Neurosci.* **9,** 3250–3271.

Graybiel A. M., Ragsdale C. W., Jr., Yoneoka E. S., and Elde R. P. (1981) An immunohistochemical study on enkephalins and other neuropeptides in the striatum of the cat with evidence that the opiate peptides are arranged to form mosaic patterns in register with the striasomal compartments visible by acetylcholinesterase staining. *Neuroscience* **6,** 377–397.

Greenamyre J. T. and Young A. B. (1989) Synaptic localization of striatal NMDA, quisqualate and kainate receptors. *Neurosci. Lett.* **101,** 133–137.

Greenamyre J. T., Penney J. B., Young A. B., D'Amato C. J., Hicks S. P., and Shoulson I. (1985) Alterations in L-[^{3}H] glutamate binding in Alzheimer's and Huntington's diseases. *Science* **227,** 1496–1499.

Gusella J. F., Wexler N. S., Conneally P. M., Naylor S. L., Anderson M. A., Tanzi R. E., Watkins P. C., Ottina K., Wallace M. R., Sakaguchi A. Y., Young A. B., Shoulson I., Bonilla E., and Martin J. B. (1983) A polymorphic marker genetically linked to Huntington's disease. *Nature* **306,** 234–238.

Hagenmeyer-Houser S. H. and Sanberg P. R. (1987) Locomotor behavior changes induced by E-17 striatal transplants in normal rats. *Pharamacol. Biochem. Behav.* **27,** 583–586.

Hayden M. R., Hewitt J., Wasmuth J. J., Kastelein J. J., Langlois S., Conneally

M., Haines J., Smith B., Hilbert C., and Allard D. (1988) A polymorphic DNA marker that represents a conserved expressed sequence in the region of the Huntington disease gene. *Am. J. Hum. Genet.* **42,** 125–131.
Hays S. E., Goodwin F. K., and Paul S. M. (1981) Cholecystokinin receptors are decreased in basal ganglia and cerebral cortex of Huntington's disease. *Brain Res.* **225,** 452–456.
Hefter H., Homberg V., Lange H. W., and Freund H. J. (1987) Impairment of rapid movement in Huntington's disease. *Brain* **110,** 585–612.
Heindel W. C., Butters N., and Salmon D. P. (1988) Impaired learning of a motor skill in patients with Huntington's disease. *Behav. Neurosci.* **102,** 141–147.
Henke H. (1979) Kainic acid binding in human caudate nucleus: Effect of Huntington's disease. *Neurosci. Lett.* **14,** 247–251.
Hiley R. C. and Bird E. D. (1974) Decreased muscarinic receptor concentration in post-mortem brain in Huntington's chorea. *Brain Res.* **80,** 355–358.
Hoffer B. J., Granholm A. C., Stevens J. O., and Olson L. (1988) Catecholamine-containing grafts in Parkinsonism: Past and present. *Clin. Res.* **36,** 189–195.
Hruska R. E. and Silbergeld E. K. (1979) Abnormal locomotion in rats after bilateral intrastriatal injection of kainic acid. *Life Sci.* **25,** 181–194.
Huettner J. E. and Bean B. P. (1988) Block of *N*-methyl-D-aspartate-activated current by the anticonvulsant MK-801: selective binding to open channels. *Proc. Natl. Acad. Sci. USA* **85,** 1307–1311.
Huntington G. (1872) On chorea. *Med. Surg. Rep. (Philadelphia)* **26,** 317.
Innis R. B. and Snyder S. H. (1980) Cholecystokinin receptor binding in brain and pancreas: regulation of pancreatic binding by cyclic and acyclic guanine nucleotides. *Eur. J. Pharmacol.* **65,** 123,124.
Innis R. B., Correa F. M. A., Uhl G. R., Schneider B., and Snyder S. H. (1979) Cholecystokinin octapeptide-like immunoreactivity: histochemical localization in rat brain. *Proc. Natl. Acad. Sci. USA* **76,** 521–525.
Isacson O., Dunnett S. B., and Bjorklund A. (1986) Graft induced behavioral recovery in an animal model of Huntington's disease. *Proc. Natl. Acad . Sci. USA* **83,** 2728 –2732.
Isacson O., Brundin P., Gage F. H., and Bjorklund A. (1985) Neural grafting in the rat model of Huntington's disease: progressive neurochemical changes after neostriatal ibotenate lesions and striatal tissue grafting. *Neuroscience* **16,** 799–817.
Isacson O., Brundin P., Kelly P., Gage F. H., and Bjorklund A. (1984) Functional neuronal replacement by grafted striatal neurons in the ibotenic acid-lesioned rat striatum. *Nature* **311,** 458–460.
Isacson O., Riche D., Hantraye P., Sofroniew M. V., and Maziere M. (1989) A primate mode of Huntington's disease; cross-species implantation of striatal precursor cells to the excitotoxically lesioned baboon caudate-putamen. *Exp. Brain Res.* **75,** 213–220.
Isacson O., Dawbarn D., Brundin P., Gage F. H., Emson P. C., and Bjorklund A. (1987a) Neural grafting in a rat model of Huntington's disease:

striosomal-like organization of striatal grafts as revealed by acetylcholinesterase histochemistry, immunocytochemistry and receptor autoradiography. *Neuroscience* **22,** 481–497.

Isacson O., Pritzel M., Dawbarn D., Brundin P., Kelly A. T., Wiklund L., Emson P. C., Gage F. H., Dunnett S. B., and Bjorklund A. (1987b) Striatal neural transplants in the ibotenic acid-lesioned rat neostriatum: cellular and functional aspects, in *Cell and Tissue Transplantation into the Adult Brain,* vol. 495 (Azmitia E. C. and Bjorklund A., eds.), 537–555 Annals of the New York Academy of Sciences, New York, pp. 537–555.

James W. E., Mefferd R. B., and Kimbell I. (1969) Early signs of Huntington's chorea. *Dis. Nerv. Syst.* **30,** 556–559.

Jervis G. A. (1963) Huntington's chorea in childhood. *Arch. Neurol.* **9,** 244–257.

Jessel T. M., Emson P. C., Paxinos G., and Cuello A. C. (1978) Topographic projections of substance P and GABA pathways in the striato- and pallido-nigral system: A biochemical and immunohistochemical study. *Brain Res.* **152,** 487–498.

Johnson J. W. and Ascher P. (1987) Glycine potentiates the NMDA response in cultured mouse brain neurons. *Nature* **325,** 529–531.

Joyce J. N., Sapp D. W., and Marshall J. F. (1986) Human striatal dopamine receptors are organized in compartments. *Proc. Natl. Acad. Sci. USA* **83,** 8002–8006.

Kanazawa I., Bird E. D., Gale J. S., Iversen L. L., Jessell T. M., Muramoto O., Spokes E.G., and Sutoo D. (1979) Substance P: decrease in substantia nigra and globus pallidus in Huntington's disease. *Adv. Neurol.* **23,** 495–504.

Kemp J. A., Foster A. C., and Wong E. H. K. (1987) Non-competitive antagonists of excitatory amino acid receptors. *Trends Neurosci.* **10,** 294–298.

Kemp J. A., Foster A. C., Leeson P. D., Priestley T., Tridgett R., Iversen L. L., and Woodruff G. N. (1988) 7-chlorokynurenic acid is a selective antagonist at the glycine modulatory site of the *N*-methyl-D-aspartate receptor complex. *Proc. Natl. Acad. Sci. USA* **85,** 6547–6550.

Kesslak J. P., Neito-Sampedro M., Globus J., and Cotman C. W. (1986) Transplants of purified astrocytes promote behavioral recovery after frontal cortex ablation. *Exp. Neurol.* **92,** 377–390.

Kim J. S., Kornhuber H. H., Holtzmuller B., Schmid Burgk W., Mergner T. and Krzepinski G. (1980) Reduction of cerebrospinal fluid glutamic acid in Huntington's chorea and in schizophrenic patients. *Arch. Psychiat. Nervenkr.* **228,** 7.

Kish S. J., Shannack K., and Hornykiewicz O. (1987) Elevated serotonin and reduced dopamine in subregionally divided Huntington's disease striatum. *Ann. Neurol.* **22,** 386–389.

Klawans H. L. and Rubovits R. (1974) The pharmacology of tardive dyskinesia and some animal models. *Excerpta Med. Int. Cong. Ser.* **359,** 355.

Klawans H. L., Geotz G. G., and Westheimer R. (1972) Pathophysiology of schizophrenia and the striatum. *Dis. Nerv. Syst.* **33,** 711–719.

Klechner N. W. and Dingledine R. (1988) Requirements for glycine in activation of NMDA receptors expressed in Xenopus Oocytes. *Science* **241,** 835–838.

Kloog Y., Nadler V., and Sokolovsky M. (1988) Mode of binding of [^{3}H]dibenzocycloalkenimine (MK-801) to the *N*-methyl-D-aspartate (NMDA) receptor and its therapeutic implication. *FEBS Lett.* **230,** 167–170.

Kochhar A., Zivin J. A., Lyden P. D., and Mazzarella V. (1988) Glutamate antagonist therapy reduces neurologic deficits produced by focal central nervous system ischemia. *Arch. Neurol.* **45,** 148–153.

Koestner A. (1973) Animal model for dyskinetic disorders. *Adv. Neurol.* **1,** 625–645.

Koh J., Peters S., and Choi D. W. (1986) Neurons containing NADPH-diaphorase are selectively resistant to quinolinate toxicity. *Science* **234,** 73–76.

Koh J. Y. and Choi D. W. (1988) Cultured striatal neurons containing NADPH-diaphorase or acetylcholinesterase are selectively resistent to injury by NMDA agonists. *Brain Res.* **446,** 374–378.

Korenyi C. and Whittier J. R. (1973) The juvenile form of Huntington's chorea: its prevalence and other observations. *Adv. Neurol.* **1,** 75–85.

Kromer L. F. and Cornbrooks L. S. (1985) Transplants of Schwann cell cultures promote axonal regeneration in adult mammalian brain. *Proc. Natl. Acad. Sci. USA* **82,** 6330–6334.

Lange H., Thorner G., Hopf A. and Schroeder K. F. (1976) Morphometric studies of the neuropathological changes in choreatic diseases. *J. Neurol. Sci* . **28,** 401–425.

Lange H. W. (1981) Quantitative changes of telencephalon, diencephalon, and mesencephalon in Huntington's chorea, postencephalitic and idiopathic parkinsonism. *Verh. Anat. Ges.* **75,** 923–925.

London E. D., Yamamura H. I., Bird E. D., and Coyle J. T. (1981) Deceased receptor-binding sites for kainic acid in brains of patients with Huntington's disease. *Biol. Psychiatr.* **16,** 155–162.

Low W. C., Lewis P. R., Bunch S. B., Dunnett S. B., Thomas S. D., Iversen L. L., Bjorklund A., and Stenevi U. (1982) Neural transplants of embryonic septal nuclei into adult rats with septohippocampal lesions: the recovery of function. *Nature* **300,** 260–262.

Lu S. Y., Giordano M., Norman A. B., Shipley M. T., and Sanberg P. R. (1990) Behavioral effects of neural transplants into the intact striatum. *Pharmacol. Biochem. Behav.* **37,** 135–148.

Mann J. J., Stanley M., Gershon S., and Rosser M. (1980) Mental symptoms in Huntington's disease and a possible primary aminergic neuron lesion. *Science* **210,** 1369–1371.

Martin J. B. (1989) Molecular genetic studies in the neuropsychiatric disorders. *Trends Neurosci.* **12,** 130–136.

Mason S. T. and Fibiger H. C. (1979) Kainic acid lesions of the striatum in rats mimic the spontaneous motor ab-normalities of Huntington's disease. *Neuropharmacology* **18,** 403–407.

Mason S. T., Sanberg P. R., and Fibiger H. C. (1978) Kainic acid lesions of the striatum dissociate amphetamine and apomorphine stereotypy: similarities to Huntington's chorea. *Science* **201,** 352–355.

Mattson B., Gottfries C. E., Roos B. E., and Winblad B. (1974) Huntington's chorea: pathology and brain amines. *Acta Psychiat. Scand.* **255,** 269–277.

McAllister J. P. (1987) Tritiated thymidine identification of embryonic neostriatal transplants, in *Cell and Tissue Transplantatian into the Adult Brain,* vol. 495 (Azmitia E. C. and Bjorklund A., eds.), Annals of the New York Academy of Sciences, New York, pp. 745–748.

McAllister J. P. and Norelle A. (1986) Tritiated thymidine analysis of neostriatal primordia transplanted into adult neostriata. *Soc. Neurosci. Abstr.* **12,** 1478.

McAllister J. P., Kaplan L., and Reynolds M. C. (1984) Morphology and connectivity of fetal neostriatal tissue transplanted into the neostriatum of adult hosts. *Anat. Rec.* **208,** 107A.

McAllister J. P., Walker P. D., and Way J. S. (1986) Evidence suggesting minimal connectivity between neostriatal grafts and host brain. *Soc. Neurosci. Abstr.* **12,** 1479.

McAllister J. P., Walker P. D., Zemanick M. C., Weber A. B., Kaplan L. I., and Reynolds M. A. (1985) Morphology of embryonic neostriatal cell suspensions transplanted into adult neostriata. *Develop. Brain Res.* **23,** 282–286.

McBean G. J. and Roberts P. J. (1985) Neurotoxicity of L-glutamate and DL-threo-3-hydroxyaspartate in the rat striatum. *J. Neurochem.* **44,** 246 –254.

McGeer P. L. and McGeer E. G. (1976a) Enzymes associated with the metabolism of catecholamines, acetylcholine and GABA in human controls and patients with Parkinson's disease and Huntington's chorea. *J. Neurochem.* **26,** 65–76.

McGeer E. G. and McGeer P. L. (1976b) Duplication of biochemical changes of Huntington's chorea by intrastriatal injections of glutamic and kainic acids. *Nature* **263,** 517–519.

McGeer P. L. and McGeer E. G. (1982) Kainic acid: the neurotoxic breakthrough. *CRC Crit. Rev. Toxicol.* **10,** 1–26.

McGeer P. I., Eccles J. C., and McGeer E. G. (eds.) (1987) *Molecular Neurobiology of the Mammalian Brain*. Plenum, New York/London.

McGeer P. L., Kimura H., and McGeer E. G. (1984) Transplantation of newborn brain tissue into adult kainic-acid-lesioned neostriatum, in *Neural Transplants* (Sladek J. R. and Gash D. M., eds.), Plenum, New York, pp. 361–371.

McGeer P. L., McGeer E. G. and Fibiger H. C. (1973) Choline acetylase and glutamic acid decarboxylase in Huntington's chorea. *Neurology* **23,** 912–917.

McGeer E. G., McGeer P. L., Hattori T., and Vincent S. R. (1979) Kainic acid neurotoxicity and Huntington's disease. *Adv. Neurol.* **23,** 577–591.

McIntosh G. E., Jameson D., and Markesbury W. R. (1978) Huntington's disease associated with Alzheimer's disease. *Ann. Neurol.* **3,** 545–548.

McKenzie G. M. and Viik K. (1975) Chemically induced choreiform activity: antagonism by GABA and EEG patterns. *Exp. Neurol.* **46,** 229–234.

McKenzie G. M., Gordon R. J., and Viik K. (1972) Some biochemical and behavioral correlates of a possible animal model of human hyperkinetic syndromes. *Brain Res.* **47,** 439–456.

Melamed E., Hefti F., and Bird E. D. (1982) Huntington's chorea is not associated with hyperactivity of dopaminergic neurons. Studies in postmortem tissues and in rats with kainic acid lesions. *Neurology* **32,** 640–644.

Meyer D. K., Beinfeld M. C., Oertel W. H., and Brownstein M. J. (1982) Origin of the cholecystokinin-containing fibers in the caudatoputamen. *Science* **215,** 187,188.

Moore R. Y. and Card J. P. (1984) Noradrenaline-containing neuron systems, in *Handbook of Chemical Neuroanatomy,* vol. 2, *Classical Transmitters in the CNS, part I* (Bjorklund A. and Hokfelt T., eds.), Elsevier, Amsterdam, pp. 123–156.

Mroz E. A., Brownstein M. J., and Leeman S. E. (1977) Evidence for substance P in the striato-nigral tract. *Brain Res.* **125,** 305–311.

Muller H. W. and Seifert W. A. (1982) A neurotrophic factor (NTF) released from primary glial cultures supports survival and fiber outgrowth of cultured hippocampal neurons. *J. Neurosci. Res.* **8,** 195–204.

Murphy D. L. and Dill R. E. (1972) Chemical stimulation of discrete brain loci as a method of producing dyskinesia models in primates. *Exp. Neurol.* **34,** 244–254.

Neill D. B., Ross J. F., and Grossman S. P. (1974) Comparison of the effects of frontal striatal and septal lesions in paradigms thought to measure incentive motivation or behavioral inhibition. *Physiol. Behav.* **13,** 297.

Neito-Sampedro M. and Cotman C. W. (1985) Growth factor induction and temporal order in CNS repair, in*Synaptic Plasticity* (Cotman C. W., ed.), pp. 407–455, Guilford Press, New York.

Nilsson O. G., Shapiro M. L., Gage F. H., Olton D. S. and Bjorklund A. (1987) Spatial learning and memory following fimbria fornix transection and grafting of fetal septal neurons to the hippocampus. *Exp. Brain Res.* **67,** 195–215.

Norman A. B., Giordano M., and Sanberg P. R. (1989a) Fetal striatal tissue grafts into excitotoxin-lesioned striatum: pharmacological and behavioral aspects. *Pharmacol. Biochem. Behav.* **34,** 139–147.

Norman A. B., Lehman M., and Sanberg P. R. (1989b) Functional effects of fetal striatal transplants. *Brain Res. Bull.* **22,** 163–172.

Norman A. B., Calderon S. F., Giordano M., and Sanberg P. R. (1988a) Striatal tissue transplants attenuate apomorphine-induced rotational behavior in rats with unilateral kainic acid lesions. *Neuropharmacology* **27,** 333–336.

Norman A. B., Calderon S. F., Giordano M., and Sanberg P. R. (1988b) A novel rotational behavior model for assessing the restructuring of striatal dopamine effector systems: Are transplants sensitive to peripherally acting drugs? in *Transplantation into the Mammalian CNS, Prog. Brain*

Res., vol. 78 (Gash D. M. and Sladek J. R., eds.), Elsevier, Amsterdam, pp. 61–67.

Norman A. B., Ford L. M., Kolmonpunporn M., and Sanberg P. R. (1990) Chronic treatment with MK-801 increases the quinolinic acid-induced loss of D-1 dopamine receptors in rat striatum. *Eur. J. Pharmacol.* **176,** 363–366.

Norman A. B., McGowan T., Calderon S. F., Giordano M., and Sanberg P. R. (1987) Attenuation of apomorphine-induced rotational behavior by fetal striatal tissue transplants in rats with unilateral striatal kainic acid lesions. *Soc. Neurosci. Abstr.* **13,** 785.

Oka H. (1980) Organization of the cortico-caudate projections. *Exp. Brain Res.* **40,** 203–208.

Okuno E., Kohler C., and Schwarcz R. (1987) Rat 3-hydroxyanthranilic acid oxygenase purification from the liver and immunocytochemical localization in the brain. *J. Neurochem.* **49,** 771–780.

Olney J. W. and de Gubareff T. (1978) Glutamate neurotoxicity and Huntington's chorea. *Nature (London)* **271,** 557–559.

Oyanagi K. and Ikuta F. A. (1987) A morphometric reevaluation of Huntington's chorea with special reference to the large neurons of the neostriatum. *Clin. Neuropathol.* **6,** 71–79.

Oyanagi K., Takeda S., Takahashi H., Ohama E., and Ikuta F. (1989) A quantitative investigation of the substantia nigra in Huntington's disease. *Ann. Neurol.* **26,** 13–19.

Perry T. L., Hansen S., and Kloster M. (1973) Huntington's chorea: deficiency of GABA in brain. *New Engl. J. Med.* **288,** 337–342.

Pinel C. (1976) Huntington's chorea. *Nursing Times* **72,** 447.

Pisa M., Sanberg P. R., and Fibiger H. C. (1980) Locomotor activity, exploration and spatial alternation learning in rats with striatal injections of kainic acid. *Physiol. Behav.* **24,** 1120.

Pisa M., Sanberg P. R., and Fibiger H. C. (1981) Striatal injections of kainic acid selectively impair serial memory performance in the rat. *Exp. Neurol.* **74,** 633–653.

Podoll K., Caspary P., Lange H. W., and Noth J. (1988) Language functions in Huntington's disease. *Brain* **111,** 1475–1503.

Pohl T. M., MacDonald M. E., Smith B., Poutska A., Volinia S., Searle S., Wasmuth J. J., Gusella J., Lehrach H., and Frischauf A. -M. (1988) Construction of a Not 1 linking library and isolation of new markers close to the Huntington's disease gene. *Nucleic Acids Res.* **16,** 9185–9197.

Pritzel M., Isacson O., Brundin P., Wiklund L., and Bjorklund A. (1986) Afferent and efferent connections of striatal grafts implanted into the ibotenic lesioned neostriatum in adult rats. *Exp. Brain Res.* **65,** 112–126.

Reed T. E. and Chandler J. H. (1958) Huntington's chorea in Michigan. I. Demography and genetics. *Am. J. Hum. Genet.* **10,** 201–225.

Reike G. K., Scarfe A. D., and Hunter J. F. (1983) L-pyroglutamic acid: a neurotoxic amino acid that produces a drug-induced model of

Huntington's disease and with a potential role in the etiology of Huntington's diease. *Soc. Neurosci. Abstr.* **6,** 269.

Reike G. K., Scarfe A. D., and Hunter J. F. (1984a) L-pyroglutamate: an alternate neurotoxin for a rodent model of Huntington's disease. *Brain Res. Bull.* **13,** 443–456.

Reike G. K., Scarfe A. D., Cannon M. S., Hunter J., and Mancall E. (1984b) L-pyroglutamic acid: a toxic amino acid that produces an animal model of Huntington's disease and with a possible etiological significance in HD. *Anat. Rec.* **208,** 147A.

Reike G. K., Smith O. B., Idusuyi J., Semenya R., Howard R., and Williams S. (1989) Chronic intrastriatal L-pyroglutamate: neuropathology and neuron sparing like Huntington's disease. *Exp. Neurol.* **104,** 147–154.

Reiner A., Albin D. L., Anderson K. D., D'Amato C. J., Penny J. B., and Young A. B. (1988) Differential loss of striatal projection neurons in Huntington's disease. *Proc. Natl. Acad. Sci. USA* **85,** 5733–5737.

Reisine T. D., Fields J. Z., Stern L. Z., Johnson P. C., Bird E. D., and Yamamura H. I. (1977) Alterations in dopaminergic receptors in Huntington's disease. *Life Sci.* **21,** 1123–1128.

Reubi J. C. and Cuenod M. (1979) Glutamate release in vitro from corticostriatal terminals. *Brain Res.* **176,** 185–188.

Reynolds I. J., Murphy S. N., and Miller R. J. (1987) [^{3}H]-labeled MK-801 binding to the excitatory amino acid receptor complex from rat brain is enhanced by glycine. *Proc. Natl. Acad. Sci. USA* **84,** 7744–7748.

Reynolds G. P., Pearson S. J., Halket J., and Sandler M. (1988) Brain quinolinic acid in Huntington's disease. *J. Neurochem.* **50,** 1959–1960.

Rezek M., Havlicek, V., Leybin L., Pinsky C., Kroeger E. A., Hughe K. R., and Friesen H. (1977) Neostriatal administration of somatostatin: differential effect of small and large doses on behavior and motor control. *Can J. Physiol. Pharmacol.* **55,** 234–242.

Rhodes K. J., Joyce J. N., Sapp D. W., and Marshall J. F. (1987) [^{3}H]hemicholinium-3 binding in rabbit striatum: correspondence with patchy acetylcholinesterase staining and a method for quantifying striatal compartments. *Brain Res.* **412,** 400–404.

Robbins C., Theilmann J., Youngman S., Haines J., Altherr J., Harper P. S., Payne C., Junker A., Wasmuth J., and Hayden M. R. (1989) Evidence from family studies that the gene causing Huntington disease is teleomeric to D4S95 and D4S90. *Am J. Hum. Genet.* **44,** 422–425.

Roberts R. C. and Difiglia M. (1989) Short- and long-term survival of large neurons in the excitotoxic lesioned rat caudate nucleus: a light and electron microscopic study. *Synapse* **3,** 363–371.

Rodda R. A. (1981) Cerebellar atrophy in Huntington's disease. *J. Neurol. Sci.* **50,** 147.

Roizin L., Kaufman M. A., Wilson W., Stellar S., and Liu J. C. (1976) Neuropathologic observations in Huntington's chorea. *Prog. Neuropathol.* **3,** 447.

Rutherford A., Garcia-Munoz, Dunnett S. B., and Arbuthott G. W. (1987) Electrophysiological demonstration of host cortical inputs to striatal grafts. *Neurosci. Lett.* **83,** 275–281.

Saint-Cyr J. A., Taylor A. E., and Lang A. E. (1988) Procedural learning and neostriatal dysfunction in man. *Brain* **111,** 941–959.

Sanberg P. R. (1980) Haloperidol-induced catalepsy is mediated by postsynaptic dopamine receptors. *Nature (London)* **284,** 472,473.

Sanberg P. R. and Coyle J. T. (1984) Scientific approaches to Huntington's disease. *CRC Crit. Rev. Clin. Neurobiol.* **1,** 1–44.

Sanberg P. R. and Creese I. (1981) Dopamine receptor stimulation and striatal kainic acid neurotoxicity. *J. Pharm. Pharmacol.* **33,** 674,675.

Sanberg P. R. and Fibiger H. C. (1979) Body weight, feeding and drinking behaviors in rats with kainic acid lesions of striatal neurons: with a note on body weight symptomology in Huntington's disease. *Exp. Neurol.* **66,** 444–466.

Sanberg P. R. and Johnston G. A. (1981) Glutamate and Huntington's disease. *Med. J. Aust.* **2,** 460–465.

Sanberg P. R., Fibiger H. C., and Mark R. F. (1981a) Body weight and dietary factors in Huntington's disease patients compared with matched controls. *Med. J. Aust.* **1,** 407–409.

Sanberg P. R., Pisa M. and Fibiger H. C. (1981b) Kainic acid injections in the striatum alter the cataleptic and locomotor effects of drugs influencing the dopaminergic and cholinergic systems. *Eur. J. Pharamacol.* **74,** 347–357.

Sanberg P. R., Hagenmeyer S. H., and Henault M. A. (1985) Automated measurement of multivariate locomotor behavior in rodents. *Neurobehav. Toxicol. Teritol.* **7,** 87–94.

Sanberg P. R., Henault M. A., and Deckel A. W. (1986) Locomotor hyperactivity: effects of multiple striatal transplants in an animal model of Huntington's disease. *Pharmacol. Biochem. Behav.* **25,** 297–300.

Sanberg P. R., Lehmann J., and Fibiger H. C. (1978) Impaired learning and memory after kainic acid lesions of the striatum: a behavioral model of Huntington's disease. *Brain Res.* **149,** 546–551.

Sanberg P. R., Pisa M., and Fibiger H. C. (1979) Avoidance operant and locomotor behavior in rats with neostriatal injections of kainic acid. *Pharmacol. Biochem. Behav.* **10,** 137–144.

Sanberg P. R., Calderon S. F., Garver D. L., and Norman A. B. (1987a) Brain tissue transplants in an animal model of Huntington's disease. *Psychopharamacol. Bull.* **23,** 476–482.

Sanberg P. R., Henault M. A., Hagenmeyer-Houser S. H., and Russell R. H. (1987b) The topography of amphetamine and scopolamine-induced hyperactivity: Toward an activity print. *Behav. Neurosci.* **101,** 131–133.

Sanberg P. R., Henault M. A., Hagenmeyer-Houser S. H., Giordano M., and Russell K. H. (1987c) Multiple transplants of fetal striatal tissue in the kainic acid model of Huntington's disease: behavioral recovery may not be related to acetylcholinesterase, in *Cell and Tissue Transplantation into the Adult Brain*, vol. 495 (Azmitia E. C. and Bjorklund A., eds.) Annals of the New York Academy of Sciences, New York, pp. 781–785.

Sanberg P. R., Calderon S. F., Giordano M., Tew J. M., and Norman A. B. (1989a) The quinolinic acid model of Huntington's disease: locomotor abnormalities. *Exp. Neurol.* **105,** 45–53.

Sanberg P. R., Giordano M., Henault M. A., Nash D. R., Ragozzino M. E., and Hagenmeyer-Houser S. H. (1989b) Intraparenchymal striatal transplants required for maintenance of behavioral recovery in an animal model of Huntington's disease. *J. Neurotransplant.* **1,** 23–31.

Sanberg P. R., Nash D. R., Calderon S. F., Giordano M., Shipley M. T., and Norman A. B. (1988) Neural transplants disrupt the blood–brain barrier and allow peripherally acting drugs to exert a centrally-mediated behavioral effect. *Exp. Neurol.* **102,** 149–152.

Sanberg P. R., Zubrycki E. M., Ragozzino M. E., Lu S. Y., Norman A. B., and Shipley M. T. (1990) NADPH-diaphorase-containing neurons and cytochrome oxidase activity following striatal quinolinic acid lesions and fetal striatal transplants. *Prog. Brain Res.* **82,** 427–431.

Schmidt R. H., Bjorklund A., and Stenevi U. (1981) Intracerebral grafting of dissociated CNS tissue suspensions: a new approach for neuronal transplantation to deep brain sites. *Brain Res.* **218,** 347–356.

Scholz O. B. and Berlemann C. (1987) Memory performance in Huntington's disease. *Int. J. Neurosci.* **35,** 155–162.

Schwarcz R. and Kohler C. (1983) Differential vulnerability of central neurons of the rat to quinolinic acid. *Neurosci. Lett.* **38,** 85–90.

Schwarcz R. and Shoulson I. (1987) Excitotoxins and Huntington's disease, in *Animal Models of Dementia: A Synaptic Neurochemical Perspective* (Coyle J. T., ed.), Alan R. Liss, New York, pp. 39–68.

Schwarcz R., Bennett J. P., and Coyle J. T. (1977) Inhibitors of GABA metabolism: implications for Huntington's disease. *Ann. Neurol.* **2,** 299.

Schwarcz R., Okuno E., and While R. J. (1989) Basal ganglia lesions in the rat: effects of quinolinic acid metabolism. *Brain Res.* **490,** 103–109.

Schwarcz R., Whetsell W. O., Jr., and Mangano R. M. (1983) Quinolinic acid: an endogenous metabolite that produces axon-sparing lesions in rat brain. *Science* **219,** 316–318.

Schwarcz R., Foster A. C., French E. D., Whetsell W. O., and Kohler C. (1984) Excitotoxic models for neurodegenerative disorders. *Life Sci.* **35,** 19–32.

Schwarcz R., Fuxe K., Agnati L. F., Hokfelt T., and Coyle J. T. (1979) Rotational behavior in rats with unilateral striatal kainic acid lesions: a behavioral model for studies of intact dopamine receptors. *Brain Res.* **170,** 485–495.

Segovia J., Meloni R., and Gale K. (1989) Effect of dopaminergic denervation and transplant-derived reinnervation on a marker of striatal GABAergic function. *Brain Res.* **493,** 185–189.

Shoulson I. (1986) Huntington's disease, in *Diseases of the Nervous System* Asbury A., McKhann G. M., and McDonald I. (eds.) Ardmore Medical Books (W. B. Saunders), Philadelphia, pp. 1258–1267.

Sirinathsinghji D. J. S., Dunnett S. B., Isacson O., Clarke D. J., Kendrick K., and Bjorklund A. (1988) Striatal grafts in rats with unilateral neostriatal lesions-II. In vivo monitoring of GABA release in globus pallidus and substantia nigra. *Neuroscience* **24,** 803–811.

Skraastad M. I., Bakker E., de Lange L.F., Vegter-van der Vis M., Klein-Breteler E. G., Ommen G. J. B., and Pearson P. L. (1989) Mapping of recombinations near the Huntington disease locus by using G8 (D4510) and newly isolated markers in the D4S10 region. *Am. J. Hum. Genet.* **44,** 560–566.

Sladek J. R. and Gash D. M. (1988) Nerve cell grafting in Parkinson's disease. *J. Neurosurg.* **170,** 485–495.

Sladek J. R., Redmond D. E., Collier T. J., Blount J. P., Elsworth J. R., Taylor J. R., and Roth R. H. (1988) Fetal dopamine neural grafts: extended reversal of methylphenyltetrahydropyridine-induced parkinsonism in monkeys. *Prog. Brain Res.* **78,** 497–506.

Smith B., Skarecky D., Bengtsson U., Magenis R. E., Carpenter N., and Wasmuth J. J. (1988) Isolation of DNA markers in the direction of the Huntington disease gene from the G8 locus. *Am. J. Hum. Genet.* **42,** 335–344.

Spokes E. G. S. (1980) Neurochemical alterations in Huntington's chorea: a study of post-mortem brain tissue. *Brain* **103,** 179–210.

Stahl W. L. and Swanson P. D. (1974) Biochemical abnormalities in Huntington's chorea brains. *Neurology* **24,** 813–819.

Staines W. A., Nagy J. I., Vincent S. R., and Fibiger H. C. (1980) Neurotransmitters contained in the striatum. *Brain Res.* **194,** 391–402.

Standefer M. J. and Dill R. E. (1978) The role of GABA in dyskinesias induced by chemical stimulation of the striatum. *Life Sci.* **21,** 1515–1519.

Starr A. (1967) A disorder of rapid eye movements in Huntington's chorea. *Brain* **90,** 545–564.

Steinbusch H. W. M. (1984) Serotonin-immunoreactive neurons and their projections in the CNS, in *Handbook of Chemical Neuroanatomy, vol. 3, Classical Transmitters and Transmitter Receptors in the CNS. Part II* (Bjorklund A., Hokfelt T., and Kuhar M. J. eds.), Elsevier, Amsterdam, pp. 68–125.

Stromberg I., Herrera-Marschitz M., Ungerstedt U., Ebendal T., and Olson L. (1985) Chronic implants of chromaffin tissue in to the dopamine-denervated striatum. Effects of NGF on graft survival fiber growth and rotational behavior. *Exp. Brain Res.* **60,** 335–349.

Toth E. and Lajtha A. (1989) Motor effects of intracaudate injection of excitatory amino acids. *Pharmacol. Biochem. Behav.* **33,** 175–179.

Tulipan N., Huang S., Whetsell W. O., and Allen G. S. (1986) Neonatal striatal grafts prevent lethal syndrome produced by bilateral intrastriatal injection of kainic acid. *Brain Res.* **377,** 163–167.

Uhlhous S. and Lange H. (1988) Striatal deficiency of L-pyroglutamic acid in Huntington's disease is accompanied by increased plasma levels. *Brain Res.* **457,** 196–199.

Van Putten T. and Menkes J. H. (1973) Huntington's disease masquerading as chronic schizophrenia. *Dis. Nerv. Syst.* **34,** 54–56.

Vincent S. R., Hokfelt T., Christensson I., and Terenius L. (1982) Immunohistochemical evidence for a dynorphin immunoreactive striatonigral pathway. *Eur. J. Pharmacol.* **85,** 251,252.
Vonsattel J-P., Myers R. H., and Stevens T. J. (1985) Neuropathological classification of Huntington's disease. *J. Neuropathol. Exp. Neurol.* **44,** 559–577.
Walker J. E. (1983) Glutamate, GABA and CNS disease: A review. *Neurochem. Res.* **8,** 521–550.
Walker P. D. and McAllister J. P. (1987) Minimal connectivity between neostriatal transplants and the host brain. *Brain Res.* **425,** 34–44.
Walker P. D., Chovanes G. I., and McAllister J. P. (1987) Identification of acetyl cholinesterase-reactive neurons and neuropil in neostriatal transplants. *J. Comp. Neurol.* **259,** 1–12.
Walsh J. P., Hull C. D., Levine M. S., Zhou F. C., and Buchwald N. A. (1986) Intracellular study of transplanted striatal neurons. *Soc. Neurosci. Abstr.* **12,** 1480.
Walsh J. P., Zhou F. C., Hull C. D., Fisher R. S., Levine M. S., and Buchwald N. A. (1988) Physiological and morphological characterization of striatal neurons transplanted into adult rat striatum. *Synapse* **2,** 37–44.
Wastek G. J. and Yamamura H. I. (1978) Biochemical characterization of the muscarinic cholinergic receptor in human brain: alterations in Huntington's disease. *Mol. Pharmacol.* **14,** 768–774.
Wastek G. J., Stern L. Z., Johnson O. C., and Yamamura H. L. (1976) Huntington's disease: regional alterations in muscarinic cholinergic receptor binding in human brain. *Life Sci.* **19,** 1033–1040.
Waters C. M., Peck R., Rossor M., Reynolds G.P., and Hunt S. P. (1988) Immunocytochemical studies on the basal ganglia and substantia nigra in Parkinson's disease and Huntington's chorea. *Neuroscience* **25,** 419–438.
Whetsell W. O., Jr. and Schwarcz R. (1989) Prolonged exposure to submicromolar concentrations of quinolinic acid causes excitotoxic damage in organotypic cultures of rat corticostriatal system. *Neurosci. Lett.* **97,** 271–275.
Wictorin K. and Bjorklund A. (1989) Connectivity of striatal grafts implanted into the ibotenic acid-lesioned striatum-II. Cortical afferents. *Neuroscience* **30,** 297–311.
Wictorin K., Ouimet C. C., and Bjorklund A. (1989a) Intrinsic organization and connectivity of intrastriatal striatal transplants in rats as revealed by DARPP-32 immunohistochemistry: specificity of connections with the lesioned host brain. *Eur. J. Neurosci.* **1,** 690–701.
Wictorin K., Clarke D. J., Bolam J. P., and Bjorklund A. (1989b) Host corticostriatal fibres establish synaptic connections with grafted striatal neurons in the ibotenic acid lesioned striatum. *Eur. J. Neurosci.* **1,** 189–195.
Wictorin K., Simerly R. B., Isacson O., Swanson L. W., and Bjorklund A. (1989c) Connectivity of striatal grafts implanted into the ibotenic acid lesioned striatum. III. Efferent projecting graft neurons and their relation to host afferents within the grafts. *Neuroscience* **30,** 313–330.
Wictorin K., Isacson O., Fischer W., Nothias F., Peschanski M., and Bjorklund

A. (1988a) Studies on host afferent imputs to fetal striatal transplants in excitotoxically lesioned striatum, in *Transplantations into the Mammalian CNS; Progress in Brain Research*, vol. 78 (Gash D.M. and Sladek J. R., eds.), Elsevier, Amsterdam, pp. 55–60.

Wictorin K., Isacson O., Fischer W., Nothias F., Peschanski M., and Bjorklund A. (1988b) Connectivity of striatal grafts implanted into the ibotenic acid lesioned striatum. I. Subcortical afferents. *Neuroscience* **27,** 547–562.

Wilson C. J., Emson P., and Feles C. (1987) Electrophysiological evidence for the formation of a corticostriatal pathway in neostriatal tissue grafts. *Soc. Neurosci. Abst.* **13,** 79.

Wong E. H. F., Knight A. R., and Ransom R. (1987) Glycine modulates [^{3}H]MK-801 binding to the NMDA receptor in rat brain. *Eur. J. Pharmacol.* **142,** 487,488.

Wong E. H. F., Kemp J. A., Priestley T., Knight A. R., Woodruff G. N., and Iversen L. L. (1986) The anticonvulsant MK-801 is a potent *N*-methyl-D-aspartate antagonist. *Proc. Natl. Acad. Sci. USA* **83,** 7104–7108.

Wong P. T.-H., Singh U. K., and McGeer E. G. (1982) Ornithine aminotransferase in Huntington's disease. *Brain Res.* **231,** 466–471.

Yang H.-Y. T., Panula P., Tang J., and Costa E. (1983) Characterization and localization of Met^5-enkephalin-Arg^6-Phe stored in various rat brain regions. *J. Neurochem.* **40,** 969–976.

Young A. B., Greenamyre J. T., Hollingsworth Z., Albin R., D'Amato C., Shoulson I., and Penny J. B. (1988) NMDA receptor losses in putamen from patients with Huntington's disease. *Science* **241,** 981–983.

Young A.B., Shoulson I., Penny J.B., Starosta-Rubinstein S., Gomez F., Travers H., Ramos M., Snodgrass S. R., Bonilla A., Moreno H., and Wexler N. (1986) Huntington's disease in Venezuela: Neurological features and functional decline. *Neurology* **36,** 244–249.

Zaczek R., Schwarcz R., and Coyle J. T. (1978) Long-term sequelae of striatal kainate lesion. *Brain Res.* **152,** 626–632.

Zemanick M. D., Walker P. D., and McAllister J. P. (1987) Quantitative analysis of dendrites from transplanted neostriatal neurons. *Brain Res.* **414,** 149–152.

Zhou F. C., and Buchwald N. (1989) Connectivities of the striatal grafts in adult rat brain: a rich afference and scant striatonigral efference. *Brain Res.* **504,** 15–30.

Zhou F. C., Hull C. D., Levine M. S., and Buchwald N. A. (1986) Host nigral fibers may grow into striatal neurons.

Note added in proof: Since the writing of this manuscript, a number of important papers have become available. Readers are referred to the following:

Emerich D. F., Zubrick E. M., Shipley M. T., Norman A. B., and Sanberg P. R. (1991) Female rats are more sensitive to the locomotor alterations following quinolinic acid-induced striatal lesions: effects of striatal transplants. *Exp. Neurol.* **111,** 369–378.

Giordano M., Ford L. M., Shipley M. T., and Sanberg P. R. (1990) Neural grafts and pharmacological intervention in a model of Huntington's disease. *Brain Res. Bull.* **25,** 453–465.

Knopman D. and Nissen M. J. (1991) Procedural learning is impaired in Huntington's disease: evidence from the serial reaction time task. *Neuropsychologia* **29,** 245–254.

Lu S. Y., Shipley M. T., Norman A. B., and Sanberg P. R. (1991) Striatal, ventral mesencephalic and cortical transplants into the intact rat striatum: a neuroanatomical study. *Exp. Neurol.* **113,** 109–130.

Majewska M. D. and Bell J. A. (1990) Ascorbic acid protects neurons from injury induced by glutamate and NMDA. *NeuroReport* **1,** 194–196.

Massman P. J., Delis D. C., and Butters N. (1990) Are all subcortical dementias alike?: Verbal learning and memory in Parkinson's and Huntington's disease patients. *J. Clin. Exp. Neuropsychol.* **12(5),** 729–744.

Meldrum B. and Garthwaite J. (1990) Excitotory amino acid neurotoxicity and neurodegenerative disease. *TIPS* **11,** 379–386.

Milasius A. M., Grinevicius K.-K. A., and Lapin I. P. (1990) Effect of quinolinic acid on wakefulness and sleep in the rabbit. *J. Neural. Transm.* [GenSect] **82,** 67–73.

Nash D. R., Kaplan S. M., Norman A. B., and Sanberg P. R. (1991) An evaluation of the possible protective effects of neonatal striatal transplants against kainic acid-induced lesions. *J. Neurol. Trans.* **2(1),** 75–79.

Norman A. B., Ford L. M., and Sanberg P. R. (1991) Differential loss of neurochemical markers following quinolinic acid-induced lesions of rat striatum. *Exp. Neurol.* **114,** 132–135.

Norman A. B., Thomas S. R., Pratt R. G., Samaratunga R. C., and Sanberg P. R. (1990) T_1 and T_2 weighted magnetic resonance imaging of excitotoxin lesions and neural transplants in rat brain *in vivo*. *Exp. Neurol.* **109,** 164–170.

Pearlman S. H., Levivier M., Collier T. J., Sladek J. R., Jr., and Gash D. M. (1991) Striatal implants protect the host striatum against quinolinic acid toxicity. *Exp. Brain Res.* **84,** 303–310.

Sanberg P. R., Zubrycki E., Ragozzino M. E., Giordano M., and Shipley M. T. (1990) Tyrosine hydroxylase-positive fibers and neurons in transplanted striatal tissue in rats with quinolinic acid lesions of the striatum. *Brain Res. Bull.* **25,** 889–894.

Wictorin K., Lagenaur C. F., Lund R. D., and Bjorklund A. (1990) Efferent projections to the host brain from intrastriatal striatal mouse-to-rat grafts: time course and tissue-type specificity as revealed by a mouse specific neuronal marker. *Eur. J. Neurosci.* **3,** 86–101.

Zubrycki E. W., Emerich D. F., and Sanberg P. R. (1990) Sex differences in regulatory changes following quinolinic acid-induced striatal lesions. *Brain Res. Bull.* **25,** 633–637.

Rodent Models of Parkinson's Disease

François B. Jolicoeur and Robert Rivest

1. Introduction

The cardinal symptoms of Parkinson's disease are rigidity, akinesia, and tremors. Secondary symptoms include postural abnormalities and neuropsychiatric disturbances such as depression, cognitive disorders, and apparent apathy (Barbeau, 1979; Schultz, 1984; Marsden et al., 1975). The basic neuropathology of Parkinson's disease involves degeneration of the heavily pigmented cells of the substantia nigra, locus ceruleus, and other brainstem nuclei (Hornykiewicz, 1972; Barbeau, 1979; Hornykiewicz and Kish, 1986). A markedly decreased concentration of dopamine (DA) and its metabolites is the main neurochemical change in the disease, although other neurotransmitters such as norepinephrine (NE) and serotonin (5-hydroxytryptamine, 5-HT) are also reduced (Hornykiewicz, 1972; Schultz, 1984; Agid and Javoy-Agid, 1985; Hornykiewicz and Kish, 1986).

Attempts to simulate Parkinson's disease in animals have been numerous. As pointed out previously (Duvoisin, 1976), Charcot was probably the first to rely on an animal model to investigate parkinsonism. Having observed that removal of the medulla, but not other brain structures, abolished nicotine-induced tremors in frogs, he hypothesized that this symptom of Parkinson's disease originated in the lower brainstem (Charcot

From: *Neuromethods, Vol. 21: Animal Models of Neurological Disease, I*
Eds: A. Boulton, G. Baker, and R. Butterworth

and Vulpian, 1862). Since then, many animal models have been proposed, and it is the purpose of this chapter to review the principal models of Parkinson's disease that have been developed using rats and mice.

Before discussing individual models, a few general comments seem in order. The often-expressed objectives of animal analogs of parkinsonism are to first replicate the disease in an animal so that the examination of the model will further our understanding of the underlying pathology of the human disease. After surveying the literature, it appears that this goal has yet to be realized. The recreation of the disease in animals has, to date, simply mimicked specific neuroanatomical, neurochemical, or neurobehavioral anomalies found in humans without extending our knowledge of the processes by which these alterations initially occur and continue to progress. Few models have addressed the constellation of symptoms. However, these shortcomings are not specific to the development of models for Parkinson's disease but probably represent a problem inherent to all animal models of neuropsychiatric disorders. The second objective is to provide an in vivo bioassay so that more effective and safer treatment strategies can be developed. It is doubtful that this objective has been met. After all, the treatment of choice in Parkinson's disease is still L-DOPA, almost 30 years after its discovery. Although this treatment constituted a major breakthrough in the management of Parkinson's disease at the time, it does not arrest the relentless progression of the disease. It ameliorates mostly the akinetic facet of the disorder and is not without important side effects. However, this does not mean that our attempts to provide an animal model of Parkinson's disease have been in vain. These endeavors have contributed enormously to the fundamental understanding of the complex interactions regulating the activity of the mesostriatal pathway as well as other dopaminergic fibers. Furthermore, the available models constitute an important data base from which improved, more comprehensive models can be established.

2. Rodent Models of Parkinson's Disease

2.1. Reserpine

By interfering with vesicular uptake and storage of amines, reserpine produces, both centrally and peripherally, prolonged depletion of concentrations of NE, DA, and serotonin. Administration of reserpine to both rats and mice induces hypokinesia, muscle rigidity, and tremors (Duvoisin and Marsden, 1974; Goldstein et al., 1975; Moss et al., 1981; Johnels, 1983; Colpaert, 1987). In rats, hypokinesia and tremors are obtained with 2.5 mg/kg, whereas rigidity first appears after administration of 10 mg/kg. Maximal intensity of all three symptoms follows administration of 40 mg/kg (Colpaert, 1987). Also, variability between and within animals appears to be minimal with this relatively higher dose of the drug (Goldstein et al., 1975; Colpaert, 1987).

Intensity of the tremors is maximal approx 40 min after injections, whereas hypokinesia and rigidity reach a peak 60–100 min following administration (Colpaert, 1987). Tremor and rigidity remain stable for approx 3 h and then gradually dissipate during the next 5 h (Colpaert, 1987). Hypokinesia endures for up to 24 h and gradually subsides by 48 h (Fischer and Heller, 1967; Colpaert, 1987).

Of the three main neurobehavioral signs induced by reserpine, muscle rigidity has been the most frequently and systematically examined. The supraspinal origin of reserpine-induced rigidity was demonstrated by the finding that it could not be produced in decerebrated animals (Morrison and Webster, 1973). More specifically, disturbances in striatal dopaminergic transmission appear to be responsible for reserpine-induced rigidity. Microapplication of the drug into the striatum but not in the nucleus accumbens induces rigidity in rats (Johnels, 1983); this can be reversed by peripheral administration of apomorphine. This effect of apomorphine is blocked by prior administration of the DA antagonist trifluoperazine into the striatum (Johnels, 1983). Moreover, microinjections of apomorphine in the striatum reverse rigidity induced by systemic administration of reserpine (Anden and Johnels, 1977). Reserpine rigidity has been associated with increased alpha and decreased gamma motoneuron activity (Steg, 1964a; Morrison and Webster, 1973). This contrasts with human parkinsonian rigidity, which results from the activation of both alpha and gamma motoneurons (Schultz, 1984).

The ability of a variety of drugs to antagonize reserpine-induced symptoms has been the subject of several studies. In the majority of these reports, rigidity was the sole symptom investigated. Anticholinergic drugs antagonize muscle rigidity (Morrison and Webster, 1973; Goldstein et al., 1975; Colpaert, 1987) but have little or no effect on hypokinesia and tremor induced by reserpine (Colpaert, 1987). In human parkinsonism, anticholinergics are of little benefit except in the early stages of the disease (Barbeau, 1979; Riederer et al., 1984). Administration of L-DOPA reverses all symptoms of reserpine, an effect that is more pronounced when the precursor is administered in combination with a DOPA decarboxylase inhibitor (Goldstein et al., 1975; Colpaert, 1987). This finding in animals parallels the beneficial effects of these substances in pharmacological management of the human disease (Barbeau, 1979; Riederer et al., 1984). Other agents stimulating dopaminergic systems—such as, methamphetamine, apomorphine, bromocriptine, and amantadine, as well as various monoamine oxidase inhibitors—were able to markedly attenuate all three facets of the reserpine syndrome (Colpaert, 1987; Goldstein et al., 1975; Gancher et al., 1990).

Nomifensine, a DA reuptake blocker, 5-HT antagonists, and histamine antagonists, as well as tricyclic antidepressants had minimal or no effects in this model (Colpaert, 1987). Surprisingly, the purported alpha-1 adrenoceptor agonist phenylephrine and both alpha-2 adrenoceptors agonists and antagonists, clonidine and yohimbine, respectively, prevented the usual tremors and rigidity but not the hypokinesia induced by reserpine (Colpaert, 1987). These drugs have no clinical usefulness in the management of Parkinson's disease.

The reserpine model of Parkinson's disease in rodents offers many similarities to but also important differences from the human disease. The three cardinal signs of Parkinson's disease are produced by the drug. Known antiparkinson agents used do block or reduce the behavioral manifestations of reserpine-treated animals. However, when comparing the reserpine model with the human disease, significant differences emerge. Reserpine induces a sudden depletion of amines without producing degeneration of catecholaminergic fibers. Also, the pervasive action of the drug on several biogenic amines, both centrally and peripherally, does not reflect the underlying pathology in Parkinson's disease. Finally, as mentioned above, adrenoceptor drugs have marked actions on reserpine's effects but, to our knowledge, have no therapeutic value in the human disease.

2.2. MPTP

The finding that humans ingesting 1-methyl-4-phenyl-1,2,3,6-tetrahydropyridine (MPTP) develop parkinson-like symptoms has prompted, in the past decade, the investigation of the effects of this toxin in animals. Although MPTP does produce both neurochemical and neurobehavioral effects in primates that are akin to Parkinson's disease, studies in rodents have, for the most part, been deceiving. In mice, MPTP has been shown in several studies to dramatically lower striatal content of DA and its metabolites (Johannessen et al., 1985; Heikkila et al., 1984). However, descriptions of any behavioral consequences of such drastic neurochemical alterations are absent from these studies. Rats appear to be impervious to the toxic action of MPTP (Murphy and Snyder, 1982; Heikkila et al., 1984; Sahgal et al., 1984; Johannessen et al., 1985). The biotransformation of MPTP to its metabolite, 1-methyl-4-phenylpyridine (MPP^+), by a type B monoamine oxidase, appears to be responsible for the toxic action on nigrostriatal neurones (Johannessen et al., 1985). This is clearly shown by the fact that the administration of an inhibitor of this enzyme prevents the neurotoxic actions of MPTP (Jarvis and Wagner, 1985). Furthermore, direct infusion of MPP^+ but not MPTP into the substantia nigra of rats results in a marked decrease in the striatal content of dopamine and its metabolites. These neurochemical changes are accompanied by a decrease in spontaneous motor activity and the emergence of muscular rigidity (Bradbury et al., 1986). Whether this treatment induced tremors was not mentioned. Clearly, the utility of MPTP treatment as a model of Parkinson's disease in rodents remains to be established.

2.3. Neuroleptics

Administration of neuroleptics frequently leads to the development of Parkinson-like neurological signs in humans. This is particularly true following prolonged treatment with doses at the high end of the therapeutic spectrum. Presumably, these side effects are mediated via blockade of striatal DA receptors. In rodents, the most obvious behavioral consequence of acute injection of a DA antagonist is the appearance of catalepsy. With chronic administration, hypokinesia is accompanied by tremors and rigidity (Maickel et al., 1974). For example, rats receiving two injections of 2 mg/kg chlorpromazine daily will display akinesia and rigidity after 7 d of administration and will show the triad of parkinson-like symptoms following 3–4 wk of daily injections (Maickel et al., 1974). The ability of antiparkinson drugs to counteract these manifestations in animals has been the subject of several studies (Simon et al., 1970; Van Woert et al., 1974; Kulkarni et al., 1980; Arnt and Christensen, 1981; Arnt et al., 1981). In most of these reports, catalepsy was the sole neurobehavioral sign investigated. Neuroleptic-induced catalepsy has been shown to be reversed by DA agonists (Simon et al., 1970). However, it is noteworthy that this symptom is only partially antagonized by L-DOPA (Derkach et al., 1974). Obviously, this contrasts to L-DOPA's effects in humans. Furthermore, anticholinergics have been shown repeatedly to be quite effective in attenuating catalepsy produced by neuroleptic administration (Van Woert et al., 1974; Kulkarni et al., 1980; Arnt and Christensen, 1981; Arnt et al., 1981). Anticholinergic agents are useful to counter the neuroleptic-induced extrapyramidal symptoms in humans, but they are of limited value in the management of idiopathic Parkinson's disease (Hornykiewicz, 1975; Barbeau, 1979; Riederer et al., 1984).

Although neuroleptics do induce Parkinson-like symptoms in humans and in animals, their usefulness in producing an animal model of Parkinson's disease is limited. Contrary to the situation with Parkinson's disease, the neurobehavioral symptoms produced by neuroleptics are not owing to a degeneration of dopaminergic fibers but are attributable to a temporary blockade of DA transmission. Second, as mentioned above, the effects of pharmacological intervention in this animal model are quite different from those seen in human parkinsonism.

2.4. Cholinomimetics

Striatal cholinergic neurons receive inhibitory afferents from mesostriatal dopaminergic fibers. Dopamine-induced inhibition of striatal acetylcholine release has been demonstrated in both in vitro and in vivo studies (DeBelleroche et al., 1982; Ajima et al., 1990).

It is generally accepted that in Parkinson's disease striatal cholinergic neurones are hyperactive because of a loss of the nigrostriatal dopaminergic inhibitory influence. In accordance with this view, administration of cholinomimetics to Parkinson's patients has been reported to exacerbate the extrapyramidal symptoms of the disease (Hornykiewicz, 1975). In rodents, both intracranial and systemic administration of various cholinergic stimulants induce tremors (Leonard, 1972; Matthews and Chiou, 1979; Dickinson and Slater, 1982). Cholinomimetics also have been reported to induce rigidity and akinesia, although these symptoms have received much less experimental attention (Dickinson and Slater, 1982). Peripheral administration of cholinergic agonists such as tremorine or its metabolite, oxotremorine, is the most frequently used procedure to induce tremors in mice; nicotine and physostigmine have also been utilized (Dandiya and Bhargava, 1968; Horst et al., 1973; Kulkarni et al., 1980). It has been shown that these tremors can be reversed by centrally acting anticholinergic agents (Nose and Kojima, 1970; Kulkarni et al., 1980), DA agonists, L-DOPA, and monoamine oxidase inhibitors (Dandiya and Bhargava, 1968; Horst et al., 1973; Kulkarni et al., 1980; Cody et al., 1986). There are also reports that cholinomimetic-induced tremors can be inhibited by a wide variety of unrelated compounds, including antihistamines, phenothiazines, and the tricyclic antidepressant imipramine (Nose and Kojima, 1970; Kulkarni et al., 1980). Furthermore, in one study, oxotremorine-induced tremors were prevented by peripheral administration of DA, despite the known inability of this amine to pass the blood-brain barrier (Horst et al., 1973). Peripheral administration of cholinomimetics to rats also induces tremors (Dandiya and Bhargava, 1968; Kulkarni et al., 1980; Hallberg and Almgren, 1987; Ray and Poddar, 1990), which have been shown recently to be blocked by central β-adrenoceptor antagonists (Hallberg and Almgren, 1987). Another method used to produce tremors in rats is the direct injection of a variety of cholinergic stimulants into the striatum (Dill et al., 1968; Matthews and Chiou, 1978,1979). For example, unilateral administration of carbachol (0.5–1.5 μg) produces contralateral bursts of forelimb tremors occurring at irregular intervals for up to 1 h after administration (Dill et al., 1968). At higher doses a "caudate stimulation behavior syndrome" results, which is described as including generalized excitement, limb rigidity, salivation, defecation, urination, stereotypy, and occasionally, convulsions. (Matthews and Chiou, 1978). Tremors following intrastriatal cholinergic stimulation as well as the above-described syndrome are reversed by anticholinergic drugs (Matthews and Chiou, 1978,1979).

In summary, the cholinergic stimulation model of Parkinson's disease in rodents appears to be of limited value for several reasons. First of all, the presumed hyperactivity of striatal cholinergic systems is not the primary neurochemical disturbance in Parkinson's disease but is rather a consequence of the progressive loss of inhibitory dopaminergic inputs. In this respect, the model does not provide an adequate neurochemical portrait of the human disease where marked decreases in DA as well as reductions in NE and serotonin are noted (Agid and Javoy-Agid, 1985). In addition, the neurobehavioral syndrome following cholinergic stimulation is dissimilar. The akinesia and rigidity that are seldom reported in the literature cannot readily be dissociated from the characteristic trembling produced by these drugs and are difficult to assess independently (personal observations). Finally, the pharmacological profile of drugs that can reverse the neurobehavioral effects of cholinergic stimulation also militates against this particular model of Parkinson's disease. As described above, cholinomimetic tremors can be antagonized by drug treatments that have minimal or no value in the management of Parkinson's disease, including the anticholinergics themselves, antihistamines, neuroleptics, tricyclic antidepressants, and the peripheral administration of DA (Barbeau, 1979; Riederer et al., 1984).

2.5. 6-Hydroxydopamine

2.5.1. Unilateral Administration

Unilateral administration of 6-hydroxydopamine (6-OHDA) in the substantia nigra, ventral tegmentum, or medial forebrain bundle of rats produces a degeneration of the nigrostriatal pathway (Ungerstedt, 1968; Ungerstedt et al., 1973). This pioneering work, which was later extended to mice (Von Voigtlander and Moore, 1973), has led to the development of what is probably the most frequently used animal model of Parkinson's disease.

Although no prominent hypokinesia or rigidity is seen in unilaterally lesioned animals at rest, the emergence of episodic head and neck tremors has been reported (Buonamici et al., 1986). Furthermore, these animals display a characteristic postural abnormality consisting of a body posture that is curved longitudinally toward the side of the lesion. This pose has been likened to the scoliotic posture frequently observed in Parkinson patients (Duvoisin, 1976).

Administration of DA-stimulating drugs to these animals results in strong whole body circling. The direction of rotations depends on the nature of the drug administered. By releasing DA from terminals of the intact nigrostriatal fiber, drugs such as amphetamine produce rotations that are ipsilateral to the lesioned side (Ungerstedt, 1971; Ungerstedt et al., 1973). On the other hand, DA agonists such as apomorphine cause circling that is contralateral to the lesioned side (Ungerstedt, 1971; Ungerstedt et al., 1973; Koller and Herbster, 1987). This effect is thought to be a result of the stimulation of striatal DA receptors that have become hypersensitive following presynaptic denervation (Melamed et al., 1982; Graham et al., 1990). L-DOPA also produces contralateral rotations, presumably by augmenting the concentration of releasable DA in remaining nerve fibers on the lesioned side (Ungerstedt et al., 1973). Anticholinergics induce ipsilateral rotations, an effect that has been attributed to removal of the inhibitory cholinergic influence in the intact side (Ungerstedt et al., 1973).

Although the rotation model has proved to be extremely valuable for studying the mechanism of action of dopaminomimetics and for investigating the complex transmitter interactions modulating the activity of the nigrostriatal DA pathway, its usefulness as an animal model of Parkinson's disease now appears somewhat restricted. First, the unilateral nature of the neuropathology and the ensuing neurochemical changes are obviously different from those present in Parkinson patients and, at best, could only be akin to hemiparkinsonism. Second, unilaterally lesioned animals do not exhibit hypokinesia and rigidity, two of the three cardinal symptoms found in parkinsonism. Third, the efficacy of drugs in inducing rotations in animals does not always parallel their ability to ameliorate the symptoms of Parkinson's disease. Amphetamine, apomorphine, and anticholinergics are very efficient for inducing rotations in lesioned animals; however, the therapeutic usefulness of these agents in the human disease is minimal (Marsden et al., 1975; Barbeau, 1979; Gancher et al., 1990).

2.5.2. Bilateral Administration of 6-OHDA

Bilateral administration of 6-OHDA in the medial forebrain bundle at the level of the lateral hypothalamic area results in widespread depletion of regional brain catecholamine contents that are accompanied by severe neurobehavioral disturbances (Ungerstedt et al., 1973; Sechzer et al., 1973; Smith et al., 1975; Rondeau et al., 1978; Ervin et al., 1977). These effects can be produced by injection of 6-OHDA in both anterolateral and posterolateral hypothalamic regions, although the magnitude of effects appears to be more pronounced after administration to the latter hypothalamic site (Smith et al., 1975; Ervin et al., 1977; Rondeau et al., 1978).

Prominent hypokinesia—which is evidenced by cataleptic manifestations, inability to initiate movement, and a strong decrease in spontaneous activity—is seen in 6-OHDA-treated animals (Smith et al., 1972; Ungerstedt et al., 1973; Smith et al., 1975; Ervin et al., 1977; Rondeau et al., 1978; Butterworth et al., 1978). Muscular rigidity is also produced by this treatment, although this effect has received relatively little attention (Rondeau et al., 1978). Recently, we have shown that tremors could also be observed in these animals (Jolicoeur et al., 1990). In addition to the above-mentioned motor deficits, bilateral 6-OHDA hypothalamic lesions also render rats severely aphagic and adipsic, to the extent that these animals must be intubated with a liquid diet in order to assure survival (Smith et al., 1972). It is noteworthy that disturbances in active avoidance responding, effects that are not attributable to motor deficits, have also been reported (Smith et al., 1975). It is tempting to speculate that this deficit corresponds to the cognitive disorders often associated with parkinsonism (Schultz, 1984). The pharmacology of hypothalamic 6-OHDA syndrome has concentrated mostly on the hypokinetic symptom. Hypokinesia is reversed temporarily by several DA agonists, including apomorphine and bromocriptine (Butterworth et al., 1978) but is not affected by amphetamine, probably because of the bilateral degeneration of dopaminergic fibers originating from the mesencephalon (Ungerstedt et al., 1973; Rondeau et al., 1978). Administration of L-DOPA alone or in combination with a peripheral decarboxylase inhibitor also reverses hypokinesia (Cashin and Sutton, 1973; Butterworth et al., 1978). On the other hand, administration of the anticholinergic trihexyphenidyl is ineffective (Butterworth et al., 1978). To date, these results indicate that this pharmacological profile corresponds closely to the pharmacotherapy of Parkinson's disease. As compared to the other models of Parkinson's disease in rodents previously reviewed, the bilateral intrahypothalamic lesioning method appears to provide a close resemblance to the human disease, both in terms of behavioral manifestations and pharmacological responsiveness. In the following sections, we present a detailed experimental protocol to induce parkinsonism in rats using the bilateral intrahypothalamic 6-OHDA lesion model. The neurochemical and neurobehavioral changes produced by this technique as well as their responsiveness to pharmacological intervention will be presented and discussed in relation to human disease.

3. Experimental Protocol

3.1. Methods

3.1.1. Animals and Lesioning Procedures

The protocol is based on the use of male hooded rats weighing 250–300 g housed under the usual laboratory conditions (temperature controlled, 12 h light/dark cycle). Animals are anesthetized with a ketamine (80 mg/kg)/xylazine (12 mg/kg) mixture intramuscularly and injected bilaterally with 4 μL of a 6-OHDA hydrobromide solution (6.5 μg/μL of distilled water containing 0.04% ascorbic acid) into the hypothalamus according to the following coordinates: A.P., 5.0 mm; L: 2.0 mm; and V, 8.0 mm using bregma and dura as points of reference (De Groot, 1959). Solutions of 6-OHDA must be prepared fresh immediately prior to each injection in order to minimize oxidation. Sham-operated animals receive isovolumetric solutions of ascorbic acid. Solutions are administered by means of 30 gage needles at a rate of 1 μL min, after which the injection needles remain in place for 4 min to allow complete diffusion. Following surgery, animals are intubated daily with 8 mL of a liquid diet containing 25 g sucrose, 1.8 mL Polyvisol vitamins, 2 eggs, 30 mL Kaopectate, 125 mL water, and 400 mL evaporated milk.

3.1.2. Testing Procedures

At 48 h after surgery, behavioral testing is initiated. Spontaneous motor activity is measured for 1 min by means of a photocell activity apparatus. The presence and intensity of catalepsy are determined by placing an animal's front paws on a horizontal wooden board 1 cm wide by 10 cm high (*see* Fig. 1). Time spent in that position, up to a maximum of 1 min, is recorded. Muscular rigidity is assessed in two tests (Fig. 1). In the grasping test, a rat is suspended by its front paws grasping a metal rod (diameter: 0.5 cm) held by the experimenter about 50 cm above the table. The time the animal remains on the bar (maximum 1

Fig. 1. Clockwise from top left: Procedures for measuring catalepsy, tremors, and muscular rigidity in the grasping and tail lift tests.

min) is noted. A prolonged grasping response has been associated with more direct measures of muscle rigidity (Steg, 1964b). In the tail rigidity test, the animal's tail is raised approx 5 cm from the table with a metal rod (diameter: 0.5 cm) positioned about 2 cm from the end of the tail. The time the tail stays on the rod is recorded (maximum 30 s). For tremors, an animal is lifted by the tail so that the hind quarters are suspended approx 8 cm above the table, with the forelimbs still resting on the surface (Fig. 1). The animal is kept in that position for a period of 10 s. When present, the intensity of body or hindleg tremors is evaluated with the following scores: 0 for absence of tremors; 1 for relatively weak and/or discontinuous tremors; and 2 for vigorous and/or continuous tremors. All tests are performed by an experimenter unaware of the treatment conditions.

3.1.3. Biochemical Analyses

If attempts to reverse the neurobehavioral symptoms are performed, care should be taken to allow sufficient time for the neurochemical effects of the drug tested to subside and for neurobehavioral symptoms to reappear prior to biochemical analyses. Animals are sacrificed and their brains rapidly removed and placed on a frozen dissection block. The nucleus accumbens, corpus striatum, hypothalamus, amygdala, and substantia nigra are excised according to the procedures outlined by Heffner et al. (1980). All regions are homogenized in 1.0*M* perchloric acid ($HC1O_4$). Homogenization volume is 400 µL except in the case of the amygdala (1.0 mL). Separation and quantification of DA and its major metabolites, 3,4-dihydroxyphenylacetic acid (DOPAC) and homovanillic acid (HVA), as well as NE, serotonin, and its metabolite 5-hydroxyindoleacetic acid (5-HIAA) are performed using high-pressure liquid chromatography (HPLC) coupled with electrochemical detection according to our previously described method (Drumheller et al., 1990).

3.2. Results and Discussion

3.2.1. Biochemical Data

Bilateral administration of 6-OHDA resulted in significant reductions in DA and its metabolites DOPAC and HVA in all regions examined except the substantia nigra, where no significant changes were found. These effects are shown in Table 1, where regional brain concentrations of DA, DOPAC, and HVA expressed in ng/mg wet wt are presented for both sham-operated and lesioned animals. As can also be seen in the table, changes in regional NE concentrations paralleled those obtained for DA. However, no changes were found in serotonin or 5-HIAA in any region examined (data not shown). Therefore, this treatment results in severe loss of DA and its main metabolites in terminal regions of both the nigrostriatal and mesolimbic DA pathways. These reductions were more pronounced in the stria-

Table 1
Regional Brain Concentrations
of NE, DA, DOPAC, and HVA Expressed in ng/mg Wet Wt

Regions	NE	DA	DOPAC	HVA
Corpus striatum				
Sham	0.08 ± 0.02	7.02 ± 0.52	1.66 ± 0.25	0.63 ± 0.09
6-OHDA	0.05 ± 0.01*	0.85 ± 0.20**	0.27 ± 0.06**	0.06 ± 0.02
N. accumbens				
Sham	0.52 ± 0.08	5.80 ± 0.56	2.62 ± 0.52	0.82 ± 0.15
6-OHDA	0.21 ± 0.10**	1.61 ± 0.52**	0.62 ± 0.14	0.15 ± 0.08**
Amygdala				
Sham	0.48 ± 0.05	0.46 ± 0.11	0.11 ± 0.005	0.08 ± 0.01
6-OHDA	0.24 ± 0.08**	0.17 ± 0.03**	0.04 ± 0.009**	0.03 ± 0.008**
Hypothalamus				
Sham	1.47 ± 0.31	0.27 ± 0.05	0.15 ± 0.015	0.06 ± 0.01
6-OHDA	0.13 ± 0.05**	0.16 ± 0.06*	0.10 ± 0.02*	0.05 ± 0.007
Substantia nigra				
Sham	0.26 ± 0.06	0.66 ± 0.15	0.30 ± 0.09	0.17 ± 0.06
6-OHDA	0.22 ± 0.05	1.72 ± 1.65	0.32 ± 0.16	0.15 ± 0.05

Values represent means ± SD of each group. Significant differences as revealed by T-tests for independent samples are indicated by asterisks (* $p < 0.05$; ** $p < 0.01$).

tum than in the accumbens, which parallels what has been reported in Parkinson patients (Agid and Javoy-Agid, 1985). Not surprisingly, pronounced reductions were also noted in the hypothalamus, the site of 6-OHDA injection. Degeneration of both nigrostriatal and mesolimbic dopaminergic systems as well as a marked decrease in hypothalamic DA concentrations has been documented in Parkinson's disease (Hornykiewicz, 1972; Agid and Javoy-Agid, 1985; Hornykiewicz and Kish, 1986). However, in contrast to Parkinson's disease, levels of DA were not affected in substantia nigra of lesioned animals. This is possibly owing to the fact that neurochemical determinations were performed too early after lesioning and that the retrograde degeneration process was not completed. However, it is interesting to note that in some animals, DA levels were markedly increased in this region. The relevance of this finding is not clear. However, accumulation of amines in the substantia nigra 4 d following hypothalamic injections of 6-OHDA has been reported by others (Willis et al., 1987). It has been proposed that local accumulation of amines might actually be responsible for some of the neurotoxic effects of 6-OHDA (Willis et al., 1983,1987).

These neurochemical findings suggest that DA reductions in terminal regions of dopaminergic fibers are sufficient, at least in rats, to induce the three cardinal symptoms of Parkinson's disease. Regional concentrations of NE were also lowered significantly in the same regions where changes in DA were found (Table 1). Similar to what has been reported in humans, these decreases in NE concentrations were less pronounced than the reductions in DA levels. Contrary to the case with Parkinson's disease, where a 50% decrease in serotonin is seen in many brain regions (Agid and Javoy-Agid, 1985), 6-OHDA administration to animals did not alter levels of this indolamine or its metabolite after 48 h.

3.2.2. Neurobehavioral Symptoms

As expected, bilateral intrahypothalamic administration of 6-OHDA resulted in a prominent hypokinesia as evidenced by a significant increase in catalepsy scores and a concomitant decrease in motor activity. Both grasping and tail suspension time were significantly increased in 6-OHDA-treated animals, indicating the presence of muscular rigidity. Finally, tremors, which were not seen in sham-operated animals, were detected in lesioned animals, with an intensity score of 46 out of a maximum possible score of 72. Therefore, bilateral administration of 6-OHDA in the medial forebrain bundle at the level of the posterolateral hypothalamus resulted in the appearance of the three principal neurological signs of Parkinson's disease: hypokinesia, rigidity, and tremors. Furthermore, bilaterally lesioned animals also adopted a characteristic hunchback position, reminiscent of the flexed posture assumed by Parkinson's patients (Duvoisin, 1976). We noted a good correlation between the intensity of hypokinesia (catalepsy and decreased motor activity) and that of muscular rigidity in lesioned animals. Tremors were reliably detected by the method described above. Most frequently, whole body tremors were observed with this technique. However, in some animals, the trembling was confined to the hind quarters or legs. Unmanipulated, lesioned animals only manifest sporadic trembling, which more closely resembles shivering movements. Our procedure has its limitations because of the subjective and qualitative nature of the scoring method. On the other hand, we have noticed in followup experiments that latency and duration of tremors during the test period were well-correlated with subjective assessment of tremor intensity and that these two parameters could be utilized to generate quantitative data. It should be mentioned that tremors were only detected in approx 65% of the treated animals. Presence or absence of tremors was unrelated to the manifestation of the other neurological signs of parkinsonism, as some animals, displaying complete hypokinesia and muscular rigidity, did not tremble. Also, the incidence of tremors was not associated with any particular neurochemical change. The dissociation of tremor from other neurological signs does not diminish the validity of the model. On the contrary, it has been argued that Parkinson's disease is not a homogeneous clinical entity but that, in fact, patients can be subdivided into two groups: one predominantly displaying tremors, and the other mainly manifesting hypokinesia and rigidity (Barbeau and Pourcher, 1982; Zetusky et al., 1985).

3.2.3. Effects of Drugs

All motor symptoms, including the hunchback posture, were reversed by the following treatments: 1 mg/kg apomorphine hydrochloride, subcutaneously, and 50 mg/kg of the peripheral decarboxylase inhibitor Ro 4-4602 (Hoffmann-LaRoche) 30 min prior to 60 mg/kg of L-DOPA, both via the intraperitoneal route. None of the deficits were counteracted by subcutaneous administration of trihexyphenidyl in doses up to 30 mg/kg. Drug effects on hypokinesia, rigidity, and tremors are summarized in Table 2.

The present findings together with the data presented above demonstrate that bilateral intrahypothalamic administration of 6-OHDA results in neurochemical and neurobehavioral manifestations that resemble those found in Parkinson's disease. Also,

Table 2
Summary of Neurobehavioral Effects Obtained with Various Treatments

Groups	Catalepsy	Motor activity	Grasping time	Tail rigidity	Tremor intensity
Sham-operated	23.1 ± 6.1	171.0 ± 24.5	64.3 ± 11.1	10.2 ± 4.3	0
6-OHDA + saline	228.0 ± 16.2^a	27.5 ± 5.0^a	363.1 ± 12.4^a	104.6 ± 5.6^a	46^a
6-OHDA + apomorphine	0^b	207.3 ± 28.4	96.8 ± 17.1^b	8.8 ± 3.6	13^b
6-OHDA + L-DOPA	0^b	268.5 ± 39.3^b	39.3 ± 14.2^b	6.7 ± 1.8^b	0^b
6-OHDA + trihexyphenidyl	219 ± 23.1^a	18.9 ± 9.0^a	358.0 ± 16.2^a	92.3 ± 7.8^a	42^a

[a]Significant differences from sham-operated animals.
[b]Significant differences from 6-OHDA animals injected with saline.

the relative ability of drugs to reverse motor symptoms created by the toxin is similar to their known therapeutic efficacy in the disease. Although further validation is needed, this model constitutes the closest approximation to the human disease in rodents.

4. Summary and Conclusion

Although several animal models of Parkinson's disease have been described in the literature, a valid and reliable model in rodents is still lacking. The administration of reserpine has been shown to induce hypokinesia, muscle rigidity, and tremors in rats and mice. However, the fact that reserpine depletes indiscriminately a multiplicity of amines in brain and in periphery raises doubts about the neurochemical validity of this model. Moreover, reserpine-induced symptoms can be reversed by nonantiparkinson agents. Administration of MPTP to mice results in prominent decreases in striatal content of DA and metabolites, but these neurochemical changes are apparently not accompanied by neurobehavioral changes. On the other hand, rats seem to be impervious to the neurotoxic action of MPTP. Neuroleptics, specifically following chronic administration, produce the three cardinal symptoms of Parkinson's disease in rodents. However, the neurochemical changes caused by these drugs differ markedly from those seen in Parkinson's disease. Furthermore, the effects of pharmacological interventions in this animal model are quite different from those in human parkinsonism. Tremors in rats can be induced by a variety of cholinomimetics, such as carbachol and oxotremorine; but akinesia and rigidity are not produced by these drugs. Furthermore, the tremors induced by cholinergic stimulation can be reversed by a variety of nonantiparkinson agents. The rotation model following unilateral lesion of the substantia nigra with 6-OHDA has proved to be a very useful tool for fundamental neuropharmacology. However, rats with unilateral nigral lesions do not display akinesia or muscular rigidity, although they do manifest sporadic bursts of head and neck tremors. Also, the efficacy of drugs in inducing rotations in animals does not always parallel their ability to ameliorate the symptoms of Parkinson's disease.

Bilateral microinjections of 6-OHDA in the medial forebrain bundle at the level of the lateral hypothalamus result in hypokinesia, muscular rigidity, and tremors in rats. This treatment also produces neurochemical changes that are similar to those seen in Parkinson's disease. Furthermore, the neurobehavioral symptoms can be reversed with known antiparkinson drugs. To date, this procedure provides the closest approximation of Parkinson's disease in rats.

References

Agid Y. and Javoy-Agid F. (1985) Peptides and Parkinson's disease. *Trends Neurosci.* **8,** 30–35.

Ajima A., Yamaguchi T., and Kato T. (1990) Modulation of acetylcholine release by D1, D2 dopamine receptors in rat striatum under freely moving conditions. *Brain Res.* **518,** 193–198.

Anden N.-E. and Johnels B. (1977) Effect of local application of apomorphine to the corpus striatum and to the nucleus accumbens on the reserpine-induced rigidity in rats. Brain Res. **133,** 386–389.

Arnt J., Christensen A. V., and Hyttel J. (1981) Differential reversal by scopolamine of effects of neuroleptics in rats: Relevance for evaluation of therapeutic and extrapyramidal side-effect potential. *Neuropharmacology* **20,** 1331–1334.

Arnt J. and Christensen A. V. (1981) Differential reversal by scopolamine and THIP of antistereotypic and cataleptic effects of neuroleptics. *Eur. J. Pharmacol.* **69,** 107–111.

Barbeau A. (1979) Parkinson's disease and its treatment. *Neurol. Neurosurg.* **1,** 1–8.

Barbeau A. and Pourcher E. (1982) New data on the genetics of Parkinson's disease. *Can. J. Neurol. Sci.* **9,** 53–60.

Bradbury A. J., Costall B., Domeney A. M., Jenner P., Kelly M. E., Marsden C. D., and Naylor R. J. (1986) 1-Methyl-4-phenylpyridine is neurotoxic to the nigrostriatal dopamine pathway. *Nature* **319,** 56–57.

Buonamici M., Maj R., Pagani F., Rossi A. C., and Khazan N. (1986) Tremor at rest episodes in unilaterally 6-OHDA-induced substantia nigra lesioned rats: EEG-EMG and behavior. *Neuropharmacology* **25,** 323–325.

Butterworth R. F., Belanger F., and Barbeau A. (1978) Hypokinesia produced by anterolateral hypothalamic 6-hydroxydopamine lesions and its reversal by some antiparkinson drugs. *Pharmacol. Biochem. Behav.* **8,** 41–45.

Cashin C. H. and Sutton S. (1973) The effect of Anti-Parkinson drugs on catalepsy induced by α-methyl-p-tyrosine in rats pretreated with intraventricular 6-hydroxydopamine. *Br. J. Pharmacol.,* **47,** 658P–659P.

Charcot J. M. and Vulpian A. (1862) Revue clinique de la paralysie agitante: III, Quelque mots concernant la physiologie pathologique de la paralysis agitante et du tremblement en general. *Gaz. Hebdomadaire* **9**, 56–59.

Cody F. W. J., MacDermott N., Matthews P. B. C., and Richardson H. C. (1986) Observations on the genesis of the stretch reflex in parkinson's disease. *Brain* **109**, 229–249.

Colpaert F. C. (1987) Pharmacological characteristics of tremor, rigidity and hypokinesia induced by reserpine in rat. *Neuropharmacology* **26**, 1431–1440.

Dandiya P. C. and Bhargava L. P. (1968) The antiparkinsonian activity of monoamine oxidase inhibitors and other agents in rats and mice. *Arch. Int. Pharmacodyn. Ther.* **176**, 157–167.

De Groot J. (1959) The rat forebrain in stereotaxic coordinates. *Trans. Roy. Soc. Acad. Sci.* **52**, 1–38.

DeBelleroche J., Coutinho-Netto J., and Bradford H. F. (1982) Dopamine inhibition of the release of endogenous acetylcholine from corpus striatum and cerebral cortex in tissue slices and synaptosomes: a presynaptic response? *J. Neurochem.* **39**, 217–222.

Derkach P., Larochelle L., Bieger D., and Hornykiewicz O. (1974) L-DOPA-chlorpromazine antagonism on running activity in mice. *Can. J. Physiol. Pharmacol.* **52**, 114–118.

Dickinson S. L. and Slater P. (1982) Effect of lesioning dopamine, noradrenaline and 5-hydroxytryptamine pathways on tremorine-induced tremor and rigidity. *Neuropharmacology* **21**, 787–794.

Dill R. E., Dorman H. L., and Nickey W. M. (1968) A simple method for recording tremors in small animals. *J. Appl. Physiol.* **24**, 598–599.

Drumheller A. D., Gagne M. A., St-Pierre S., and Jolicoeur F. B. (1990) Effects of neurotensin on regional brain concentrations of dopamine, serotonin and their main metabolites. *Neuropeptides* **15**, 169–178.

Duvoisin R. C. and Marsden C. D. (1974) Reversal of reserpine-induced bradykinesia by a-methyldopa: new light on its modus operandi. *Brain. Res.* **71**, 178–182.

Duvoisin R. C. (1976) Parkinsonism: Animal analogues of the human disorder. *Assoc. Res. Nerv. Ment. Dis.* **55**, 293–303.

Ervin G. N., Fink J. S., Young R. C., and Smith G. P. (1977) Different behavioral responses to L-DOPA after anterolateral or posterolateral hypothalamic injections of 6-hydroxydopamine. *Brain Res.* **132**, 507–520.

Fischer E. and Heller B . (1967) Pharmacology of the mechanism of certain effects of reserpine in the rat. *Nature* **216**, 1221–1222.

Gancher S. T., Woodward W. R., Gliessman P., Boucher B., and Nutt J. G. (1990) The short-duration response to apomorphine: Implications for the mechanism of dopaminergic effects in parkinsonism. *Ann. Neurol.* **27**, 660–665.

Goldstein J. M., Barnett A., and Malick J. B. (1975) The evaluation of antiparkinson drugs on reserpine-induced rigidity in rats. *Eur. J. Pharmacol.* **33**, 183–188.

Graham W. C., Crossman A. R., and Woodruff G. N. (1990) Autoradiographic studies in animal models of hemi-parkinsonism reveal dopamine D2 but not D1 receptor supersensitivity. I. 6-OHDA lesions of ascending mesencephalic dopaminergic pathways in the rat. *Brain Res.* **514**, 93–102.

Hallberg H. and Almgren O. (1987) Modulation of oxotremorine-induced tremor by central B-adrenoceptors. *Acta. Physiol. Scand.* **129**, 407–413.

Heffner T. G., Hartman J. A., and Seiden L. S. (1980) A rapid method for dissection of the rat brain. *Pharmacol. Biochem. Behav.* **13**, 453–456.

Heikkila R. E., Cabbat F. S., Manzino L., and Duvoisin R.C. (1984) Effects of 1-methyl-4-phenyl-1,2,5,6-tetrahydropyridine on neostriatal dopamine in mice. *Neuropharmacology* **23**, 711–713.

Hornykiewicz O. (1972) Neurochemistry of parkinsonism, in *Handbook of Neurochemistry*, vol. 7 (A. Lajtha, ed.), pp. 465–501, Plenum, New York.

Hornykiewicz O. (1975) Parkinsonism induced by dopaminergic antagonists. *Adv. Neurol.* **9**, 155–164.

Hornykiewicz O. and Kish S. J. (1986) Biochemical pathophysiology of parkinson's disease. *Adv. Neurol.* **45**, 19–34.

Horst W. D., Pool W. R., and Spiegel H. E. (1973) Correlation between brain dopamine levels and L-DOPA activity in anti-parkinson tests. *Eur. J. Pharmacol.* **21**, 337–342.

Jarvis M. F. and Wagner G. C. (1985) Neurochemical and functional consequences following 1-methyl-4-phenyl-1,2,5,6-tetrahydropyridine (MPTP) and metamphetamine. *Life Sci.* **36**, 249–254.

Johannessen J. N., Chiueh C. C., Burns R. S., and Markey S. P. (1985) Differences in the metabolism of MPTP in the rodent and primate parallel differences in sensitivity to its neurotoxic effects. *Life Sci.* **36**, 219–224.

Johnels B. (1983) Reserpine-induced rigidity in rats: Drug effects on muscle tone from corpus striatum and nucleus accumbens . *Pharmacol. Biochem. Behav.* **19**, 463–470.

Jolicoeur F. B., Rivest R., and Drumheller A. (1991) Hypokinesia, rigidity and tremor induced by hypothalamic 6-OHDA lesions in the rat. *Brain Res. Bull.* 26, 317–320

Koller W. C. and Herbster G. (1987) Terguride, a mixed dopamine agonist-antagonist, in animal models of parkinson's disease. *Neurology* **37**, 723–727.

Kulkarni S. K., Arzi A., and Kaul P. N. (1980) Modification of drug-induced catatonia and tremors by quipazine in rats and mice. *Jpn. J. Pharmacol.* **30**, 129–135 .

Leonard B. E. (1972) Anti-tremorine effects of some mono- and diacyloxytropanes. *Arch. Int. Pharmacodyn. Ther.* **196**, 93–97.

Maickel R. P., Braunstein M. C., McGlynn M., Snodgrass W. R., and Webb R. W. (1974) Behavioral, biochemical, and pharmacological effects of chronic dosage of phenothiazine tranquilizers in rats. *Adv. Biochem. Psychopharmacol.* **9**, 593–602.

Marsden C. D., Duvoisin R. C., Jenner P., Parkes J. D., Pycock C., and Tarsy D. (1975) Relationship between animal models and clinical parkinsonism. *Adv. Neurol.* **9**, 165–175.

Matthews R. T. and Chiou C. Y. (1978) Cholinergic stimulation of the caudate nucleus in rats: A model of parkinson's disease. *Neuropharmacology* **17**, 879–882.

Matthews R. T. and Chiou C. Y. (1979) Effects of diethylcholine in two animal models of parkinsonism tremors. *Eur. J. Pharmacol.* **56**, 159–162.

Melamed E., Hefti F., and Wurtman R. J. (1982) Compensatory mechanisms in the nigrostriatal dopaminergic system in parkinson's disease: Studies in an animal model. *Isr. J. Med. Sci.* **18**, 159–163.

Morrison A. B. and Webster R. A. (1973) Drug-induced experimental parkinsonism. *Neuropharmacology* **12**, 715–724.

Moss D. E., McMaster S. B ., and Rogers J. (1981) Tetrahydrocannabinol potentiates reserpine-induced hypokinesia. *Pharmacol. Biochem. Behav.* **15**, 779–783 .

Murphy K. M. M. and Snyder S. H. (1982) Heterogeneity of adenosine A1 receptor binding in brain tissue. *Mol. Pharmacol.* **22**, 250–257.

Nose T. and Kojima M. (1970) A simple screening method for antiparkinsonian drugs in mice. *Eur. J. Pharmacol.* **10**, 83–86.

Ray S. K. and Poddar M. K. (1990) Interaction of central serotonin and dopamine in the regulation of carbaryl-induced tremor. *Eur. J. Pharmacol.* **181**, 159–166.

Riederer P., Reynolds G. P., and Jellinger K. (1984) The pharmacology of parkinson's disease: *L*-DOPA and beyond. *Trends Pharmacol. Sci.* **5**, 25–27.

Rondeau D. B., Jolicoeur F. B., Belanger F., and Barbeau A. (1978) Differential behavioral activities from anterior and posterior hypothalamic lesions in the rat. *Pharmacol. Biochem. Behav.* **9**, 43–47.

Sahgal A., Andrews J. S., Biggins J. A., Candy J. M., Edwardson J. A., Keith A. B., and Turner J. D. (1984) *N*-Methyl-4-phenyl-1,2,3,6-tetrahydroxypyridine (MPTP) affects locomotor activity without producing a nigrostriatal lesion in the rat. *Neurosci. Lett.* **48**, 179–184.

Schultz W. (1984) Minireview: Recent physiological and pathophysiological aspects of parkinsonian movement disorders. *Life Sci.* **34**, 2213–2223.

Sechzer J. A., Ervin G. N., and Smith G. P. (1973) Loss of visual placing in rats after lateral hypothalamic microinjections of 6-hydroxydopamine. *Exp. Neurol.* **41**, 723–737.

Simon P., Malatray J., and Boissier J. R. (1970) Antagonism by amantadine of prochlorpemazine-induced catalepsy. *J. Pharm. Pharmacol.* **22**, 546,547.

Smith G. P., Strohmayer A. J., and Reis D. J. (1972) Effect of lateral hypothalamic injections of 6-hydroxydopamine on food and water intake in rats. *Nature* **235**, 27–29.

Smith G. P., Levin B. E., and Ervin G. N. (1975) Loss of active avoidance responding after lateral hypothalamic injections of 6-hydroxydopamine. *Brain Res.* **88**, 483–498.

Steg G. (1964a) X-rigidity in reserpinized rats. *Experientia* **20**, 79–80.

Steg G. (1964b) Efferent muscle innervation and rigidity. *Acta Physiol. Scand.* **61: (Suppl. 225)**, 5–53.

Ungerstedt U. (1968) 6-Hydroxydopamine induced degeneration of central monoamine neurons. *Eur. J. Pharmacol.* 5, 107–110.

Ungerstedt U. (1971) Striatal dopamine release after amphetamine or nerve degeneration revealed by rotational behavior. *Acta Physiol. Scand.* **367: (Suppl. 83)**, 49–68.

Ungerstedt U., Avemo A., Avemo E., Ljungberg T., and Ranje C. (1973) Animal models of parkinsonism. *Adv. Neurol.* **3**, 257–270.

Van Woert M. H., Sethy V. H., and Ambani L. M. (1974) Effect of phenothiazines on central cholinergic activity. *Adv. Biochem. Psychopharmacol.* **9**, 707–715.

Von Voigtlander P. F. and Moore K. E. (1973) Turning behavior in mice with unilateral 6-hydroxydopamine lesions in the striatum: Effects of apomorphine, L-DOPA, amantadine, amphetamine and other psychomotor stimulants. *Neuropharmacology* **12**, 451–462.

Willis G. L., Smith G. C., and McGrath B. P. (1983) Deficits in locomotor behaviour and motor performance after central 6-hydroxydopamine or peripheral L-DOPA. *Brain Res.* **266**, 279–286.

Willis G. L., Sleeman M., Pavey G. M., and Smith G. C. (1987) Further studies on the neurochemical specificity of 6-hydroxydopamine as compared to radiofrequency lesions. *Brain Res.* **403**, 15–21.

Zetusky W. J., Jankovic J., and Pirozzolo F. J. (1985) The heterogeneity of parkinson's disease: Clinical and prognostic implications. *Neurology* **35**, 522–526.

Primate Models of Parkinson's Disease

Paul J. Bédard, René Boucher,
Baltazar Gomez-Mancilla, and Pierre Blanchette

1. Introduction

An adequate animal model of a disease should reproduce the cardinal features of its human counterpart. Parkinson's disease is characterized by akinesia, rigidity, and tremor. Attempts to reproduce Parkinsonian symptoms can be divided in three broad categories according to the method used. Earlier studies were based mostly on destruction of certain brain structures by surgical lesions with or without the help of stereotaxy. Then, with development of our knowledge of neurotransmitters, Parkinsonian features could be reproduced by drugs that affect more or less selectively certain biogenic amines. More recently, the serendipitous discovery of 1 methyl-4-phenyl-1,2,3,6-tetrahydropyridine (MPTP) provided us with the best model of Parkinson's disease to date. We will review these models,which have evolved over a period of over fifty years and have provided us with powerful tools to understand the pathophysiology of Parkinson's disease and to test new therapeutic approaches.

1.1. Surgical Lesions

With the development of neurosurgery in the first half of this century, experimental lesions were placed in most structures of the monkey brain in an attempt to elucidate their physiology as well as their role in certain symptoms and signs. Thus it was observed that lesions placed in the dentate nucleus or the superior cerebellar peduncle were sometimes followed

From: *Neuromethods, Vol. 21: Animal Models of Neurological Disease, I*
Eds: A. Boulton, G. Baker, and R. Butterworth

by the appearance of rhythmic oscillations and bursts of 4–7 Hz present with maintenance of posture or at rest (Ferraro and Barrera, 1936; Mettler, 1946; Walker and Botterell, 1937; Carpenter, 1958). Lesions placed more rostrally with the help of stereotaxic surgery induced a similar but more sustained postural or resting tremor. Such lesions were thought to involve predominantly the superior cerebellar peduncle and the red nucleus in which it travels at the midbrain level (Carpenter, 1956; Mettler and Whittier, 1947; Peterson et al., 1949; Ward et al., 1948). A more sustained postural and resting tremor accompanied by hypokinesia of the same limb was obtained by placing the lesion slightly more ventrally to involve in addition to the red nucleus and superior cerebellar peduncle the ventromedial tegmental area of the mesencephalon (Poirier, 1960; Poirier et al., 1969a; Goldstein et al., 1969) (Fig. 1). Interestingly, such lesions were accompanied by almost complete degeneration of the ipsilateral substantia nigra (contralateral to the tremor and hypokinesia). The substantia nigra was known to be the site of neuropathological changes in Parkinson's disease (Tretiakoff, 1919; Hassler, 1938) and had been shown by histofluorescence to be the origin of a dopaminergic projection to the ipsilateral striatum (Andén et al., 1965; Dahlström and Fuxe, 1964). Moreover, at about the same time, it was found that the most consistent neurochemical alteration in Parkinson's disease was a marked decrease in the level of dopamine in the striatum. Following these reports, Poirier and Sourkes (1965) measured levels of dopamine and its metabolites in the striatum of their experimental animals and found them to be extremely low on the side of the lesion contralateral to the tremor and hypokinesia. The same authors also observed a decrease of 5-hydroxytryptamine (5-HT) concentrations in the ipsilateral striatum that originates in the dorsal raphé nucleus and that they believed to contribute to the Parkinsonian signs.

Poirier and Sourkes emphasized the presence of tremor and hypokinesia but noted that rigidity was not present in their animals. On the contrary, the affected limbs were generally hypotonic. A partial answer to this discrepancy was provided in later experiments where the same group (Péchadre et al., 1976) found

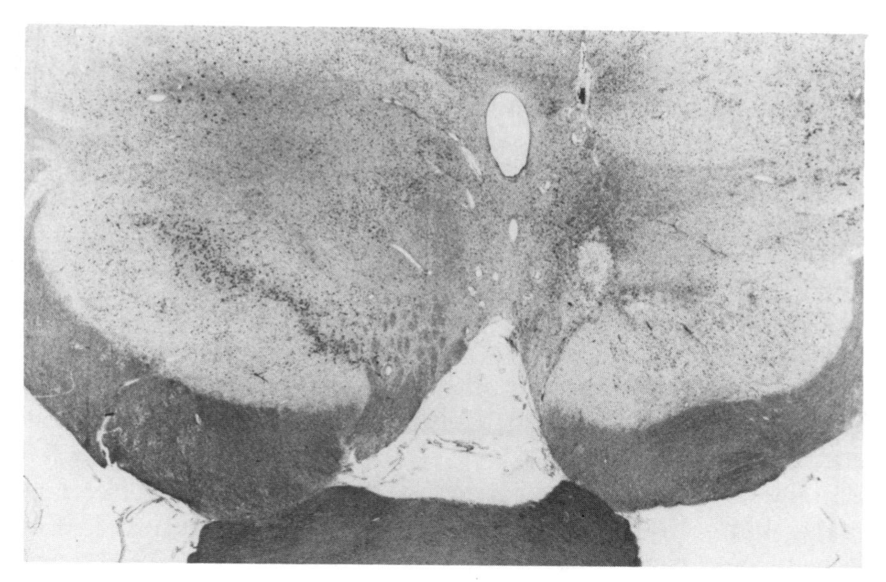

Fig. 1. Microphotograph showing a unilateral ventromedial tegmental lesion of the midbrain (right) with resulting almost complete degeneration of the substantia nigra pars compacta on that side. Kluver Barrera stain.

that typical rigidity with a cogwheel phenomenon could be observed together with tremor and akinesia after bilateral ventral tegmental lesions of the midbrain that were placed more rostrally and spared the parvocellular division of the red nucleus.

Surprisingly, Poirier and Sourkes (1965) observed that when the lesion involved the nigrostriatal pathway but spared the cerebellar or rubral connections, hypokinesia, but not tremor, was seen. They thus concluded that this type of Parkinsonian-like tremor required an association of a lesion of the nigrostriatal dopaminergic pathway and of the corresponding rubro-olivo-cerebello-rubral loop (Poirier et al., 1969,1972). They investigated this point further by using a combination of lesions and drugs.

2. Surgical Lesions and Pharmacological Agents

In the course of their investigation of Parkinsonian-like tremor induced by a ventromedial tegmental lesion in the monkey, Poirier and Sourkes (1966) studied a series of drugs that were known to affect brain neurotransmitters in one way or another. They identified a drug called harmaline, which was known to be a reversible monoamine oxidase (MAO) inhibitor, and which could increase the amplitude of the lesion-induced tremor and even trigger its appearance in cases where the lesion had seemed unsuccessful. By careful analysis of several types of lesions, this group (Larochelle et al., 1970) came to the conclusion that harmaline could induce Parkinsonian-like tremor in animals provided they had a lesion in any of the structures constituting the rubro-olivo-cerebello-rubral loop, i.e., the red nucleus, the rubro-olivary pathway (central tegmental tract) the inferior olivary nucleus, the dentate nucleus, and the superior cerebellar peduncle.

The exact mechanism of action of harmaline was not clear and could not be explained by its role as a MAO inhibitor. However, Llinas et al. (1973), Lamarre and Mercier (1971), and deMontigny and Lamarre (1973), later showed that harmaline has the curious property of triggering rhythmic firing in the inferior olivary nucleus.

In an attempt to ascertain the role of the lesion of the nigrostriatal pathway and of the rubro-olivo-cerebellar loop, Bédard et al. (1970) and Larochelle et al. (1971) administered to monkeys already bearing a lesion of the *loop* but displaying no sustained tremor alpha-methyl-para-tyrosine (AMPT) (1.5 g over 24 h), an inhibitor of tyrosine hydroxylase which depletes stocks of catecholamines. In such a preparation they observed typical Parkinsonian tremor coupled to profound akinesia. A similar effect was obtained by using (instead of AMPT) reserpine, a depletor of monoamines, or thiopropazate, a receptor blocker. Thus they concluded that akinesia resulted mostly from the absence of brain dopamine, whereas tremor necessitated in addition some degree of involvement of the rubro-olivo-cerebello-rubral loop (Larochelle et al., 1970).

3. MPTP-Induced Parkinsonism

The discovery that 1-methyl-4-phenyl-1,2,3,6-tetra-hydropyridine (MPTP) could, after systemic administration, induce a typical Parkinsonian syndrome in humans (Langston et al.,

1983) and subhuman primates (Burns et al., 1983) represented a major breakthrough in research on Parkinson's disease. It opened the way to a new hypothesis to explain the etiology of Parkinson's disease by exogenous or endogenous toxins, and the search still continues to pinpoint the possible toxic agents involved in the natural disease.

MPTP can reproduce in monkeys almost exactly the Parkinsonian syndrome (Burns et al., 1983) and such animals as well as their human counterparts respond to the same therapeutic agents, even including common side effects such as dyskinesia (Langston and Ballard, 1984; Bédard et al., 1986). It has thus become an indispensable tool with which to test new therapeutic agents for their efficacy and for their potential to induce side effects. We will review in detail our experience with this agent in monkeys. We have used almost exclusively cynomolgus *(Macaca fascicularis)* monkeys of 2–4 kg. Almost all primate species are sensitive to MPTP (Langston et al., 1984) although some species such as the Marmoset seem to recover more easily, do not show an increased density of dopamine (DA) receptors after denervation, and do not develop dyskinesia after treatment (Temlett et al., 1988).

3.1. Method of Administration

Although MPTP salts are available, we have always used the base. It can be administered easily intravenously or subcutaneously with equivalent efficacy. Despite earlier claims to the contrary, MPTP base is easily soluble in saline. The powder is weighed under a fume hood, the operator wearing a chemical-proof mask and rubber gloves. It is dissolved (1 mg/mL) in saline, and we try to calculate the exact amount necessary for an experiment, leaving no residue. The degree of toxicity of MPTP by inhalation or skin contact is not known but it is volatile when in solution and it is excreted through feces and urine of the animals during the first 48 h. The powder is first placed in a bottle that is then closed with a rubber cover through which the exact amount of saline is injected.

For the injection, the animals are placed in a chair and the solution is given intravenously in the leg vein or subcutaneously.

The animals are then placed in their cages in a special room equipped with separate ventilation. During the first 48 h, only the caregivers, wearing special clothing, a chemical-proof mask and gloves, are allowed in the room. Excrement from the animals is collected and disposed of separately.

3.2. Dosage

Some investigators use a fixed schedule of 0.3 mg/kg/d for 5 days. We have found that the response to MPTP is extremely variable and we have therefore adopted a more flexible dosage. We start with 0.3 mg/kg and wait three days. In some animals this dose is sufficient and they become maximally akinetic. If not, we repeat the same dose every three to four days. After three trials the dose is increased to 0.6 mg/kg and, if not sufficient, to 0.9 mg/kg. In some animals we have had to give more than 40 mg in cumulative doses. There seems to be a threshold effect and some animals apparently will not respond to a dose of 0.3 mg/kg no matter how many times it is repeated. We have not been able to predict reliably which animals will respond better than others. Age does not seem to be a good criterion.

3.3. Acute Syndrome

Most animals display immediately after each injection and for 10–30 min a syndrome characterized by agitation, ataxia, myoclonus, hallucinations, lingual dyskinesia, and sometimes, vomiting. The mechanism of this syndrome is not clear, but it may involve sudden release of DA or 5-HT. At any rate, the intensity or the presence of this acute syndrome in our experience bears no relationship to the later development of the Parkinsonian syndrome.

3.4. Initial Phase of Parkinsonian Syndrome

A minority of animals become akinetic progressively in the first hours following the first dose of MPTP. Most require repeated administration, and akinesia becomes evident in the days following the last dose. It is generally heralded by aphagia and adypsia followed by immobility and a stooped posture. This period is critical and many animals need intensive care to make it through. As soon as it is realized that the animal does not eat or drink, the animal should be assisted by putting food and fluids into the mouth. If it does not swallow, it must be fed through an esophageal tube with a syringe. We normally use a commercial infant milk formula 50–100 mL once or twice a day, to which we add ranitidine (50 mg, 1/4 tablet crushed and suspended in the milk), since we have observed that many animals suffer from stomach ulcers. The animal must be moved several times a day and placed on paper cushions to avoid pressure sores.

If after a few days the animal shows no sign of recuperation and loses weight, there is no choice but to start treatment with L-DOPA (Sinemet 100/25; 1/2–1 tablet twice a day) or DA agonists (e.g., bromocriptine 5 mg/kg/d). This allows the animal to move and feed, and if treatment is started early enough there is prompt improvement of the animal's condition and it starts to gain weight. One should avoid doses that induce excessive stereotyped behavior that interfere with feeding and consume energy.

3.5. When Should the Experiment Start?

If the aim of the experiment is to test antiparkinsonian agents, it should be remembered that after the initial phase, there is generally some degree of functional recovery, so that within a few weeks most animals are capable of feeding without assistance and move about in their cages, albeit more slowly. We have found that after 6–8 weeks, the syndrome is stable and there will be little if any further recovery. We thus generally wait 2 months after the last dose of MPTP before starting the experiment.

Moreover, the process of adaptation to sudden denervation takes a few weeks to develop, since animals treated with L-DOPA early after MPTP develop only lingual dyskinesia, whereas if one waits a few weeks they develop typical limb dyskinesia.

3.6. How Severe Must the Lesion Be?

Based on our experience of over 100 MPTP-treated monkeys studied in the last few years, we believe that a visible Parkinsonian syndrome requires at least a 90% loss of DA in the striatum. We have found (Di Paolo et al., 1986; Falardeau et al., 1988a,b) that the loss in the nucleus accumbens is generally less complete and of the order of 60%. Upregulation of D-1 and D-2 receptors also requires that degree of denervation. Animals with levels of DA around 25% of control in the striatum look almost normal after 2 months, their DA receptors are not increased and they do not develop dyskinesia after L-DOPA.

3.7. How Do You Monitor the Loss of Dopamine?

As we mentioned in the preceding paragraph, the clinical syndrome is a useful tool. However if one has to be more precise in the assessment before sacrifice, measurement of homovanillic acid (HVA) in the cerebrospinal fluid (CSF) may give a good idea of the degree of DA loss (Burns et al., 1983; DiPaolo et al., 1986) in the days after the injection. Even months after MPTP, animals that have recovered almost to normal still show a marked decreased of HVA in the CSF.

3.8. How Do You Administer Drugs Orally?

In several types of experiments, antiparkinsonian agents have to be administered daily and even several times daily. Although administration through an esophageal tube is probably the best way to ascertain ingestion, it is cumbersome, requires capturing the animal, and probably causes distress. It should therefore be used only when the animal will not ingest spontaneously. Drugs that are water soluble can be mixed in orange juice, which most primates will drink avidly. However, some drugs such as L-DOPA or bromocriptine have a poor water solubility and when suspended will accumulate at the bottom of the bottle, leading to erratic dosage. We have found that it is convenient to have such drugs (powdered) in the tip of a banana and to offer the banana early in the morning in the fasting state. The animal will bite the top off with the drug inside.

3.9. How Do You Evaluate the Response to Pharmacological Agents?

Since the Parkinsonian syndrome is essentially motor, the response to drugs will have to be evaluated in terms of motor parameters. We have evaluated *tremor* by electromyography (EMG) to assess the frequency and amplitude as well as by accelerometry (Gomez-Mancilla et al., in press). These techniques are objective and provide a good record of the tremor. They are useful in a paradigm where the assessment is performed in the same experiment before and after the drug since it is difficult to install the needles or the accelerometer with exactly the same parameters in separate experiments. In recent studies, we used these methods to have a permanent record of the tremor; but to assess the quantitative effect of a drug we use a simple method, which consists of watching the animal through a one-way screen and recording with a stopwatch the cumulative time with and without tremor in the experimental period.

Rigidity is felt by the observer during passive movements, but manipulation makes the animals nervous and they contract their muscles voluntarily, making the assessment very difficult. The cogwheel phenomenon can be recorded by EMG (Di Paolo et al., 1986).

Akinesia is the fundamental Parkinsonian deficit. Ideally, its assessment should involve precise measures of reaction time and movement time, which in monkeys must involve training the animals to manipulate a rather complex system for reward.

Unfortunately, these procedures are too cumbersome for most pharmacological experiments, which require frequent evaluation in relatively large numbers of animals. We therefore developed simple methods of assessment that do not require capturing the animals and taking them out of their cages for each assessment.

First, our cages are equipped with photocells linked to a computer that allow, through interruption of the beams, evaluation of gross locomotor activity. We found that after MPTP there is a marked decrease in locomotion, which can be restored by DA agonists (Rouillard et al., 1990). The computer displays a count every 15 min and we generally use cumulative 24 h counts, although we can also compare, for instance, the 3 h preceding and the 3 h following injection of a pharmacological agent.

The speed and precision of arm movements are evaluated by presenting the animals each morning in the fasting state with a wooden board with sixteen holes in which we have placed small pieces of bananas. A normal monkey can collect all pieces in less than 15 s with almost no training. We can evaluate with a stopwatch the increase in time necessary to perform the task after MPTP and the effect of treatment.

Finally, we use a disability score that is inspired by the types of scoring systems used in humans but that, with one exception, involves only decisions in terms of yes or no and that does not require handling of the animals. The following components are scored: posture: normal = 0, flexed = 1, crouched = 2; mobility: active = 1; passive = 2; climbing: present = 1; absent = 0; gait: normal = 0; abnormal = 1; eating: present = 0, absent = 1; social interactions: present = 0, absent = 1; grooming: present = 0, absent = 1; tremor: present = 1, absent = 0. Assessment of these items gives a score of 0 for normal monkeys and 10 as maximal disability. We have found the scoring system convenient for following a group of animals before and after experimental treatment where they can be evaluated as frequently as every 30 min. Elements of motor activity such as locomotion, which are related to more anterior structures, can be assessed in parallel with indices related to the putamen, such as picking up objects with the hand. We thus obtain a rather complete picture of the Parkinsonian state.

4. Dyskinesia

Many of the animals treated with L-DOPA or DA agonists develop (after a few days to a few weeks peak dose) dyskinesia of the face and limbs. These are scored with a so-called abnormal involuntary movements scale (AIMS) similar to the one used in humans. Dyskinetic movements are scored on a scale of 0 (none) to 3 (severe) in the face, neck, trunk, and each of the four limbs. Video sequences are also taken to document the type of movements: chorea, dystonia, myoclonus, and so on.

It is important to note that when a new drug is tested for its potential to induce dyskinesia it should be administered daily for at least 4 weeks to animals having received no L-DOPA or any other DA agonists. We have observed that when dyskinesias appeared after L-DOPA, they can be reproduced by DA agonists that would not have induced them when given alone.

4.1. Unilateral MPTP Model

This technique was pioneered by Bankiewicz et al. (1986) and consists of administering MPTP unilaterally by intracarotid infusion so that only one side of the telencephalon is affected. In a typical protocol (Brücke et al., 1988), the carotid artery was catheterized under ketamine anesthesia under fluoroscopic control and MPTP (HCl) dissolved in physiological saline at a concentration of 0.0325 mg/mL and infused at a rate of 4 mL/min during 23 min for a total dose of 3 mg.

The advantages of the technique are numerous. First, since the lesion is unilateral, the animal can still move around and, more importantly, can eat and drink. The normal intact side can be used as a control for behavioral or biochemical studies (Joyce et al., 1986).

On the other hand, it cannot be denied that the procedure is more complicated since it requires fluoroscopic control. Moreover, the lesion is not always strictly unilateral and variable degrees of denervation can occur on the so-called intact side. Finally, even when the lesion appears restricted to the side of the infusion, it appears that such sudden and severe denervation on one side has repercussions on the so-called intact side. For instance, Graham et al. (1990) have reported bilateral receptor changes after a unilateral lesion.

4.2. Is It Necessary to Use a Monkey Model of Parkinsonism to Assess New Drugs?

It is true that we have at our disposal excellent models of Parkinsonism in rodents (*see* preceding chapter), which have contributed enormously to our progress in the understanding of this disease. In many instances the response to drugs in terms of locomotor activity is the same in rats and monkeys, notably with DA D-2 agonists. However, rats do not display certain signs, such as tremor, and it is difficult to assess limb movements. Moreover, rats do not develop dyskinesia after DA agonists except stereotyped behavior, which is also seen in normal animals after high doses. On the contrary, Parkinsonian monkeys develop abnormal movements of a choreic or dystonic type identical to those seen in humans.

Finally, for inexplicable reasons, certain drugs that are effective in rats have little effect in monkeys. A typical example is the D-1 agonist SKF 38393, which has dramatic effects on circling in the 6-hydroxydopamine (6-OHDA) rat model (Arnt and Hyttel, 1984; Robertson and Robertson, 1986; Rouillard and Bédard, 1988) but has no effect in monkeys (Close et al., 1985; Nomoto et al., 1986; Barone et al., 1987). When this drug was tried in humans (Braun et al., 1987) it was also found to be ineffective.

We thus believe that, despite the additional cost and work involved, no new antiparkinsonian agents should be tried in patients before being tested in monkeys for their therapeutic effects on the various components of the Parkinsonian syndrome and for their potential to induce side effects such as dyskinesia.

References

Andén N. E., Dahlström A., Fuxe K., and Larsson K. (1965) Further evidence for the presence of nigro-neostriatal dopamine neurons in the rat. *Am. J. Anat.*, **116**, 329–333.

Arnt J. and Hyttel J. (1984) Differential inhibition by dopamine D-1 and D-2 agonists of circling behavior induced by dopamine agonists in rats with unilateral 6-hydroxydopamine lesion. *Eur. J. Pharmacol.*, **102**, 349.

Bankiewicz K. S., Oldfield E. H., Chiueh C. C., Doppman J. L., Jacobowitz D. M., and Kopin, I. J. (1986) Hemiparkinsonism in monkeys after unilateral internal carotid artery infusion of 1-methyl-4-phenyl-1,2,3,6-tetrahydropyridine (MPTP). *Life Sci.*, **39**, 7–16.

Barone P., Bankiewicz K. S., Corsini G. U., Kopin I. J., and Chase T. N. (1987) Dopaminergic mechanisms in hemiparkinsonian monkeys. *Neurology*, **37**, 1592–1595.

Bédard P. J., Di Paolo T., Falardeau P., and Boucher R. (1986a) Chronic treatment with Levodopa, but not bromocriptine, induces dyskinesia in MPTP-parkinsonian monkeys. Correlation with [^{3}H]spiperone binding. *Brain Res.*, **379**, 294.

Bédard P. J., Larochelle L., Poirier L. J., and Sourkes T. L. (1970) Reversible effect of L-DOPA on tremor and catatonia induced by α-methyl-*p*-tyrosine. *Can. J. Physiol. Pharmacol.*, **48**, 82–84.

Braun A., Fabbrini G., Mouradian M. M., Serrati C., Barone P., and Chase T. N. (1987) Selective D-1 dopamine receptor agonist treatment of Parkinson's disease. *J. Neural Transm.*, **68**, 41–50.

Brücke T., Bankiewicz K., Harvey-White J., and Kopin I. (1983) The partial dopamine receptor agonist terguride in the MPTP-induced hemiparkinsonian monkey model. *Eur. J. Pharmacol.*, **13**, 148(3), 445–448.

Burns R. S., Chiueh C. C., Marky S. P., Ebert M. H., Jacobowitz D. M., and Kopin I. J. (1983) A primate model of parkinsonism: Selective destruction of dopaminergic neurons in the pars compacta of the substantia nigra by 1-methyl-4-phenyl-1,2,3,6-tetrahydropyridine. *Proc. Natl. Acad. Sci.,USA,* **80,** 4546.

Carpenter M. B. (1966) A study of the red nucleus in the rhesus monkey. Anatomic degeneration and physiologic effects resulting from localized lesions of the red nucleus. *J. Comp. Neurol.,* **105,** 195–249.

Carpenter M. B. (1958) The neuroanatomical basis of dyskinesia. In: *Pathogenesis and Treatment of Parkinsonism* (W.S. Fields, ed.), Charles Thomas, Springfield, IL, pp. 50–85.

Close S. P., Marriott A. S., and Pay S. (1985) Failure of SKF 38393 to relieve parkinsonian symptoms induced by 1-methyl-4-phenyl-1,2,3,6-tetrahydropyridine in the marmoset. *Br. J. Pharmacol.,* **85,** 320–322.

Dahlström A. and Fuxe K. (1964) Evidence for the existence of monoamine-containing neurons in the central nervous system. I. Demonstration of monoamines in the cell bodies of brain stem neurons. *Acta Physiol. Scand.,* 62 (suppl.), **232,** 1–55.

de Montigny C. and Lamarre Y. (1973) Rhythmic activity induced by harmaline in the olivo-cerebello-bulbar system of the cat. *Brain Res.,* **53,** 81–95.

Di Paolo T., Bédard P. J., Daigle M., and Boucher R. (1986) Long-term effects of MPTP on central and peripheral catecholamine and indoleamine concentrations in monkeys. *Brain Res.,* **379,** 286.

Falardeau P., Bédard P. J., and Di Paolo T. (1988a) Relation between brain dopamine loss and D_2 dopamine receptor density in MPTP monkeys. *Neurosci. Lett.,* **86,** 225–229.

Falardeau P., Bouchard S., Bédard P. J., Boucher R., and Di Paolo, T. (1988b) Behavioral and biochemical effect of chronic treatment with D_1 and/ or D_2 dopamine agonist in MPTP monkeys. *Eur. J. Pharmacol.,* **150,** 59–66.

Ferraro A. and Barrera S. E. (1936) The effects of lesions of the superior cerebellar peduncle in the Macacus rhesus monkey. *Bull. Neur. Inst. NY* **5,** 165–179.

Goldstein M., Anagnoste B., Battista A. F., Owen W. S., and Nakatari S. (1969) Studies of amines in the striatum in monkeys with nigral lesions. *J. Neurochem.,* **16,** 645–653.

Gomez-Mancilla B., Boucher R., and Bédard P. J. (1991) Effect of clonidine and atropine on rest tremor in the MPTP monkey model of parkinsonism. *Clin. Neuropharmacoal.* (in press).

Graham W. C., Clarke C. E., Boyce S., Sambrook M. A., Crossman A. R., and Woodruff G. N. (1990) Autoradiographic studies in animal models of hemi-parkinsonism reveal dopamine D_2 but not D_1 receptor supersensitivity. II Unilateral intra-carotid infusion of MPTP in the monkey (Macaca fascicularis). *Brain Res.,* **514,** 103–110.

Hassler R. (1938) Zur pathologie der Paralysis agitans und des postenzephalitischen Parkinsonismus. *J. Psych. Neurol. (Leipzig)* **48,** 5–6, 387–476.

Joyce J. N., Marshall J. F., Bankiewicz K. S., Kopin I. J., and Jacobwitz D. M. (1986) Hemiparkinsonism in a monkey after unilateral internal carotid artery infusion of 1-methyl-4-phenyl-1,2,3,6-tetrahydropyridine (MPTP) is associated with regional ipsilateral changes in striatal dopamine D_2 receptor density. *Brain Res.*, **382,** 360–364.

Lamarre Y. and Mercier L. A. (1971) Neurophysiological studies of harmaline-induced tremor in the cat. *Can. J. Physiol. Pharmacol.*, **49,** 1049–1058.

Langston J. W. and Ballard P. A. (1984) Parkinsonism induced by 1-methyl-4-phenyl-1,2,3,6-tetrahydropyridine: implications for treatment and the pathophysiology of Parkinson's disease. *Can. J. Neurol. Sci.*, **11,** 160–165.

Langston J. W., Ballard P., Tetrud J. W., and Irwin I. (1983) Chronic parkinsonism in humans due to a product of meperidine-analog synthesis. *Science,* **219,** 979,980.

Langston J. W., Forno L. S., Robert C. S., and Irwin I. (1984) Selective nigral toxicity after systemic administration of 1-methyl-4-phenyl-1,2,5,6-tetrahydropyridine (MPTP) in the squirrel monkey. *Brain Res.*, **292,** 390–394.

Larochelle L., Bédard P. J., Boucher R., and Poirier L. J. (1970) The rubro-olivo-cerebello-rubral loop and postural tremor in the monkey. *J. Neurol. Sci.*, **11,** 53–64.

Larochelle L., Bédard P., Poirier L. J., and Sourkes T. L. (1971) Correlative neuroanatomical and neuropharmacological study of tremor and catatonia in the monkey. *Neuropharmacology* **10,** 273–288.

Llinas R. and Wolkind R. A. (1973) The olivo-cerebellar system: functional, properties as revealed by harmaline-induced tremor. *Exp. Brain Res.*, **18,** 69–87.

Mettler F. A. (1946) The experimental production of static tremor. *Fed. Proc.*, **5,** 72 (abstract).

Mettler F. A., and Whittier J. R. (1947) The experimental production of abnormal involuntary movements in primates. *Trans. Am. Neurol. Assoc.*, **72,** 96–101; 190–192.

Nomoto M., Stahl S., Jenner P., and Marsden C. D. (1986) Alterations in motor behavior produced by the isomers of 3-PPP in the MPTP-treated marmoset. *Eur. J. Pharmacol.*, **121,** 123.

Péchadre J. C., Larochelle L., and Poirier L. J. (1976) Parkinsonian akinesia, rigidity and tremor in the monkey. *J. Neurol. Sci.*, **28,** 147–157.

Peterson E. W., Magoun H. W., McCulloch W. S., and Lindssley D. B. (1949) Production of postural tremor. *J. Neurophysiol.*, **12,** 371–384.

Poirier L. J. (1960) Experimental and histological study of midbrain dyskinesias. *J. Neurophysiol.*, **23,** 534–551.

Poirier L. J. and Sourkes T. L. (1965) Influence of the substantia nigra on the catecholamine content of the striatum. *Brain,* **88,** 181,182.

Poirier L. J., Bédard P. J., Langelier P., Larochelle L., Parent A., and Roberge A. G. (1972) Les circuits neuronaux implisqués dans la physiopathologie des syndromes parkinsoniens. *Rev. Neurol.*, **127,** 37–50.

Poirier L. J., Bouvier G., Bédard P. J., Bocuer R., Larochelle L., Olivier A., and Singh P. (1969a) Essai sur les circuits neuronaux impliques dans le tremblement postural et l'hypokinesie. *Rev. Neurol.*, **120,** 15–40.

Poirier L. J., Sourkes T. L., Bouvier G., Boucher R., and Carabin S. (1969) Striatal amines, experimental tremor and the effect of harmaline in the monkey. *Brain*, **89,** 37–52.

Robertson G. S. and Robertson H. A. (1986) Synergistic effects of D_1 and D_2 dopamine agonists on turning behaviour in rats. *Brain Res.*, **384,** 387–390.

Rouillard C. and Bédard P. J. (1988) Specific D-l and D-2 dopamine agonists have synergistic effects in the 6-hydroxydopamine circling model in the rat. *Neuropharmacology*, **27,** 1257–1264.

Rouillard C., Bédard P. J., and Di Paolo T. (1990) Effects of chronic treatment of MPTP monkeys with bromocriptine alone or in combination with SKF 38393. *Eur. J. Pharmacol.*, **185,** 209–215.

Temlett J. A., Chang P. N., Oertel W. H., Jenner P., and Marsden C. D. (1988) The D_1 receptor partial agonist CY-208243 exhibits antiparkinsonian activity in the MPTP-treated marmoset. *Eur. J. Pharmacol.* **156,** 197–201.

Tretiakoff C. C. (1919) Contribution à l'étude de l'anatomie pathologique du locus niger de Soemmering avec quelques déductions relatives à la pathogénie des troubles du tonus musculaire de la maladie de Parkinson. Thèse Paris.

Walker A. E., and Botterell, E. H. (1937) The syndrome of the superior cerebellar peduncle in the monkey. *Brain*, **60,** 329–353.

Ward A. A. Jr., McCulloch W. S., and Magoun H. W. (1948) Production of an alternating tremor at rest in monkeys. *J. Neurophysiol.*, **11,** 317–330.

Genetic Dysmyelination Models

A Key to the Mechanisms and Regulation of Myelination

Gregory Konat and Richard C. Wiggins

1. Introduction

The myelin sheath is a highly organized membranous structure that wraps tightly around an axon and provides for the energy-efficient, high velocity propagation of nerve impulses—known as saltatory conduction—without depolarization of adjacent nerve fibers (*see* chapter by Miller). The myelin sheath is a prerequisite for the development of complex nervous function, as indicated by the appearance of increasing amounts of compact myelin in the nervous systems of the higher vertebrates, with evolutionary development, and by the disease sequelae associated with demyelinating disorders of the nervous system.

Morphologically, the myelin sheath is an extension of the plasma membrane of the oligodendrogial cell in the central nervous system (CNS) or the Schwann cell in the peripheral nervous system (PNS). Glial cell processes engulf the axon, extend around it spirally, form junctional specializations with the axon, and eventually extrude cytoplasm to form the mature myelin lamellae and paranodal architecture. The composition of the membrane also modified in the course of the process, and the matured sheath contains specific proportions of certain lipids and proteins (*see* chapter by Miller). It should be emphasized at this point that, although the major myelin proteins and their genes are quite well known, the sheath contains a host of minor undefined and unidentified proteins that may be essential

From: *Neuromethods, Vol. 21: Animal Models of Neurological Disease, I*
Eds: A. Boulton, G. Baker, and R. Butterworth © 1992 The Humana Press Inc.

for normal extension and/or maturation (i.e., establishing proper myelin sheath–axon relationship and compaction) of the organelle. The biosynthesis of the myelin sheath is a relatively late stage in the development of the nervous system (Wiggins et al., 1988), and the whole process, collectively referred to as "myelination," implies the existence of a myriad of genes that encode for the structural components of the myelin sheath as well as for regulatory elements of various steps in its development. Although these developmental processes remain largely unknown (Wiggins et al., 1988), it is clear that a complex sequence of events is featured in the specialization and elaboration of glial membrane around previously bare axons and that these events are precisely regulated and synchronized by corresponding genetic factors. The respective genes are not organized spatially on a chromosome(s), but rather are randomly scattered throughout the genome. For example, mouse genes encoding for major myelin proteins map as follows: proteolipid protein (PLP) on chromosome X (Willard and Riordan,1985), basic protein (BP) on chromosome 18 (Roach et al., 1985), myelin-associated glycoprotein (MAG) on chromosome 6 (D'Eustachio et al., 1988), and major peripheral myelin P_0 glycoprotein on chromosome 1 (Kuhn et al., 1990). Studies on regulatory mechanisms of the synchronous and temporary expression of these genes are still at a rudimentary stage.

2. Myelin Abnormalities

The formation of myelin is an essential element of brain development. A number of genetic (Baumann, 1980; Hogan and Greenfield, 1984) and environrnental factors (Wiggins 1982,1988; chapter by Miller) may perturb the process and result in *dysmyelination,* i.e., qualitatively and/or quantitatively abnormal myelinogenesis, which in turn leads to serious neurological problems. *Hypomyelination* is a more restricted term pertaining to decreased, or retarded production of myelin. Consequently, *amyelination* would refer to a complete lack of myelin in a particular anatomical structure. *Demyelination,* on the other hand, denotes loss of already existing myelin sheath in a course of a pathological process, which may be followed by *remyelination,* i.e., secondary ensheathment of denuded axons. Studies of these abnormalities have already contributed to the elucidation of basic features of myelinogenesis and will be instrumental in further endeavors to decipher molecular mechanisms of the process, such as the nature of the relationship between the axon and the oligodendrocyte leading to the initiation myelinogenesis, the differentiation of oligodendrocyte plasma membrane into myelin, the assembly of molecules into the growing sheath, the process of myelin compaction, the function of various myelin sheath domains, factors facilitating stability of the sheath, and so on.

3. Genetic Dysmyelination

The value of mutation in understanding genetic organization and regulation is well known, and in the case of myelination, too, mutations affecting the process are of paramount value in the study of molecular mechanisms of myelin formation, maintenance, and function. In these animal models the disturbance is selective and limited to the myelin-forming cells, whereas the axons appear to be spared. Furthermore, the availability of several mutations affecting different genes may facilitate the dissection of the complex processes of myelinogenesis. In general, any condition producing dysmyelination of the developing nervous system can be either specific to myelinogenesis (primary effect) or secondary to some other, more general, developmental abnormality (pleiotropic effect).

There are several known single gene mutations (most of them in the mouse) that selectively impair myelination. This heterogenous group of genetic disorders is characterized by widespread lesions of the white matter of the CNS (leukodystrophies) or in the peripheral nerve fibers. These dysmyelinating mutations exhibit clear-cut defects leading to neurological and behavioral abnormalities associated with histological pathologies, and can either be restricted to the myelin sheath or result in a more complex phenotype. Characterization of these genetic defects and the resultant dysmyelination provides valuable data that can be applied to determining the "blueprint" for myelin development, structure, and function. It can be envisaged that some of the mutations primarily affecting myelination may affect the biosynthesis of structural elements of the myelin sheath, whereas others may perturb regulatory mechanisms of the process.

The emphasis of this review will be on mutations that primarily affect myelination in experimental animals. Since the most extensively studied mutants have already been reviewed excellently by other authors (Baumann, 1980; Hogan and Greenfield,1984), we will only highlight the most characteristic features of these mutants and update them with recent findings.

Murine rodents, i.e., mice and rats, provide the most convenient experimental mammals with which to study inherited dysgenesis of myelin. The normal pattern of brain myelination in murine rodents indicates that little myelin protein expression occurs prenatally and that myelin protein synthesis and accumulation increase significantly in the brain at about 10 days and reaches a maximum at about 21 days *post partum* (Norton and Poduslo, 1973; Konat and Clausen, 1976). Myelination begins earlier in the peripheral nervous system, although myelin protein synthesis and accumulation do not begin until 2–3 days postnatally in the developing sciatic nerve (Wiggins et al., 1975; Wiggins and Morell, 1980). Furthermore, the genetic analysis of these animals is greatly facilitated by small body size, short lifespan, and high reproductive rates, and, consequently, most of the dysmyelinating mutations have been isolated and characterized in these species.

3.1. PLP Gene Mutations

The structure of major myelin proteins appears to be remarkably conserved during the mammalian evolution, which is consistent with their highly specialized functions. PLP, an integral protein of the CNS myelin, spans the membrane bilayer severalfold (Stoffel et al., 1984; Hudson et al., 1989a) and is thought to participate in the formation of minor dense (interperiod) line, hence contributing to the compaction process. The 15 kb-long PLP gene is located on the X chromosome (Willard and Riordan, 1985) and contains 7 exons (Macklin et al., 1987). Some of the primary transcripts are alternatively spliced to produce a second form of PLP, i.e., intermediate protein (IP), also called DM-20 (Morello et al., 1986; Simons et al., 1987; Hudson et al., 1987). Because the PLP gene is on the X chromosome, recessive mutations at this locus affect only male animals; however, retardation of early myelination and myelin mosaicism can be observed in heterozygous females (Bartlett and Skoff, 1986; Benjamins et al., 1986,1989; Duncan et al.,1987a). Jimpy and myelin synthesis-deficient mice, myelin-deficient rat, shaking pups, Pelizaeus-Merzbacher disease, and—probably also—paralytic tremor rabbit leukodystrophies result from mutations in the PLP gene.

3.1.1. Jimpy Mouse

The jimpy *(jp)* disorder (Phillips, 1954; reviewed by Hogan and Greenfield, 1984) is a recessive, sex-linked, dysmyelinating mutation in the mouse. The viability of newborn *Y/jp* animals is not affected by the mutation prior to the CNS myelination. The neurological manifestations of the mutation are concomitant with the onset of myelination and include axial body tremor followed by severe tonic seizures that cause the death of mutants between 30–35 days of age.

The myelin content of the entire CNS is reduced to less than 10% of control, whereas the PNS myelin appears completely normal. The mutant CNS contains only half the normal number of oligodendrocytes, which appear to have a distorted cell cycle (Knapp and Skoff, 1987). Ultrastructural abnormalities of the CNS myelin sheath are conspicuous. The CNS myelin deficiency is paralleled by drastic reductions of myelin-specific lipids and the activities of several enzymes involved in their biosynthesis (cf., Hogan and Greenfield, 1984; Ganser et al., 1988). The levels of myelin proteins, BP, MAG, and Wolfgram protein (WP) are reduced to only several percent of the control, whereas PLP is undetectable by an immunoblotting technique (Yanagisawa and Quarles,1986). Abnormal isoforms of BP in jimpy were recently reported (Fannon and Moscarello, 1990).

The genetic defect in jimpy has recently been deciphered (Dautigny et al., 1986; Nave et al., 1986; Moriguchi et al., 1987). There is a single base transition (A to G) at the splice acceptor site of intron 4 that activates a cryptic splicing site and results in the deletion of exon 5 (74 nucleotides) from the message RNA and in a reading frame shift (Nave et al., 1987). These, in turn, generate a 243 amino acid PLP polypeptide with an altered, missense and unusually cysteine-rich carboxyl terminus (residues 206–243)(Hudson et al., 1987). The abnormal protein cannot be posttranslationally processed through the Golgi apparatus and is retained in the rough endoplasmic reticulum (RER) (Roussel et al., 1987). The accumulation of the faulty PLP in RER may be responsible for the oligodendrocyte death in jimpy. The presence of PLP-reactive material in rare, loosely arranged myelin lamellae (Roussel et al., 1987) may indicate the existence of an alternate intracellular pathway for the maturation of PLP to myelin membrane. On the other hand, Duncan et al. (1989) found jimpy myelin to be PLP-negative. The discrepancy between these two studies may reflect a difference in the specificity of antibodies used by these authors. In any case, the truncated jimpy PLP, if actually present in myelin, cannot fulfill its function and results in a formation of an abnormal interperiod line (Duncan et al., 1989).

3.1.2. Myelin Synthesis-Deficient Mouse

A sex-linked mutation closely related to jimpy, dubbed myelin synthesis-deficiency *(msd)* was reported by Meier and MacPike in 1970 (reviewed by Hogan and Greenfield, 1984). This mutation has been shown to be allelic to jimpy and is now denoted by the jp^{msd} symbol. Neurological symptoms in this disorder closely resemble those of jimpy but are quantitatively lesser. The affected animals die at about 3 weeks of age.

The histological and biochemical abnormalities in the *msd* mutant are very similar to those in jimpy, although quantitatively less severe (for lipid composition, *see* Ganser et al., 1988). The affected males produce twice as much myelin as jimpy. Residual amounts of normal size PLP and IP can be detected in *msd* brain by Western blot analysis (Gardinier and Macklin, 1988), althoug these proteins are present in an inverted proportion to that in controls, i.e., IP is more abundant than PLP.

The molecular defect was recently defined as a point mutation, a C to T transition in exon 6 of the PLP gene. This single base change results in the substitution of valine for alanine at position 242 in both PLP and its alternatively spliced isoform IP (Gencic and Hudson, 1990). This alanine residue is in a hydrophobic domain of PLP and IP, and this domain is thought to reside on the extracellular side of the oligodendrocyte plasma membrane and participate in the formation of minor dense line in compact myelin (Hudson et al., 1989a). The amino acid substitution at this position results not only in the loss of the biological activity of the protein but also arrests normal maturation of the oligodendrocytes.

3.1.3. Myelin-Deficient Rat

Myelin deficiency *(md)* is a hereditary X-linked dysmyelination described in a Wistar strain rat by Csiza and de Lahunta (1979) (reviewed by Hogan and Greenfield, 1984). Clinically the mutation first manifests in the second week of life as generalized tremor later followed by seizures and severe ataxia. Pups expire within a month after birth. Myelination is extremely deficient in the CNS, whereas the PNS myelin appears to be normal. The oligodendrocytes are drastically reduced in number in older animals and the cells are morphologically abnormal (vacuolated, immature) (Dentinger et al., 1985; Kahn et al., 1986; Jackson and Duncan, 1988). *md* Axons appear to produce normal myelination signals (Duncan et al., 1988). Sporadic CNS myelin sheaths are devoid of PLP and have an abnormally condensed interperiod line (Duncan et al., 1987b) as well as an abnormal axoglial junction (Rosenbluth, 1987). Total phospholipids and cholesterol are substantially reduced; however, cerebrosides and sulfatides are the most severely deficient lipids in the brain. The glycolipid-synthesizing enzymes are also deficient in the mutant CNS.

The level of PLP and BP messages in the mutant brain is reduced to 10–20% of control, although only BP can be detected by immunolabeling (Zeller et al., 1989; Dentinger et al., 1982; Duncan et al., 1987b; Yanagisawa et al., 1986). The molecular basis for the mutation has been identified as an A-to-C transversion in exon 3 of the PLP gene that causes a substitution of threonine by proline at position 75 (Boison and Stoffel, 1989; Simons and Riordan, 1990). The helix-breaking proline probably induces a dramatic structural change in the second α-helical transmembrane domain (Stoffel et al., 1984; Hudson et al., 1989b) and, consequently, impairs proper conformational integration of the polypeptide into the lipid bilayer.

3.1.4. Shaking Pups

Shaking pups is a recessive, sex-linked trait in Springer-Spaniel dogs associated with severe hypomyelination in the CNS (Griffiths et al., 1981a,b; Duncan et al., 1983). Clinical manifestations develop as gross tremors in the second week *post partum* and do not comport any peripheral neuropathy (Griffiths et al., 1981b). The lifespan of the affected males is 3–4 mo. Histologically, the CNS myelin is poorly compacted and the oligodendrocytes are conspicuously abnormal and reduced in number (Duncan et al., 1983).

The myelin yield is reduced to less than 3% of control and the membrane is drastically deficient in PLP (Yanagisawa et al., 1987) and BP (Inuzuka et al., 1986). MAG is less severely affected but displays a higher apparent molecular weight (Inuzuka et al., 1986), resembling the feature found in quaking and trembler *(vide infra).*

The molecular basis for the mutation turned out to be a point mutation resulting in the substitution of histidine by proline at position 36 of the PLP polypeptide (Nadon et al., 1988). Such mutation changes the conformation of the α-helix structure of the first transmembrane segment (cf., *md* rat).

3.1.5. Pelizaeus-Merzbacher Disease

Pelizaeus-Merzbacher disease (PMD) is a clinically and pathologically heterogenous group of leukodystrophies in humans. It is beyond the scope of this chapter to review this disorder, although recently discovered features of the molecular genetics in some PM kindreds are worth mentioning.

In three PMD pedigrees displaying X-linked inheritance, the genotypic flaws accountable for the neuropathological phenotype have been unraveled in the PLP gene. Single point mutations were identified; one family has a C-to-T transition, resulting in a substitution of 215-serine for proline (Gencic et al., 1989); another family has the same transition at nucleotide 40 of the second exon leading to proline-to-leucine transversion (Trofatter et al., 1989); and another family has a T-to-C transition, resulting in an arginine-to-tryptophan change at residue 167 (Hudson et al., 1989b). Besides the scholarly interest, these findings may provide for a rapid prenatal diagnosis, and eventually, for the development of new medical therapies as well as for better classification of various clinical conditions grouped under PMD.

3.2. BP Gene Mutations

BP is a peripheral membrane protein in both PNS and CNS myelin. In the mouse the BP gene is located on mouse chromosome 18 (Roach et al., 1985). The gene spans 32 kb and contains 7 exons that are differentially spliced to produce mRNA messages coding for 5 distinct BP polypeptides (Takahashi et al., 1985; deFerra et al., 1985; Newman et al., 1987). The polypeptides are posttranslationally modified, further augmenting their heterogeneity. BP is localized in the major dense line of myelin (Omlin et al., 1982) and appears to be quintessential for the compaction and stability of the CNS myelin sheath (Matthieu et al.,1984,1986). Mutations in the BP gene result in shiverer and myelin-deficient mouse phenotypes.

3.2.1. Shiverer Mouse

Shiverer *(shi)* is an autosomal, recessive mutation isolated by Biddle et al. (1973). The affected locus maps on chromosome 18. The clinical symptoms include shaking voluntary movements that appear by the onset of myelination. The intensity of symptoms increases, progressively leading to tonic seizures, paralysis, and finally death between 50–100 days *post partum.*

The pathohistology and biochemistry of the shiverer mouse is reviewed by Hogan and Greenfield (1984). Briefly, the mutant is virtually devoid of CNS myelin, whereas the PNS myelin appears qualitatively and quantitatively normal. The cardinal feature of the residual CNS myelin is a lack of the major dense line. The biochemical correlate of the hypomyelination is the decrease in the concentration of galactolipids and cholesterol in the shiverer brain (Ganser et al., 1988). Detectable BP content is less than 3% of control, PLP is reduced to a lesser degree, whereas WP is increased in the shiverer brain. Both CNS and PNS myelin lack BP, with the exception of P2 in the PNS, which is normal.

Detailed mapping experiments show that 20 kb from the 3' end of the BP gene, containing exons 3 through 7 (almost two-thirds of the gene), is deleted from the shiverer haplotype

(Kimura et al., 1985; Roach et al., 1985; Molineaux et al., 1986), resulting in a truncated protein. The abnormal protein is not compatible with its original function in the sheath, which appears to be the production of major dense line, and, thus, the genetic defect prevents the formation of compact myelin. The compaction of the PNS myelin seems not to be exclusively associated with BP as its morphology is virtually normal in the mutant.

The shiverer phenotype can be experimentally induced in transgenic experiments using antisense BP constructs (Katsuki et al., 1988). Consequently, endowment of shiverer mice with the wild-type BP gene restores normal myelin formation and behavior in the transgenic shiverer mouse (Readhead et al., 1987). Introduction of a single 14 kDa BP type gene appears to be sufficient for the formation of major dense line and compaction of the myelin sheath (Kimura et al., 1989).

3.2.2. Myelin-Deficient Mouse

The myelin deficient *(mld)*, an autosomal, recessive mutation in mice is allelic to shiverer mutant and displays a similar phenotype characterized by a dramatic myelin deficit in the CNS and trembling locomotion (Doolittle and Schweikart, 1977; reviewed by Hogan and Greenfield, 1984). The peripheral myelin appears to be normal. The affected tissues have reduced lipid concentration (Ganser et al., 1988). Myelin-deficient mouse has higher expression of BP than does shiverer. The amounts of MAG and Wolfgram protein, however, are increased as compared to control animals.

The BP gene in *mdl* has a more complicated structure than the gene in shiverer. The *mdl* mutants carry two tandemly duplicated BP genes (Akowitz et al., 1987; Okano et al., 1987; Popko et al., 1987). The upstream copy of the gene has a large inversion spanning exons 3–7, whereas the downstream copy has an intact organization (Okano et al., 1988; Popko et al., 1988). The two genes are expressed independently, although the upstream gene is transcribed faster and produces antisense RNA complementary to exons 3–7 (Okano et al., 1988; Popko et al., 1988). Although the overall transcription rate of the BP gene is normal and the nuclear concentration of sense transcript is relatively high in the *mdl* mutant, the concentration of mature mRNA in the cytoplasm is extremely low (Popko et al., 1988; Roch et al., 1989; Tosic et al., 1990). This finding infers that the sense and antisense transcripts form a duplex RNA in the nucleus, and, hence, arrest normal posttranscriptional processing of BP-coding RNA and/or its transport to the cytoplasm (Tosic et al., 1990). The normal phenotype can be restored by introducing wild-type BP transgene into *mld* (Popko et al., 1987).

3.3. Mutations of Unknown Genetic Defect

The other well established dysmyelinating phenotypes in experimental animals include quaking, twitcher, trembler, and dystrophic mice, CBB hamster, and paralytic tremor rabbit. The dysmyelination in trembler mouse and in paralytic tremor rabbit seems to result from a stymied lipid metabolism and from a defective PLP gene, respectively. The molecular pathobiology of quaking, trembler, and dystrophic mice as well as of CBB hamster remains elusive.

3.3.1. Quaking Mouse

The quaking mutant was originally described by Sidman et al. (1964) (reviewed Hogan and Greenfield, 1984). The trait is a recessively inherited autosomal mutation. The quaking locus *(qk)* is situated on chromosome 17. The appearance of neurological symptoms coincides with the onset of myelination and includes postural and voluntary action tremor and convulsive seizures. The symptoms progressively reach their maximum intensity after 4 wk of age.

The quaking mutation causes a pronounced myelin deficit throughout the CNS (myelin concentration less than 30% of control) and, to a lesser degree, in the PNS. In the CNS, the cardinal ultrastructural features are (1) a marked reduction in the myelin lamellae relative to axon diameter, and (2) a lack of compaction of the myelin sheath with prominent cytoplasmic retention in the spiral cell process. The quaking brain contains normal or excessive numbers of oligodendrocytes. Nerve grafting experiments show that the primary defect leading to myelin deficiency in the quaking mutant resides in the Schwann and oligodendroglial cells (cf. Hogan and Greenfield, 1984; Poltarak and Freed, 1988).

Several studies on quaking have reported abnormalities in both cerebral lipids and lipid metabolizing enzymes (cf. Hogan and Greenfield, 1984; Ganser et al., 1988). The depletion in myelin-specific lipids has, however, been attributed to a defective intracellular transport mechanism or to the defects in the mechanism of the final myelin assembly, rather than to a lipid-specific genetic error (Deshmukh and Bear, 1977). The quaking brain is practically devoid of acylgalactosylceramides, an early marker of oligodendrocyte differentiation (Theret et al., 1988) and is severely depleted in monosialogangliosides, which are also qualitatively abnormal (Iwamori et al., 1985).

Although the content of several myelin-specific proteins has been found to be decreased in the quaking mouse, there are indications of a possible defect in the mechanisms of myelin assembly, rather than abnormal synthesis of myelin components. A specific developmental pattern for the individual myelin proteins has been observed in the quaking brain. Thus, the expression of the BP gene is only initially delayed (Roth et al., 1985; Konat et al., 1988), although, even in adult mutant brain, the 14 kDa BP isoform is conspicuously decreased, indicating the immaturity of quaking myelin (Inuzuka et al., 1987). The synthesis of PLP is more severely reduced (Sorg et al., 1986; Konat et al., 1988) and its posttranslational acylation with long-chain fatty acids appears to be impaired (Konat et al., 1986). PLP seems not to be assembled properly into the myelin sheath in the mutant brain and this disorder presumably leads to its increased turnover (Konat et al., 1987a).

MAG is also decreased in quaking, but its genetic expression (the message level) is enhanced several-fold and appears to be progressively dysregulated (Frail and Braun, 1985; Konat et al., 1988). The splicing pattern of the primay transcript is also abnormal (Frail and Braun, 1985; Fujita et al., 1988). Furthermore, MAG exhibits a higher apparent molecular weight when compared to the control animals (Matthieu et al., 1974; Inuzuka et al., 1987), which is probably a result of an abnormal processing of the *N*-linked oligosaccharide moiety of the protein in quaking oligodendrocytes (Konat et al., 1987b). In the PNS, the lack of normal myelin compaction seems to be owing to inability of the Schwann cells to remove MAG from their mesaxon membrane (Trapp, 1988). The defective compaction has also been linked to a deficiency in NCAM-180, a cell adhesion glycoprotein which is structurally and functionally related to MAG (Bhat and Silberberg, 1988).

A sixfold increase in total calcium concentration along with a 30% elevation of calcium-activated neutral proteinase (CANP) activity has been reported in quaking brain (Banik et al., 1987). Furthermore, the assembly of CANP into quaking myelin seems to be impaired (Banik et al.,1987).

In addition to defective myelinogenesis, the quaking mice suffer from aberrant spermatogenesis, leading to azoospermia. Thus, the locus affected by the quaking mutation important in the development of at least two types of cells; myelin-forming cells and spermatids.

In conclusion, the quaking mutation that is expressed largely as an arrest in myelinogenesis is not genetically linked to any of the major myelin proteins and maps on chromosome 17 within the *t*-complex, a region that harbors several genes involved in ontogeny. Thus, the mutation signals the existence of an important myelination-related genomic locus that is presently undefined and unidentified. The product of this locus, which is defective in quaking, may play a regulatory role in the process of myelinogenesis by activating the expression of specific proteins and enzymes required for the assembly of myelin membrane. Alternatively, the product of this locus may be an undefined and uncharacterized minor protein that is necessary for normal extension and maturation of the sheath. The presence of dysmyelination in both the CNS and PNS also indicates that the affected gene normally regulates myelin formation by both Schwann and oligodendrocyte cells.

3.3.2. Twitcher Mouse

The twitcher mouse *(tw)* is an autosomal recessive mutation of the mouse of yet unknown genetic linkage (Duchen et al., 1980; reviewed by Hogan and Greenfield, 1984). The clinical and morphological features resemble those of human globoid cell leukodystrophy (Krabbe's disease). Neurological symptoms are progressive muscular weakness and wasting of trunk and limbs that appear at about 30 days of age and lead to death by the third month. Both the CNS and PNS display a developmental sequence of normal early myelination, hypomyelination, and demyelination. The progressive loss of myelin is associated with the presence of inclusions in the perikarya of myelin-forming cells. Lipid composition of optic and trigeminal nerves is abnormal (Ganser et al., 1988). The primary defect in twitcher has been identified as a deficiency in a lysosomal enzyme, galactosylceramidase. This metabolic dysfunction leads to the accumulation of a toxic metabolite, psychosine (galactosylsphingosine) (Igisu and Suzuki, 1984; Shinoda et al., 1987) and ultimately to myelin loss and cell death (Nagara et al., 1986; Tanaka et al., 1988,1989). The neurological symptoms can be alleviated by supplying the deficient enzyme through allogeneic transplantation of hematopoietic cells (Ichioka et al., 1987; Suzuki et al., 1988; Hoogerbrugge et al., 1988a,b; Kondo et al., 1988). Twitcher myelin is also deficient in PLP-acylation activities (Yoshimura et al., 1989).

3.3.3. Trembler Mouse

The trembler mutation *(Tr)* in mice displays an autosomal inheritance (Falconer, 1951; reviewed by Hogan and Greenfield, 1984). The defective gene maps on chromosome 11. The pathology features convulsions in young animals and a generalized tremor in the adults. This chronic hypertrophic neuropathy is restricted to the PNS and causes a severe hypomyelination associated with segmental demyelination. The phenotype results from a primary disorder of Schwann cells, which show greatly increased proliferation and fail to produce compact myelin. Histology of the CNS is normal. The mutation is semidominant as it also manifests in heterozygotes, although the extent of hypomyelination is less than in *Tr/Tr* homozygotes (Fryxell, 1983). An allelic mutation, Tr^{j}, was recently described by Henry and Sidman (1988). The homozygotic Tr^{j} animals are also severely hypomyelinated, but whereas *Tr* animals are viable, the Tr^{j} animals die within 2–3 weeks *post partum*. The content of major lipid classes is reduced in trembler sciatic nerve to less than half, whereas the cholesterol esters are increased approximately five-fold as compared to the control tissue (Yao and Bourre, 1985; Heape et al.,1986a,b; Juguelin et al., 1986; White et al., 1986; Ganser et al., 1988). The reduction in total lipid is accompanied by even greater deficiency in myelin-specific proteins. Trembler myelin is virtually devoid of P_0 glycoprotein and BP, whereas the high-molecular weight proteins are increased relative to control myelin.

In spite of severe deficiency in myelin specific proteins, the concentration of MAG is increased by approximately 30% in the sciatic nerves of adult mutants (Inuzuka et al., 1985). Furthermore, MAG in the sciatic nerves has higher than normal apparent molecular weight, while MAG in the brain of the mutant is normal.

The reduced concentration of P_0 and BP in trembler sciatic nerves (Garbay et al., 1988) correlates with reduced level of corresponding mRNAs, although the posttranscriptional processing of the messages appears to be normal (Garby et al., 1989). An excessive production of multilayered basal lamina around the Schwann cells has been implicated as the cause for dysmyelination (Koenig et al., 1989; Ferzaz et al., 1989).

3.3.4. Dystrophic

The dystrophic mutation, dystrophia muscularis *(dy)*, appears to have a primary effect on the PNS, although there is striking and extensive muscular involvement (Michelson et al., 1955; reviewed by Hogan and Greenfield, 1984). The trait is recessive and the affected locus is on chromosome 10. Clinically the mutation manifests as progressive paresis and atrophy of axial and limb muscles. The spinal roots and cranial nerves of affected mice are practically naked, lacking both myelin and glial wrappings. The unique feature in the affected nerves is an almost total lack of basal lamina around poorly differentiated Schwann cells. The CNS myelination is normal. The defect is all the more curious since axons that are practically devoid of myelin (amyelination) in the lumbosacral roots appear to be normally myelinated as they continue into the sciatic nerve (Wiggins and Morell, 1978). Furthermore, the Schwann cells in the dystrophic spinal roots regain their ability to myelinate axons when transplanted into more distal regions of the nerves of normal or mutant animals (Aguayo et al., 1982). All major PNS myelin proteins are proportionally reduced in the affected lumbosacral spinal roots (John and Purdom, 1984) and BP shows a difference in microheterogeneity of basic isomers (John, 1990). Several nonmyelin proteins are augmented (Davis, 1987) and lipid composition is altered as compared to control (Ganser et al., 1988).

The nerve graft experiments and the conspicuous deficiency in the basal lamina and collagen (John and Purdom, 1984) tend to indicate that the primary defect resides in the contiguous extracellular matrix in the spinal roots of the mutant and may be of axonal origin.

3.3.5. CBB Hamster

The trembling mutant hamster CBB is an autosomal, recessive trait (Nunoya et al., 1985). The only neurological abnormalities manifest as a fine axial tremor that commences at about the 14th day and disappears around 100 days of age—no clinical symptoms can be detected in older animals. The axons are poorly myelinated throughout the CNS but the peripheral nerves appear normal (Kunishita et al., 1986). The yield of CNS myelin is reduced by 70% in younger animals and still by 60% in adult animals as compared to controls. Smaller axons (less than 0.6 μm) are exclusively unmyelinated, whereas the thickness and lamellar structure of myelin sheath of those larger myelinated axons is normal. Myelin seems to be normally compacted and no alteration in its chemical composition has been detected. The primary defect seems to involve the oligodendrocyte-axon recognition process, especially in smaller fibers.

3.3.6. PT Rabbit

Paralytic tremor *(pt)* rabbit mutation is a sex-linked recessive trait that produces a deficiency of CNS myelin (Osetowska and Leszawski, 1975). Clinical symptoms appear in the 2nd week of life as rapid rhythmic tremor followed by spastic paresis. In most of the affected animals the clinical conditions improve, although some rabbits develop a severe progressive disease and expire within 3–4 mo. The aberration in the sheath formation is evident in all structures of the CNS (Taraszewska, 1979,1983; Taraszewska and Zelman,1985; Zelman and Taraszewska, 1984). The oligodendrocytes are slightly increased in number (Taraszewska, 1983). Myelin-specific lipids, and especially cerebrosides and sulfatides, are significantly reduced in the mutant brain (Domanska-Janik et al., 1986). The recovery of myelin from affected brains is reduced to 20–30% of control and the membrane is deficient in galactosphingolipids and the most basic BP isoform, whereas the ganglioside content is increased (Domanska-Janik et al., 1988).

The mapping of the *pt* locus on the X chromosome suggests involvement of the PLP gene. The mutation, however, seems to affect the PLP molecule at less vulnerable point(s), generating more benign phenotypic alterations than in the previously discussed mutants. Alternatively, the mutation may be in the regulatory sequences of the gene, resulting in an attenuated response of the oligodendrocytes to the signals initiating and/or sustaining normal myelination.

4. Mechanisms of Genetic Dysmyelination

The highly organized and regulated developmental events collectively described as myelinogenesis can be affected at different levels by genetic defects, as indicated in the characterizations of the specific mutations *(vide supra)*. There are several proven and/or hypothetical molecular mechanisms by which various genomic abnormalities may result in dysmyelinating phenotypes:

Defects in genes encoding for myelin proteins. There are several mutations that affect the protein building blocks of the myelin membrane, such that there may be either an absence of the protein or the presence of an abnormal form of the protein that precludes myelin formation. Failure to produce BP is observed in the *md* rat as a result of the production of antisense RNA from a duplicated and partly inverted BP gene. A large deletion in the BP gene results in truncated polypeptides in shiverer. PLP with a vastly altered carboxy terminus in jimpy results from distorted splicing concomitant with a frame shift caused by a point mutation. PLP species in *msd* mouse, shaking pups, and PMD have single amino acid changes in different parts of the polypeptide caused by point mutations in the coding sequence of the PLP gene. The accumulation of aberrant protein may impair post-translational processing of not only its polypeptides, but also of other proteins, resulting in cell death. Such a situation exists in the jimpy mutation.

Characterized mutations affecting these genes all occur in the genes themselves, i.e., in the coding sequences or in the splicing sequences. Out of nine known mutations (seven for PLP and two for BP), none seems to affect regulatory elements of these structural genes. It can be envisaged that mutations in promoter or enhancer regions of the genes as well as in translation regulating sequences of the respective mRNAs could result in attenuated (or amplified, which may be equally detrimental) production of otherwise normal proteins; *pt* rabbit may turn out to be this type of mutation.

Besides the major proteins of purified myelin, it is possible that other minor and unidentified proteins are involved in the full spectrum of events during myelinogenesis. Certain of the dysmyelinating mutants that have not yet been characterized may be caused by mutations affecting genes for minor myelin proteins. Such a possibility cannot be ruled out for the quaking mutant.

Defects in genes involved in metabolism of myelin constituents. Myelin is a lipid-rich membrane. Although there are no *sensu stricto* myelin-specific lipids, some species are specifically enriched in myelin. Hence, a defect in the synthesis of any of these lipids would perturb myelination, whereas the deficiency could be circumvented in other cells. So far, there are no dysmyelinating mutations known to be caused by deficient lipid synthesis. In Twitcher, which features abnormal lipid metabolism, the dysmyelination results from a deficit in lipid catabolism leading to the accumulation of the toxic metabolite, psychosine.

Processing of primary transcripts, translation of mRNA as well as posttranslational processing of myelin proteins, including both chemical modifications of the polypeptides and intracellular translocation, represent other points at which myelination could be disrupted. Abnormalities in quaking MAG resulting from distorted splicing and/or glycosylation may underlie dysmyelination in this mutant.

Defects in genes of intercellular signaling. Myelinogenesis is thought to involve numerous trophic factors produced by axons, myelin-forming cells (and maybe also other glial cells) as well as specific recognition molecules displayed on the surface of these cells. Attenuation or lack of proper signals may cause dysmyelination. Dystrophic mouse may represent such a mechanism of dysmyelination, since the flaw appears to reside in the lack of signals from certain parts of axons (dorsal and ventral roots, in particular). An abnormality in axonal signaling has also been implicated in the CBB hamster.

Similarly, it is possible that defective recognition sites on myelin-forming cells may not permit a proper response to axonal signals. Dysmyelination in trembler may be due to impaired recognition of axonal signals as a result of excessive formation of basal lamina around the Schwann cells. In the absence of any clear understanding of the quaking mutations, a defect in glial receptors would account for the observed dysmyelination.

Defects in genes regulating myelination. The numerous separate events in myelinogenesis require precisely controlled expression of genes coding for structural proteins of myelin as well as lipid-synthesizing enzymes. The genetic expression must be controlled by as yet unknown regulatory factors, and a mutation in such a regulatory gene would be expected to produce a spectrum of dysmyelinating conditions, even though normal proteins and lipids are available to make myelin. Perhaps the dysmyelination observed in quaking can be explained by such a mutation.

5. Conclusions and Applications

Cellular and molecular events involved in myelinogenesis remain largely unknown. As in other cellular and developmental processes, the interacting cells represent a highly complex cybernetic system that features a plethora of precisely controlled spatial and temporary events at both the inter- and intracellular level. The availability of dysmyelinating mutants combined with advanced molecular biology methodologies provide powerful tools to discern the regulatory mechanisms of the assembly and maintenance of myelin. The elucidation of the developmental processes leading to myelination will provide the quantum leap to understand the molecular mechanisms of not only myelinogenesis but also of other morphogenic processes.

Because myelin is strongly conserved phylogenetically, studies in experimental animals are likely to reveal mechanisms of myelination in humans. This information may be of paramount medical value, as lacking a basic understanding of the mechanisms of myelinogenesis limits the number of available hypotheses to spur investigation of myelin pathologies. This is potentially of great importance to the human condition in view of the importance of myelination to the developing nervous system and to neural repair processes. Above all it would be highly pertinent to the pathology of myelin sheath as it would increase our understanding of demyelinating conditions and especially the most frequent and most debilitating, multiple sclerosis (MS). Once the mechanisms of myelin sheath formation are defined, therapies can be developed to facilitate recovery of neurological conditions by promoting the preservation of the myelin sheath during exacerbations (and/or even prevent exacerbations from occurring), and eventually by potentiating remyelination by stimulating quiescent oligodendrocytes that have been deactivated in the course of the disease.

Acknowledgments

The authors wish to express their gratitude to Sheldon Miller for helpful suggestions and comments. This work was partially supported by PHS grant NS13799 and National MS Society grant RG 1917.

References

Akowitz A. A., Barbarese E., Scheld K., and Carson J. H. (1987) Structure and expression of myelin basic protein gene sequences in the *mld* mutant mouse: Reiteration and rearrangement of the MBP gene. *Genetics* **116,** 447–464.

Aguayo A., Perkins S., and Bray G. M. (1982) Cell interaction in nerves of dystrophic mice. In: *Disorders of the Motor Unit* (Schotland, D. L., ed.), Wiley, New York, pp. 37–50.

Banik N. L., Chakrabarti A. K., Gantt G., and Hogan E. L. (1987) Distribution of calcium-activated neutral proteinase activity in quaking mouse brain: a subcellular study. *Brain Res.* **435,** 57–62.

Bartlett W. P. and Skoff R. P. (1986) Expression of the Jimpy gene in the spinal cords of heterozygous female mice. An early myelin deficit followed by compensation. *J Neurosci.* **6,** 2802–2812.

Baumann N. (1980) *Neurological Mutations Affecting Myelination,* INSERM Symposium No. **14,** Elsevier/North-Holland, Amsterdam.

Benjamins J. A., Studzinski D. M., and Skoff R. P. (1986) Biochemical correlates of myelination in brain and spinal cord of mice heterozygous for the jimpy gene. *J. Neurochem.* **47,** 1857–1863.

Benjamins J. A., Studzinski D. M., Skoff R. P., Nedelkoska L., Carrey E. A., and Dyer C. A. (1989) Proteolipid protein in mice heterozygous for the jimpy gene. *J. Neurochem.* **53,** 279–286.

Bhat S. and Silberberg D. H. (1988) NCAM-180, the largest component of the neural cell adhesion molecule, is reduced in dysmyelinating quaking mutant mouse brain. *Brain Res.* **452,** 373–377.

Biddle F., March E., and Miller J. R. (1973) Research news, *Mouse News Lett.* **48,** 24.

Boison D. and Stoffel W. (1989) Myelin-deficient rat: a point mutation in exon III (A-C, Thr75-Pro) of the myelin proteolipid protein causes dysmyelination and oligodendrocyte death. *EMBO J.* **8,** 3295–3302.

Csiza C. K., and de Lahunta A. (1979) Myelin deficiency (md), a neurologic mutant in the Wistar rat. *Am. J. Pathol.* **95,** 215–222.

Dautigny A., Mattei M. G., Morello D., Alliel P. M., Pham-Dinh D., Amar L. Arnaud D., Simon D., Mattei J. F., Guenet J. L., Jolles P., and Avner P. (1986) The structural gene coding for myelin-associated proteolipid protein is mutated in Jimpy mice. *Nature* **321,** 867–869.

Davis H. L. (1987) Sciatic nerve protein composition in normal and dystrophic C57BL/6J mice. *Neurosci. Lett.* **75,** 95–100.

deFerra F., Engh H., Hudson L., Kamholz J., Puckett C., Molineaux S., and Lazzarini R. A. (1985) Alternative splicing accounts for the four forms of myelin basic protein. *Cell* **43,** 721–727.

Dentinger M. P., Barron K.D., and Csiza C. K. (1982) Ultrastructure of the central nervous system in a myelin deficient rat. *J. Neurocytol.* **11,** 671–691.

Dentinger M. P., Barron K. D., and Csiza C. K. (1985) Glial and axonal development in optic nerve of myelin deficient rat mutant. *Brain Res.* **344,** 255–266.

Deshmukh D. S. and Bear W. D. (1977) The distribution and biosynthesis of the myelin-glycolipids in the subcellular fractions of brains of quaking and normal mice during development. *J. Neurochem.* **28,** 987–993.

D'Eustachio P., Colman D. R., and Salzer J. L. (1988) Chromosomal location of the mouse gene that encodes the myelin-associated glycoproteins. *J. Neurochem.* **50,** 589–593.

Domanska-Janik K., Wikiel H., Zelman I., and Strosznajder J. (1986) Brain lipids of a myelin-deficient rabbit mutant during development. *Neurochem. Pathol.* **4,** 135–151.

Domanska-Janik K., Gajkowska B., de Nechaud B., and Bourre J. M. (1988) Myelin composition and activities of CNPase and Na^+, K^+ -ATPase in hypomyelinated "pt" mutant rabbit. *J. Neurochem.* **50,** 122–130.

Doolittle D. P. and Schweikart K. M. (1977) Myelin deficient, a new neurological mutant in the mouse. *J. Hered.* **68,** 331–332.

Duchen L. W., Eicher E. M., Jacobs J. M., Scaravilli F., and Texeira F. (1980) Hereditary leukodystrophy in the mouse: the new mutation, twitcher. *Brain* **103,** 695–710.

Duncan I. D., Griffiths I. R., and Munz M. (1983) Shaking pups: a disorder of central myelination in the spaniel dog. III. Quantitative aspects of glia and myelin in the spinal cord and optic nerve. *Neuropathol. Appl. Neurobiol.* **9,** 355–368.

Duncan I. D., Hammang J. P., and Gilmore S. A. (1988) Schwann cell myelination of the myelin deficient rat spinal cord following X-irradiation. *Glia* **1,** 233–239.

Duncan I. D., Hammang J. P., Goda S., and Quarles R. H. (1989) Myelination in the jimpy mouse in the absence of proteolipid protein. *Glia* **2,** 148–154.

Duncan I. D., Hammang J. P., and Jackson K. F. (1987a) Myelin mosaicism in female heterozygotes of the canine shaking pups and myelin-deficient rat mutants. *Brain Res.* **402,** 168–172.

Duncan I. D., Hammang J. P., and Trapp B. D. (1987b) Abnormal compact myelin in the myelin-deficient rat: absence of proteolipid protein correlates with a defect in the interperiod line. *Proc. Natl. Acad. Sci USA* **84,** 6287–6291.

Falconer D. S. (1951) Two new mutants, "Trembler" and "Reeler" with neurological actions in the mouse (MUS. MUSCULUS L). *J. Genet.* **50,** 192–201.

Fannon A. M. and Moscarello M. A. (1990) Myelin basic protein is affected by reduced synthesis of myelin proteolipid protein in the jimpy mouse. *Biochem. J.* **268,** 105–110.

Ferzaz B., Koenig H., Ressouches A. (1989) Axon regeneration in the Trembler mouse, with mutation affecting Schwann cells. *C. R. Acad. Sci. [III]* **309,** 377–382.

Frail D. E. and Braun P. E. (1985) Abnormal expression of the myelin-associated glycoprotein in the central nervous system of dysmyelinating mutant mice. *J. Neurochem.* 45, 1071–1075.

Fujita N., Sato S., Kurihara T., Inuzuka T., Takahashi Y. and Miyataka T. (1988) Developmentally regulated alternative splicing of brain myelin-associated glycoprotein mRNA is lacking in the quaking mouse. *FEBS Lett.* **232,** 323–327.

Fryxell K. F. (1983) Biochemical and Genetic Studies of Peripheral Myelination in Normal Development and in the Mouse Mutant Trembler. Ph.D. Dissertation, California Institute of Technology, Pasadena.

Ganser A. L., Kerner A. -L. Brown B. J., Davisson M.T., and Kirschner D. A. (1988) A survey of neurological mutant mice. I. Lipid composition of myelinated tissue in known myelin mutants. *Dev. Neurosci.* **10,** 99–122.

Garbay B., Domec C., Fournier M., and Bonnet J. (1989) Developmental expression of the P_0 glycoprotein and basic protein mRNA in normal and trembler mutant mice. *J. Neurochem.* **53,** 907–911.

Garbay B., Fournier M., Sallafranque M. L., Muller S., Boiron F., Heape A., Cassagne C., and Bonnet J. (1988) P_0, MBP, histone and DNA levels in sciatic nerve: postnatal accumulation studies in normal and Trembler mice. *Neurochem. Pathol.* **8,** 91–107.

Gardinier M. V. and Macklin W. B. (1988) Myelin proteolipid protein gene expression in jimpy and jimpy^{msd} mice. *J. Neurochem.* **51,** 360–369.

Gencic S., Abuelo D., Ambler M., and Hudson L. D. (1989) Pelizaeus-Merzbacher disease: an X-linked neurologic disorder of myelin metabolism with a novel mutation in the gene encoding proteolipid protein. *Am. J. Hum. Genet.* **45,** 435–442.

Gencic S. and Hudson L. D. (1990) Conservative amino acid substitution in the myelin proteolipid protein of jimpy^{mds} mice. *J. Neurosci.* **10,** 117–124.

Griffiths I. R., Duncan I. D., and McCulloch M. (1981a) Shaking pups: a disorder of central myelination in the spaniel dog. II. Ultrastructural observations on the white matter of the cervical spinal cord. *J. Neurocytol.* **10,** 847–858.

Griffiths I. R., Duncan I. D., McCulloch M., and Harvey M. J. A. (1981b) Shaking pups: a disorder of central myelination in the spaniel dog. I. Clinical, genetic and light microscopical observations *J. Neurol. Sci.* **50,** 423–433.

Heape A., Juguelin H., Fabre M., Boiron F., and Cassagne C. (1986a) A quantitative developmental study of the peripheral nerve lipid composition during myelinogenesis in normal and Trembler mice. *Devel. Brain Res.* **25,** 181–189.

Heape A., Juguelin H., Fabre M., Boiron F., Garbay B., Fournier M., Bonnet J., and Cassagne C. (1986b) Correlation between the morphology and the lipid and protein compositions in the peripheral nervous system of individual 8-day-old normal and trembler mice. *Devel. Brain Res.* **25,** 173–180.

Henry E. W. and Sidman R. L. (1988) Long lives for homozygous Trembler mutant mice despite virtual absence of peripheral nerve myelin. *Science* **241,** 344–346.

Hogan E. L. and Greenfield S. (1984) Animal models of genetic disorders of myelin, In: *Myelin* (Morell, P., ed.), Plenum, New York, pp. 489–534.

Hoogerbrugge P. M., Poorthuia B. J., Romme A. E., van de Kamp J. J., Wagemaker G., and van Bekkum D. W. (1988a) Effect of bone marrow transplantation on enzyme levels and clinical course in the neurologically affected twitcher mouse. *J. Clin. Invest.* **81,** 1790–1794.

Hoogerbrugge P. M., Suzuki K., Suzuki K., Poorthuia B. J., Kobayashi T., Wagemaker G., and van Bekkum D. W. (1988b) Donor-derived cells in the central nervous system of twitcher mice after bone marrow transplantation. *Science* **239,** 1035–1038.

Hudson L. D., Berndt J. -A., Puckett C., Kozak C. A., and Lazzarini R. A. (1987) Aberrant splicing of proteolipid protein mRNA in the dysmyelinating jimpy mutant mouse. *Proc. Natl. Acad. Sci. USA* **84,** 1454–1458.

Hudson L. D., Friedrich V., Behar T., Dubois-Dalcq M., and Lazzarini R. A. (1989a) The initial events in myelin synthesis: Orientation of proteolipid protein in the plasma membrane of cultured oligodendrocytes. *J. Cell Biol.* **109,** 717–727.

Hudson L. D., Puckett C., Berndt J., Chan J., and Gencic S. (1989b) Mutation of proteolipid protein gene PLP in a human X chromosome-linked myelin disorder. *Proc. Natl. Acad. Sci. USA* **86,** 8128–8131.

Ichioka T., Kishimoto Y., Brennan S., Santos G. W., and Yeager A. M. (1987) Hematopoietic cell transplantation in murine globoid cell leukodystrophy (the twitcher mouse): Effects on levels of galactosylceramidase, psychosine, and galactocerebrosides. *Proc. Natl. Acad. Sci. USA* **84,** 4259–4263.

Igisu H. and Suzuki K. (1984) Progressive accumulation of toxic metabolite in a genetic leukodystrophy. *Science* **224,** 753–755.

Inuzuka T., Duncan I. D., and Quarles R. H. (1986) Myelin proteins in the CNS of 'shaking pups'. *Devel. Brain Res.* **27,** 43–50.

Inuzuka T., Johnson D., and Quarles R. H. (1987) Myelin-associated glycoprotein in the central and peripheral nervous system of quaking mice. *J. Neurochem.* **49,** 597–602.

Inuzuka T., Quarles R. H., Heath J., and Trapp B. D. (1985) Myelin-associated glycoprotein and other proteins in trembler mice. *J. Neurochem.* **44,** 793–797.

Iwamori M., Harpin M. -L., Lachapelle F., and Baumann N. (1985) Brain gangliosides of quaking and shiverer mutants: qualitative and quantitative changes of monosialogangliosides in the quaking brain. *J. Neurochem.* **45,** 73–78.

Jackson K. F. and Duncan I. D. (1988) Cell kinetics and cell death in the optic nerve of the myelin deficient rat. *J. Neurocytol.* **17,** 657–670.

John H. A. (1990) Microheterogeneity of myelin basic proteins in the partially myelinated spinal roots of the Bar Harbor 129 ReJ muscular dystrophic mouse. *Neurosci. Lett.* **109,** 321–324.

John H. A. and Purdom I. F. (1984) Myelin proteins and collagen in the spinal roots and sciatic nerves of muscular dystrophic mice. *J. Neurol. Sci.* **65,** 69–80.

Juguelin H., Heape A., Boiron F., and Cassagne C. (1986) A quantitative developmental study of neutral lipids during myelinogenesis in the peripheral nervous system of normal and Trembler mice. *Devel. Brain Res.* **25,** 249–252.

Kahn S., Tansey F. A., and Cammer W. (1986) Biochemical and immunocytochemical evidence for a deficiency of normal interfascicular oligodendroglia in the CNS of the dysmyelinating mutant (md) rat. *J. Neurochem.* **47,** 1061–1065.

Katsuki M., Sato M., Kimura M., Yokoyama M., Kobayashi K., and Nomura T. (1988) Conversion of normal behavior to shiverer by myelin basic protein antisense cDNA in transgenic mice. *Science* **241,** 593–595.

Kimura M., Inoko H., Katsuki M., Ando A., Sato, T., Hirose T., Inaymama S., Takahashima H., Takamutsu K., Mikoshiba K., Tsukada Y., and Watanabe I. (1985) Molecular genetic analysis of myelin-deficient mice: Shiverer mutant mice show deletion in gene(s) coding for myelin basic protein. *J. Neurochem.* **44,** 692–696.

Kimura M., Sato M., Akatsuka A., Nozawa-Kimura S., Takahashi R., Yokoyama M., Nomura T., and Katsuki M. (1989) Restoration of myelin formation by a single type of myelin basic protein in transgenic shiverer mice. *Proc. Natl. Acad. Sci. USA* **86,** 5661–5665.

Knapp P. E. and Skoff R. P. (1987) A defect in the cell cycle of neuroglia in the myelin deficient jimpy mouse. *Brain Res.* **432,** 301–306.

Koenig J., Hantaz-Ambroise D., De La Porte S., Do Thi N. A., Bourre J. M., La Chapelle F., and Koenig H. L. (1989) In vitro evidence for a neurite growth-promoting activity in Trembler mouse serum. *Int. J. Devel. Neurosci.* **7,** 281–294.

Konat G. and Clausen J. (1976) Triethyllead-induced hypomyelination in the developing rat forebrain. *Exp. Neurol.* **50,** 124–133.

Konat G., Gantt G., Singh I., and Hogan E. L. (1986) Synthesis and acylation of myelin proteolipid protein in quaking mouse brain. *Metab. Brain Dis.* **1,** 241–247.

Konat G., Gantt G., and Hogan E. L. (1987a) Increased turnover of myelin proteolipid protein in quaking mouse brain. *Metab. Brain Dis.* **2,** 113–116.

Konat G., Hogan E. L., Leskawa K. C., Gantt G., and Singh I. (1987b) Abnormal glycosylation of myelin-associated glycoprotein in quaking mouse brain. *Neurochem. Int.* **10,** 555–558.

Konat G., Trojanowska M., Gantt G., and Hogan E. L. (1988) Expression of myelin protein genes in quaking mouse brain. *J. Neurosci. Res.* **20,** 19–22.

Kondo A., Hoogerbrugge P. M., Suzuki K., Poorthuia B. J., van Bekkum D. W., and Suzuki K. (1988) Pathology of the peripheral nerve in the twitcher mouse following bone marrow transplantation. *Brain Res.* **460,** 178–183.

Kuhn R., Pravtcheva D., Ruddle F., and Lemke G. (1990) The gene encoding peripheral myelin protein zero is located on mouse chromosome 1. *J. Neurosci.* **10,** 205–209.

Kunishita T., Tabira T., Umezawa H., Mizutani M., and Katsuie Y. (1986) A new myelin-deficient mutant hamster: Biochemical and morphological studies. *J. Neurochem.* **46,** 105–111.

Macklin W. B., Campagnoni C. W., Deininger P. L., and Gardinier M. V. (1987) Structure and expression of the mouse myelin proteolipid protein gene. *J. Neurosci. Res.* **18,** 383–394.

Matthieu J. -M., Brady R. O., and Quarles R. H. (1974) Anomalies of myelin-associated glycoprotein in "quaking" mice. *J. Neurochem.* **22,** 291–296.

Matthieu J. -M., Omlin F. X., Ginalski-Winkelmann H., and Cooper B. J. (1984) Myelination in the CNS of *mld* mutant mice: Comparison between composition and structure. *Devel. Brain Res.* **13,** 149–158.

Matthieu J. -M., Roch J. -M., Omlin F. X., Rambaldi I., Alamazan G., and Braun P. E. (1986) Myelin instability and oligodendrocyte metabolism in myelin-deficient mutant mice. *J. Cell Biol.* **103,** 2673–2682.

Meier H. and MacPike A. D. (1970) A neurological mutation (msd) of the mouse causing a deficiency of myelin synthesis. *Exp. Brain Res.* **10,** 512–525.

Michelson A. M., Russell E. S., and Pinckney J. (1955) Dystrophia muscularis: A hereditary primary myopathy in the house mouse. *Proc. Natl. Acad. Sci. USA* **41,** 1079–1081.

Molineaux S., Engh H., deFerra F., Hudson L., and Lazzarini R. A. (1986) Recombination within the myelin basic protein gene created the dysmyelinating shiverer mouse mutation. *Proc. Natl. Acad. Sci. USA* **83,** 7542–7546.

Morello D., Dautigny A., Pham-Dinh D., and Jolles P. (1986) Myelin proteolipid protein (PLP and DM-20) transcripts are deleted in jimpy mutant mice. *EMBO J.* **5**, 3489–3493.

Moriguchi A., Ikenaka K., Furuichi T., Okano H., Iwasaki Y., and Mikoshiba K. (1987) The fifth exon of myelin proteolipid protein-coding gene is not utilized in the brain of jimpy mutant mice. *Gene* **55**, 333–337.

Nadon N., Duncan I. D., and Hudson L. D. (1988) Molecular analysis of the shaking pup mutation. *Soc. Neurosci. Abstr.* **14**, 829.

Nave K. -A., Bloom F. E., and Milner R. J. (1987) A single nucleotide difference in the gene for myelin proteolipid protein defines the *jimpy* mutation in mouse. *J. Neurochem.* **49**, 1873–1877.

Nave K. -A., Lai C., Bloom F. E., and Milner R. J. (1986) Jimpy mutant mouse: A 74-base deletion in the mRNA for myelin proteolipid protein and evidence for a primary defect in RNA splicing. *Proc. Natl. Acad. Sci. USA* **83**, 9264–9268.

Nagara H., Ogawa H., Sato Y., Kobayashi T., and Suzuki K. (1986) The twitcher mouse: degeneration of oligodendrocytes in vitro. *Devel. Brain Res.* **26**, 79–84.

Newman S., Kitamura K., and Campagnoni A. T. (1987) Identification of a cDNA coding for a fifth form of myelin basic protein in mouse. *Proc. Natl. Acad. Sci. USA* **84**, 886–890.

Norton W. T. and Poduslo S. E. (1973) Myelination in rat brain: Changes in myelin composition during brain maturation. *J. Neurochem.* **21**, 759–773.

Nunoya T., Tajima M., Mizutani M., and Umezawa H. (1985) A new mutant strain of Syrian hamster with myelin deficiency. *Acta. Neuropathol.* (Berl.) **65**, 305–312.

Okano H., Miura M., Moriguchi A., Ikenaka K., Tsukada Y., and Mikoshiba K. (1987) Inefficient transcription of the myelin basic protein gene possibly causes hypomyelination in myelin-deficient mutant mouse. *J. Neurochem.* **48**, 470–476.

Okano H., Ikenaka K., and Mikoshiba K. (1988) Recombination within the upstream gene of duplicated myelin basic protein genes of *myelin deficient shi^{mld}* mouse results in the production of antisense RNA. *EMBO J.* **7**, 3407–3412.

Omlin F. X., Webster H. deF, Palkovits C. G., and Cohen S. R. (1982) Immunocytochemical localization of basic protein in major dense line regions of central and peripheral myelin. *J. Cell Biol.* **95**, 242–248.

Osetowska E. and Leszawski F. (1975) Prolegomena do badan doswiadczalnych nad choroba dziedziczna ukladu nerwowego na modelu krolika pt. *Neuropathol. Pol.* **13**, 61–70.

Phillips J. S. R. (1954) Jimpy, A new total sex-linked gene in the house mouse. *Arch. Indukt. Abstammungs-Vererbungsl.* **86**, 322.

Poltorak M. and Freed W. J. (1988) Cerebral allografts in brain of quaking mice. *Exp. Brain Res.* **71**, 163–170.

Popko B., Puckett C., Lai E., Shine H. D., Readhead C., Takahashi N., Hunt S. W. III, Sidman R. L., and Hood L. (1987) Myelin-deficient mice:

expression of myelin basic protein and generation of mice with varying levels of myelin. *Cell* **48,** 713–721.

Popko B., Puckett C., and Hood L. (1988) A novel mutation in myelin-deficient mice results in unstable myelin basic protein gene transcripts. *Neuron* **1,** 221–225.

Readhead C., Popko B., Takahashi N., Shine H. D., Saavedra R. A., Sidman R. L., and Hood L. (1987) Expression of a myelin basic protein gene in transgenic shiverer mice: correction of the dysmyelinating phenotype. *Cell* **48,** 703–712.

Roach A., Takahashi N., Pravtcheva D., Ruddle F., and Hood L. (1985) Chromosomal mapping of mouse myelin basic protein gene and structure and transcription of the partially deleted gene in shiverer mutant mice. *Cell* **42,** 149–155.

Roch J. -M., Tosic M., Roach A., and Matthieu J. -M. (1989) The duplicated myelin basic protein gene in *mld* mutant mice does not impair transcription. *Brain Res.* **477,** 292–299.

Rosenbluth J. (1987)Abnormal axoglial junctions in the myelin-deficient rat mutant. *J. Neurocytol.* **10,** 497–509.

Roth H. J., Hunkeler M. J., and Campagnoni A. T. (1985) Expression of myelin basic protein genes in several dysmyelinating mouse mutants during early postnatal brain development. *J. Neurochem.* **45,** 572–580.

Roussel G., Neskovic N. M., Trifilieff E., Artault J. -C., and Nussbaum J. -L. (1987) Arrest of proteolipid transport through the Golgi apparatus in Jimpy brain. *J. Neurocytol.* **10,** 195–204.

Shinoda H., Kobayashi T., Katayama M., Goto I., and Nagara H. (1987) Accumulation of galactosylsphingosine (psychosine) in the twitcher mouse: determination by HPLC. *J. Neurochem.* **49,** 92–99.

Sidman R. L., Dickie M. M., and Appel S. H. (1964) Mutant mice (quaking and jimpy) with deficient myelination in the central nervous system. *Science* **144,** 309–311.

Simons R., Alon N., and Riordan J. R. (1987) Human myelin DM-20 proteolipid protein deletion defined by cDNA sequence. *Biochem. Biophys. Res. Comm.* **146,** 666–671.

Simons R. and Riordan J. R. (1990) The myelin-deficient rat has a single base substitution in the third exon of the myelin proteolipid protein gene. *J. Neurochem.* **54,** 1079–1081.

Sorg B. J. A., Agrawal D., Agrawal H. C., and Campagnoni A. T. (1986) Expression of myelin proteolipid protein and basic protein in normal and dysmyelinating mutant mice. *J. Neurochem.* **46,** 379–387.

Stoffel W., Hillen H., and Giersiefen H. (1984) Structure and molecular arrangement of proteolipid protein of central nervous system myelin. *Proc. Natl. Acad. Sci. USA* **81,** 5012–5016.

Suzuki K., Hoogerbrugge P. M., Poorthuia B. J., Bekkum D. W., and Suzuki K. (1988) The twitcher mouse. Central nervous system pathology after bone marrow transplantation. *Lab. Invest.* **58,** 302–309.

Takahashi N., Roach A., Teplow D. B., Prusiner S. B., and Hood L. (1985)

Cloning and characterization of the myelin basic protein gene from mouse: one gene can encode both 14 kd and 18.5 kd MBPs by alternate use of exons. *Cell* **42,** 139–148.

Tanaka K., Nagara H., Kobayashi T., and Goto I. (1988) The twitcher mouse: accumulation of galactosylsphingosine and pathology of the sciatic nerve. *Brain Res.* **454,** 340–346.

Tanaka K., Nagara H., Kobayashi T., Goto I., and Suzuki K. (1989) The twitcher mouse: attenuated processes of Schwann cells in unmyelinated fibers. *Brain Res.* **503,** 160–162.

Taraszewska A. (1979) Ultrastructural changes in the spinal cord of "pt" rabbit in the symptomatic period of the disease. *Neuropathol. Pol.* **17,** 19–38.

Taraszewska A. (1983) Evaluation of disturbances in myelin sheath formation in "pt" rabbit, electron microscopic study of spinal cord. *Neuropathol. Pol.* **21,** 327–342.

Taraszewska A. and Zelman I. B. (1985) Morphometric studies on myelin development in "pt" rabbit brain. 1. Optic nerve. *Neuropathol. Pol.* **23,** 219–227.

Theret N., Boulenguer P., Fournet B., Fruchart J. C., Bourre J., and Delbart C. (1988) Acylgalactosylceramides in developing dysmyelinating mutant mice. *J. Neurochem.* **50,** 883–888.

Tosic M., Roach A., de Rivaz J. -C., Dolivo M., and Matthieu J. -M. (1990) Post-transcriptional events are responsible for low expression of myelin basic protein in myelin deficient mice: role of natural antisense RNA. *EMBO J.* **9,** 401–406.

Trapp B. D. (1988) Distribution of the myelin-associated glycoprotein and P0 protein during myelin compaction in quaking mouse peripheral nerve. *J. Cell Biol.* **107,** 675–685.

Trofatter J. A., Dlouhy S. R., DeMyer W., and Conneally P. M. (1989) Pelizaeus-Merzbacher disease: tight linkage to proteolipid protein gene exon variant. *Proc. Natl. Acad. Sci. USA* **86,** 9427–9430.

White F. V., Burroni D., Ceccarini C., Matthieu J. -M., Manetti R., and Constantino-Ceccarini E. (1986) Trembler mouse Schwann cells in culture: anomalies in the synthesis of lipids and proteins. *Brain Res.* **381,** 85–92.

Wiggins R. C. (1982) Myelin development and nutritional insufficiency. *Brain Res. Rev.* **4,** 151–175.

Wiggins R. C. (1988) Myelination: A critical stage in development. *Neurotoxicology,* **7,** 103–120.

Wiggins R. C. and Morell P. (1978) Myelin of the peripheral nerve of the dystrophic mouse. *J. Neurochem.* **31,** 1101–1105.

Wiggins R. C. and Morell P. (1980) Phosphorylation and fucosylation of myelin protein in vitro by sciatic nerve from developing rats. *J. Neurochem.* **34,** 627–634.

Wiggins R. C., Benjamins J.A., and Morell P. (1975) Appearance of myelin proteins in rat sciatic nerve during development. *Brain Res.* **89,** 99–106.

Wiggins R. C., Chongjie G., Delaney C., and Samorajski T. (1988) Development of axonal-oligodendroglial relationships and junctions during myelination of the optic nerve. *Int. J. Devel. Neurosci.* **6,** 233–243.

Willard H. F. and Riordan J. R. (1985) Assignment of the gene for myelin proteolipid protein to the X chromosome: Implications for X-linked myelin disorders. *Science* **230,** 940–942.

Yanagisawa K., Duncan I. D., Hammang J. P., and Quarles R. H. (1986) Myelin-deficient rat: Analysis of myelin proteins. *J. Neurochem.* **47,** 1901–1907.

Yanagisawa K. and Quarles R. H. (1986) Jimpy mice: Quantitation of myelin-associated glycoprotein and other proteins. *J. Neurochem.* **47,** 322–325.

Yanagisawa K., Moller J. R., Duncan I. D., and Quarles R. H. (1987) Disproportional expression of proteolipid protein and DM-20 in the X-linked, dysmyelinating shaking pup mutant. *J. Neurochem.* **49,** 1912–1917.

Yao J. K. and Bourre J. M. (1985) Metabolic alterations of endoneural lipids in developing Trembler nerve. *Brain Res.* **325,** 21–27.

Yoshimura T., Kobayashi T., Mitsuo K., and Goto I. (1989) Decreased fatty acylation of myelin proteolipid protein in the twitcher mouse. *J. Neurochem.* **52,** 836–841.

Zeller N. K., Dubois-Dalcq M., and Lazzarini R. A. (1989) Myelin protein expression in the myelin-deficient rat brain and cultured oligodendrocytes. *J. Mol. Neurosci.* **1,** 139–149.

Zelman I. B. and Taraszewska A. (1984) Pathology of myelin in "pt" rabbit. *Neuropathol. Pol.* **22,** 205–218.

Nongenetic Animal Models of Myelin Disorders

Sheldon L. Miller

1. Introduction

This chapter concerns animal models of a diverse group of insults that can directly perturb myelin or the cells that synthesize myelin in the central nervous system (CNS) and peripheral nervous system (PNS). Section 2 briefly relates the developmental sequence of events culminating in myelin deposition and the biochemical characteristics of myelin in the CNS and PNS. Diseases or other insults, e.g., chemical, which appear to affect myelin primarily do so either because of vulnerable periods in development leading to maturation of myelinating cells, including the myelination process, or because of unique aspects of myelin composition. Section 3 concerns general considerations in the selection and evaluation of an animal model.

Sections 4–7 concern different categories of insults. Each section, to varying degrees, includes discussion on:

1. Abnormal myelin deposition and/or demyelination which can occur in humans as a result of the insult;
2. Animal models of the human myelin disorders; and
3. Experimental techniques involved in establishing and evaluating the animal model.

The task of detailing all the insults that have been known to perturb myelin and animal models that have been or might be

From: *Neuromethods, Vol. 21: Animal Models of Neurological Disease, I*
Eds: A. Boulton, G. Baker, and R. Butterworth

used to investigate these insults would be formidable. Therefore, the biological insults reviewed are representative and the numerous references provide much of the experimental details. In this regard, "Myelin" (Morell, 1984) is especially useful as a comprehensive reference.

To avoid any ambiguity in the terminology used, it is important to define certain terms used in this chapter. *Dysmyelination* refers collectively to the synthesis of myelin that differs from the normal composition. *Hypomyelination* is used in its usual sense to indicate decreased myelin deposition. *Abnormal myelin* refers to hypomyelination and/or dysmyelination. An *insult* indicates any exogenous factor, chemical or biological, that interferes with normal biological processes. *Myelin disorders* include the formation of abnormal myelin or demyelination irrespective of the cause. *Developmental myelination* refers to the initial and rapid phase of myelination that occurs in the development of the CNS and PNS but not to the post-developmental myelination that occurs at a slower rate for an extended period of time.

2. Myelin and Myelination: An Overview

2.1. Central Nervous System (CNS)

2.1.1. Developmental Events Culminating in Myelin Deposition

A proposed outline of the developmental steps resulting in a functional myelin membrane in the CNS is given in Fig. 1. Until the use of tissue culture, little was known about the development of the oligodendrocyte, the myelinating cell in the CNS, from its undifferentiated progenitor. Although many of the details of this developmental sequence remain unclear, it appears that extracellular factors and cell–cell interactions may play an important role in this development (e.g., McMorris and Dubois-Dalcq, 1988; Raff, 1989). Migration and proliferation of the progenitor cell have also been postulated to occur under the influence of extracellular agents. Oligodendrocyte synthesis of unique or highly enriched myelin components has been found

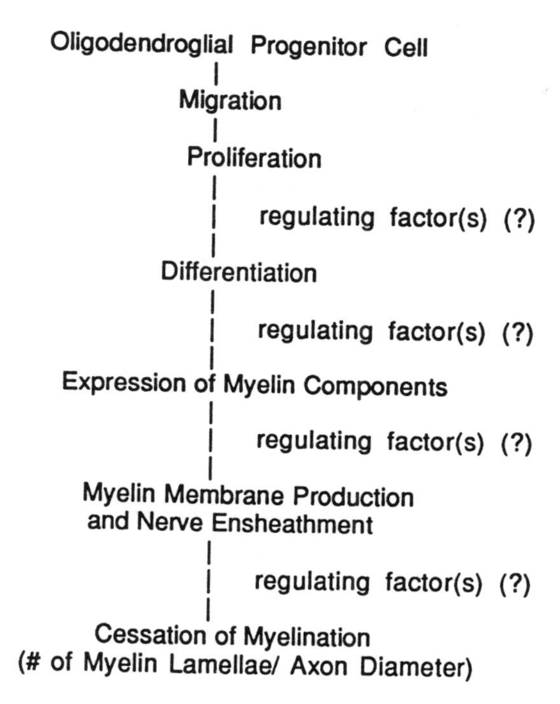

Fig. 1. Developmental steps, culminating in myelin deposition, which may be susceptible to an insult (from Miller, 1990 by permission of Wiley-Liss, a division of John Wiley & Sons, Inc.).

to begin distinctly in advance of the commencement of the assembly of these components into myelin membrane (e.g., Sternberger et al., 1978; Roussel and Nussbaum, 1981; Nishizawa et al., 1981).

During myelination, a cellular process from the oligodendrocyte is extended toward a portion of a nerve axon. At some point along this process (possibly at or near the point where the process extends out from the cell body) the composition of the plasma membrane is remodeled to reflect the lipid and protein composition of myelin *(see below)* that differs considerably from that of the oligodendrocyte plasma membrane. The extended process encounters the axon and begins to wrap around a portion of the axon. As the "leading edge" of the process completes the encirclement of the axon, it continues around the axon intercalating between the "trailing edge" of the process and the axolemma. The wrapping of this process around the axon continues until a multilayered structure is formed. As this multilayered structure is formed, the cytoplasm of the process is displaced and the juxtapositional cytoplasmic sides of the remodeled membrane fuse. This sequence of events results in the compact multilamellar structure characteristic of myelin. A single oligodendrocyte can extend from 10 to 40 processes, each myelinating a portion of different axons (Wood and Bunge, 1984). During the period of developmental myelination in brain, the oligodendrocyte requires considerable amounts of metabolic energy and biochemical precursors to synthesize the proteins and lipids necessary to extensively modify the composition of the oligodendroglial plasma membrane to the composition found in myelin. During the peak period of developmental myelination in rat brain, the oligodendrocyte may synthesize in one day an amount of myelin that is equal to twice the weight of the oligodendrocyte cell body (Norton and Cammer, 1984a,b).

As presented, the process of myelination is the last in a series of interdependent developmental events, each important in the normal accumulation of myelin. Thus, abnormal myelin deposition could potentially be the result of an insult to one or more of several different stages of development. This underscores the necessity of understanding in the human and the representative animal model general aspects of developmental events that culminate in myelin deposition.

2.1.2. Temporal Considerations of Events Leading to Myelin Deposition in the Human

Of the developmental sequence of events leading to myelin deposition in the human, only the last step, the process of myelination, has been extensively studied as a function of age. The histological data relating to developmental myelination are based primarily on the extent of myelin found at autopsy in pre- and postnatal brains (e.g., Rorke and Riggs, 1969; Gilles et al., 1983; Brody et al., 1987). These studies indicate that considerable myelination takes place *in utero* in the CNS, although the extent of myelination varies, with some of the myelinated tracts showing only a paucity of myelin, whereas other tracts show considerable myelin deposition. Of the areas of the brain that have little or no myelin deposition at birth, most are completely myelinated by the end of the second postnatal year, although some discrete areas of the brain remain with little or no myelin. Because of the large portion of the CNS that shows some myelin deposition at birth, the onset of myelination is often characterized as predominantly an *in utero* event, and thus oligodendroglial development and maturation is predominantly an *in utero* event. However, by analogy with the asynchrony of oligodendrocyte development and myelination known to occur in various regions of rat brain, some oligodendrocyte maturation probably takes place postnatally in some brain regions concurrently with myelination occurring in other brain regions (Wood and Bunge, 1984).

As a general rule, the onset of myelination in developing brain occurs in a caudal to rostral direction, although numerous exceptions can be noted (Gilles et al., 1983). The rate at which myelin deposition occurs in the CNS is difficult to ascertain. The process of developmental myelination of the CNS appears to be a protracted process starting *in utero* and lasting through the second year of life, at which time almost all nerve tracts have been myelinated to an extent that approaches the levels of myelin found in the adult. Myelin accumulation may continue to occur at a slower rate possibly into the third decade of life (Rorke and Riggs, 1969). The observation that developmental myelination extends through the pre- and postnatal periods is a misleading assessment of the rate at which an oligodendrocyte myelinates. Gilles et al. (1983) and Brody et al. (1987) noted that:

1. The onset and rate of myelin accumulation varies among the different brain regions;
2. The rate of myelin accumulation within a brain region varies among the various tracts;
3. For a given tract the onset and rate of myelination do not contain a group of developmentally synchronous myelinating axons; and
4. The data are consistent with the concept that for a given axon myelination proceeds from the region of the perikaryon to the distal end.

Thus, attempts to grossly determine the rate of myelination can be obscured by the heterogeneity of events. Gilles et al. (1983) attempted to estimate the rate of myelination by determining, at a defined anatomical site, the rate of myelin accumulation from the period at which it is initially detected to the time that it approaches mature levels. They concluded that:

1. At a defined anatomical site, the period of time necessary for myelin accumulation to approach levels found in the adult brain is measured in weeks;
2. Myelination consists of a rapid phase of accumulation followed by a slower rate as myelin deposition approaches levels found in the adult; and
3. The initial rapid rate of myelination in the human may approach the initial rapid rate of myelination observed in rat CNS *(see below)*.

2.1.3. Comparative Aspects of Myelin Deposition in the Rat Brain

Rat and mouse, which appear to be very similar in development with respect to oligodendrocyte development, myelination, and the composition of myelin, are the most commonly used animals in studying the developmental events leading to myelin deposition. The anatomical sequence in which myelination occurs in the rat appears to be similar to that of the human, and in the brain exhibits a general caudal to rostral course of myelination.

A major difference between these rodents and humans in the developmental events leading to myelin deposition is the period of time relative to birth during which these events occur. As noted above, brain oligodendrocyte maturation and the onset and period of rapid myelination in many brain regions begin *in utero* in humans; however, in the rat and the mouse these events occur primarily postnatally in the brain. In the rat brain, most oligodendrocytes complete a proliferation phase between approximately 1–2 weeks after birth which is followed by a period of a very rapid rate of myelination (16–30 days after birth). The rate of myelination increases to a maximum at about 20 days of age and then declines to a low level that appears to be maintained throughout the life of the rat (Norton and Poduslo, 1973a).

2.1.4. Biochemical Characteristics of Myelin

Although the structural features of myelin resemble the general features of other membranes, i.e., a lipid bilayer and the inclusion of transmembrane and extrinsic proteins, myelin is an atypical membrane in several aspects of its biochemical makeup, and it is these unusual aspects that could preferentially predispose it to insult. Myelin is composed of approximately 1/3 protein and 2/3 lipid, as compared to most other membranes, which are approximately 1/3 lipid and 2/3 protein. The diversity of proteins in myelin is relatively small compared to the membrane from which it is derived, oligodendroglial plasma membrane. Three proteins account for ~80% of the mass of myelin proteins: Folch-Lees proteolipid protein, PLP, ~50%; myelin basic proteins, MBP, ~20%; and 2',3'-cyclic nucleotide-3'-phosphohydrolase, CNPase, ~10%. PLP, a transmembrane protein, appears to be specific for CNS myelin. MBP, probably a nonintegral (peripheral) protein, appears to be specific for CNS and PNS myelin and is comprised of several structurally homologous proteins that vary from 14 to 21.5 kDa (Campagnoni et al., 1989). In rat and mouse, the 14 and 18.5 kDa forms of MBP are quantitatively dominant, whereas in humans the 18.5 kDa protein is the dominant species. CNPase catalyzes the hydrolysis of the 3' ester bond in 2',3'-cyclic nucleotide monophosphate and has a greater catalytic rate when the base is adenine (3',5'-cyclic adenosine monophosphate does not serve as a substrate). 2',3'-Cyclic nucleotide monophosphate is probably not the natural substrate of CNPase since it has not been found as a normally occurring constituent of the body. Despite the efforts of many individuals, the natural substrate for this compound has not been determined. Although CNPase is found in other tissues in small amounts, it is highly enriched in CNS myelin. The remaining proteins in myelin, although constituting a small proportion of the total protein, could play an important role in either the myelination process, normal functioning of myelin, or the metabolic maintenance of myelin. Two proteins, DM-20 (intermediate protein) and myelin-associated glycoprotein (MAG), have been the focus of a number of studies. DM-20, occasionally grouped with PLP in myelin studies, appears to resemble PLP in its primary structure (Agrawal and Hartman, 1980). MAG, present in CNS and PNS myelin, has been postulated to play a role in the recognition of the axon by myelin or oligodendrocyte (Sternberger, 1979).

Although myelin does not contain lipids that are unique to this membrane, several of the lipid components are highly enriched in myelin as compared to other membranes in the body. Preeminent among these lipids are galactosylceramide (cerebroside) and sulfatide (the sulfated form of galactosylceramide). These two lipids together constitute approximately 30% of the total lipid of myelin by weight and may have an important structural role in the stabilization of myelin. It has also been suggested that they may also play a function in developmental regulation and recognition. The glycerophospholipids of myelin are enriched in plasmalogen (a glycerophospholipid with a long chain alkene bound to the 1 position of glycerol by a vinyl ether linkage). About 30% of the glycerophospholipids are present as plasmalogens and are found predominantly in the ethanolamine phosphoglycerides. Approximately 80% of the ethanolamine phosphoglycerides are present as plasmalogens, whereas only small amounts of serine and choline phosphoglycerides are present as plasmalogens. Glycerophospholipid in which an alkane is bound to the C-1 carbon of glycerol through an ether linkage is also present in myelin, although these are present in much smaller quantities than the plasmalogens. Other differences in the lipid composition, though not as pronounced, also exist between myelin and other membranes, e.g., *see* Norton and Cammer (1984a,b).

The relative composition of whole brain myelin changes as a function of age (e.g., in rat—Smith, 1973; Horrocks, 1973; Banik and Smith, 1977: in human—Eng et al., 1968; Fishman et al., 1975; Söderberg et al., 1990). These quantitative changes involve lipids and proteins and occur in both human and rat. It is not known whether these differences reflect differences in the turnover rate of the myelin components or changes in the synthesis or transport of these components as the animal matures. In addition to age-related changes, regional differences in the relative composition of myelin also occur (e.g., Smith and Sedgewick, 1975; Sapirstein et al., 1978).

2.2. Peripheral Nervous System (PNS)

2.2.1. Developmental Aspects

In the rat, the onset of myelin deposition in the PNS precedes that in the CNS, and, like the CNS, myelination in the PNS is not a synchronous event. These differences are reflected in the timing and rate of myelination among different PNS nerve axons (e.g., Fraher, 1976; Fraher et al., 1988).

In the human, as in the rat, PNS myelination begins *in utero*, is not synchronous among the various peripheral nerves, and continues postnatally. In the sural nerve, myelination begins at 18 weeks gestational age but does not reach adult values for all the nerve fibers until 5 years of age. PNS myelination in the human and rat has been reviewed by Webster and Favilla (1984).

Myelination in the PNS is accomplished by the Schwann cell. Unlike the oligodendrocyte, each internodal segment of myelin is deposited by a separate myelin-producing Schwann cell. Although compact myelin appears very similar in the PNS and the CNS, there are differences in the myelin and the cellular dynamics of events leading to myelin deposition. (A description of the cellular events in the CNS and PNS is given in the review by Raine (1984).)

2.2.2. Biochemical Aspects

The relative composition of the major lipid classes of PNS myelin differs quantitatively from those in CNS myelin; in addition, the relative composition of the various fatty acid moieties differs in PNS myelin phosphoglycerides and sphingolipids as compared to CNS myelin. With respect to minor lipid constituents, rat and human PNS myelin contain fewer gangliosides. Human PNS myelin contains a glycosphingolipid that is not a ganglioside and is not found in the CNS. This glycosphingolipid has a carbohydrate moiety that is immunologically crossreactive with a portion of the carbohydrate found in human CNS and PNS myelin-associated glycoprotein (Chou et al., 1985). The carbohydrate portion of this glycolipid has been implicated as an antigen in a peripheral neuropathy that is believed to be immunologically mediated (*see* Section 7.).

Several proteins are common to PNS and CNS myelin (including MAG, CNPase, and MBP). Although quantitative differences exist between PNS and CNS myelin in the relative amount of the proteins that are common to both, more profound is the qualitative difference between the protein composition of CNS and PNS myelin. PNS myelin does not contain PLP, but does contains a 30,000 molecular weight, hydrophobic, glycosylated protein, P_0, which constitutes about 50% of the total PNS myelin protein. In addition, PNS myelin contains a basic protein, termed P_2, mol wt of ~14,000, whose primary structure is unrelated to any of the MBP proteins found in the CNS. Two other proteins termed X and Y (19 and 23 kDa, respectively) are also found in the PNS but not the CNS.

3. Experimental Considerations in Animal Models

Establishing a nongenetic animal model of a human myelin disorder can be based on 3 different circumstances.

1. The myelin-related deficit is believed to be fairly well described, but the etiology and mechanism by which the deficit is produced are unknown.
2. The etiology is known, but the myelin-related deficit has not been well characterized and the mechanism is unknown.
3. The etiology is known and the myelin-related deficit is believed to be fairly well described, but the mechanism by which the myelin-related deficit is produced is unknown.

3.1. Biochemical Considerations

Although the biochemical composition of myelin shows a striking similarity among mammalian species, qualitative and quantitative differences exist in lipid and protein components that can be important depending on the myelin disorder studied in the animal model and the techniques used to evaluate it. For instance, the relative concentration of individual gangliosides in myelin varies considerably among mammalian species

(Cochran et al., 1982). Qualitative differences exist in the carbohydrate of MAG, the major glycoprotein of CNS myelin, between human and rat (Miller et al., 1987). Although the amino acid sequence of myelin proteins is highly conserved during evolution, certain amino acid deletions and substitutions exist in myelin proteins among the various mammalian species. These variations, which are often demonstrated by the lack of crossreactivity of monoclonal antibodies to the same protein of different species (e.g., Miller et al., 1989), could be important in animal models that involve an autoimmune mechanism (*see* Section 7.).

3.2. Developmental Considerations

If it can be assumed that the general sequence of events from oligodendroglial precursor to myelin deposition found to occur in certain regions of rodent brain can be generalized to include all of the CNS in both rat and human, then, based on the myelination pattern in the human CNS, oligodendrocyte maturation and myelination is continually occurring in discrete areas of the CNS from early gestation to the third postnatal year of life. The proposed developmental steps depicted in Fig. 1, each of which represents a critical step leading to the normal deposition of myelin, are potentially sites of heightened vulnerability (Dobbing and Smart, 1973) with a greatly increased probability that a chemical or biological insult will lead to abnormal myelin deposition. One implication of this line of reasoning is that, to the extent that an animal model is to reflect the anatomical specificity and mechanism of a human disorder with an onset during human brain development, consideration should be given to the developmental stage of brain development in the animal model.

Rat and mouse are the most commonly used animal models for the study of abnormal myelin deposition. As described above, the development of oligodendrocytes and the process of myelination in the CNS is, to a much greater extent, a postnatal event in rodents rather than in humans. Thus, anatomical areas that may undergo oligodendrocyte development or myelination in the human *in utero*, may occur postnatally in the rat and mouse. Therefore, in comparing a particular animal model with the human disorder, consideration may need to be given to any maternal modulation of the insult during *in utero* development.

Despite temporal differences in human development as compared to rat and mouse, these animal models offer some distinct advantages.

1. Since these models have been studied extensively, considerable knowledge has been accrued on the biochemical composition, metabolism, morphology, and immunology of myelin and developmental events relating to the deposition of myelin.
2. As described above, these animals undergo a period of very rapid myelination in brain in which considerable myelin deposition occurs in a period of only 2 weeks, making this a convenient model with which to study the effects of an insult on the process of myelination.
3. In vivo studies on the susceptibility of oligodendroglial development to insult are easier to accomplish since the major portion of their development takes place postnatally.
4. Since considerable deposition of myelin occurs in the spinal cord of rat *in utero* (Banik and Smith, 1977), this region of the CNS can be used as a general model to assess the effects of *in utero* insults.

The complication of temporal asynchrony in the events leading to myelin deposition among the various brain regions (and to some extent even within these brain regions) is approached in several ways in designing experiments to examine the effect of an insult on maturation of the oligodendrocyte and the process of myelination. One approach has been to carry out the experiment in vivo and to analyze the results on tissue derived from one or more brain regions. These methods have included: use of light and electron microscopy combined with standard histological techniques (quantitative survey of cell types present) (e.g., Robain and Ponsot, 1978) and/or [^{3}H]thymidine (to label dividing cells) (e.g., Lai et al., 1980); immunocytochemistry to identify the appearance of enriched or specific myelin antigens in oligodendrocytes (e.g., Sternberger, 1979); and quantitation of myelin formed using nonradioactive (e.g., Wiggins and Fuller,

1978) or radioactive methods (e.g., Wiggins and Fuller, 1979). However, until the availability of monospecific antisera, and more recently, monoclonal antibodies, to antigens specific for oligodendrocytes, their precursors, and myelin-enriched and myelin-specific antigens, little was known about the cellular events leading to myelin deposition. Most experiments concerned with the effects of insults on myelin deposition determined the effects of the insult on the metabolism involved in myelin synthesis during developmental myelination or the net accumulation or composition of myelin in whole brain after this period.

3.3. Adult Considerations

The nature of the insult initiating demyelination in the adult human is only understood in a limited number of instances (e.g., in the CNS, post-rabies immunization encephalomyelitis; in the PNS, diptheric neuritis); however, even in these cases the exact mechanism is not well understood. Demyelination is often found to be associated with oligodendroglial cell death, but it is uncertain whether the insult first affects the oligodendrocyte that subsequently results in demyelination or whether the myelin is the primary target of the insult resulting in demyelination and subsequently death of the oligodendrocyte. The selection of the animal model will depend on the agent or putative agent and the suspected mechanism by which demyelination occurs. For instance, if an immune mechanism is suspected of mediating demyelination, then the host response of the immune system to the putative antigen should be similar for the animal model and human. If demyelination, is initiated by specific chemicals or biologically related molecules, e.g., vaccine or bacterial toxin, then the metabolic similarities (i.e., metabolic transformations and rates) between human and the animal model will be of importance. If the agent is to be delivered systemically to the CNS, then permeability across the blood–brain barrier may need to be considered. However, the blood–brain barrier is not always a consideration; some viruses, e.g., rabies and herpes simplex, gain access to the brain via retrograde transport along peripheral nerve.

3.4. Parameters To Be Measured

In initial experiments to determine whether an insult during development will result in abnormal myelin deposition, a reasonable first step is the quantitation of the net accumulation and composition of myelin in whole brain after the peak period of myelination. Differences between the net accumulation of experimental and control animal myelin can be measured by: weighing the total amount of lyophilized myelin recovered; extraction of the myelin lipids (e.g., Folch et al., 1957) and determination of either lipid weight, or total lipid phosphorus (Eng and Noble, 1968) or cholesterol; or determination of total myelin protein (Lowry et al., 1951; or Smith et al., 1985). Accurate comparison of the experimental and control myelin yields based on a myelin lipid or protein (or a component of one of these fractions) depends on that component or fraction of myelin being unaltered in experimental animals. Because of biological differences such as variability in myelin deposition among animals of the same age, even among littermates (e.g., Wiggins et al., 1976; Norton and Poduslo, 1973a), a considerable number of animals may be required to reveal significant differences in decreased deposition of myelin, especially if these differences are of the order of 10–20%. Brain and body weights of experimental and control animals are also useful parameters to determine *(see below)*.

To detect changes in composition, a relatively rapid preliminary survey can be accomplished by determining the ratio of total protein to some lipid parameter in myelin samples. If isolated myelin is to be stored before analysis, we routinely separate the isolated myelin from the isolation medium by centrifugation, wash the myelin several times in cold deionized water, and lyophilize it before storage at –70°C. When lipid and protein analyses are to be done on the same myelin sample, it is convenient to take two separate aliquots of the suspended myelin before lyophilization, since we have found it difficult to obtain a uniform suspension of myelin in water after lyophilization. For the measurement of protein, aliquots of myelin can be taken after dissolving the lyophilized myelin in 1% SDS. In lieu of the Folin-Ciocalteau reagent, which is often used for protein analysis (Lowry et al., 1951), we have found that the use of the bicinchoninic acid method of protein quantitation (Smith et al., 1985) to be more advantageous because of: sensitivity; linearity over a wide range of protein concentrations; and less interference by substances used in the isolation and solubilization of biological material.

To quantitate individual myelin components that may have been affected by an insult to the animal, the compositional changes in individual myelin lipid classes and myelin proteins need to be determined. Methods to quantitate the individual class of lipids have recently been described in Volume 7 (1988) of the Neuromethods series. To determine the percent composition of individual myelin proteins with respect to total protein, we separate the proteins using gradient SDS-PAGE (Miller et al., 1989), stain the gel in 1% Fast Green dye (Greenfield et al., 1971) for 2 hours at 37°C or overnight at room temperature, destain the background, scan the gel, and calculate the relative protein concentration of individual proteins using the sum of the area under all the absorption peaks as representing total protein. The advantage of using Fast Green dye in lieu of one of the more commonly used Coomassie dyes is that the background can be destained without loss of dye bound to the quantitatively minor proteins. The disadvantage of Fast Green dye is that several days of destaining is generally required to completely destain the background. The identification of a compositional deficit is preliminary to determining the metabolic step that results in this deficit, e.g., alteration of synthesis, incorporation, and turnover of a myelin component and, in the case of proteins, posttranslational modification.

In evaluating the biochemical composition of myelin derived from animals in which an experimental insult occurred during the period of development, the composition may differ from the age-matched controls accompanied by a smaller quantity of myelin recovered from the experimental animals. One explanation for these observations is that the myelin in the experimental animal reflects the myelin of a younger animal, i.e., the insult has resulted in a developmental delay. This appears to be the case in some types of undernutrition (e.g., Wiggins et al., 1976). In a severe undernutrition paradigm, body and brain weights for the experimental animals were also less than controls, and myelin isolated from the experimental animals corresponded in composition to that of younger control animals. However, decreases in body weight, brain weight, and recovery of myelin from experimental animals with respect to controls do not necessarily reflect slower development. In developmental delay in experimental animals, the composition of myelin should change with animal maturation and correspond to the myelin composition of a younger control animal.

An important step in evaluating abnormal myelin composition resulting from an insult is the isolation of the abnormal myelin. Techniques for isolating myelin involve centrifugation on a continuous or discontinuous gradient taking advantage of the low density of myelin, which is less than that of organelles and other membrane components of the cell (e.g., Laatsch et al., 1962; Agrawal et al., 1970; Greenfield et al., 1971; Norton and Poduslo, 1973b; Bourre et al., 1980). The method of Norton and Poduslo (1973b), or some variation of it, is probably the most widely used method for the isolation of myelin and utilizes a 0.32*M*/0.85*M* sucrose gradient with the myelin accumulating at the interface and the remaining cell debris in or at the bottom of the 0.85*M* sucrose. This myelin fraction has been shown to consist of at least 2 fractions; one of these fractions is regarded as representative of myelin and another is referred to in the literature as "myelin-like" (Agrawal et al., 1970), "SN4" (Waehneldt, 1978), or "light myelin" (Sapirstein et al., 1978). The proportion of these lighter membranes relative to the total myelin isolated decreases with age (Norton and Poduslo, 1973a). Although this myelin-like fraction contains myelin-enriched or myelin-specific markers, its composition and density differ from that of myelin, and evidence has been presented that this membrane may be a precursor to myelin (e.g., Agrawal et al., 1974; Benjamins et al., 1976; Miller et al., 1989); however, this has not been proven definitively. Isolation of myelin from experimental animals, in which an insult has resulted in either dysmyelination or demyelination, can yield a myelin-derived fraction with a density differing from that of myelin from the control animals. The composition of the myelin-derived fraction resulting from the insult does not appear to be related to the myelin-like fraction.

In the isolation of myelin from animals in which an insult may have resulted in dysmyelination or demyelination, use of a discontinuous gradient technique developed for the isolation of normal myelin may be inadequate since abnormally constituted myelin could either be lost during the isolation procedure or accumulate with the normal myelin fraction. Loss of the abnormally constituted myelin would lead to the conclusion that hypomyelination of normally constituted myelin was the effect of the insult. Isolation of insult-induced, abnormally constituted myelin together with myelin of normal composition would result in an erroneous observation that a single membrane fraction of altered composition results from the insult (i.e., a composition that is the weighted average of myelin and the abnormal myelin-derived membrane).

A more cautious approach to myelin isolation under such experimental conditions would be to isolate the myelin by centrifugation on a continuous density gradient (e.g., Greenfield et al., 1971). Successive aliquots of this gradient could then be screened for putative myelin-derived fraction using assay methods for myelin-enriched or myelin-specific components, such as cerebroside, sulfatide, myelin basic protein, and proteolipid protein and compared with a gradient isolation from control animals. Myelin components in the gradient fraction can be monitored by standard methods including: chemical assay, e.g., protein analysis; use of myelin-specific antibodies, e.g., radioimmunoassay; or in vivo labeling of myelin components with radioactive precursor(s), e.g., [^{3}H]galactose to label galactosylseramide and sulfatide. Putative myelin-derived fractions can then be analyzed for their protein and lipid composition.

Another approach to determining myelin fractions of abnormal density resulting from an insult is to use a "double label technique" (Wiggins et al., 1976). A lipid or protein precursor radiolabeled with ^{3}H is injected into experimental animals and the same precursor radiolabeled with ^{14}C is injected into control animals. Animals are sacrificed, brains removed, homogenized together, and separated on a continuous gradient. The presence of protein or lipid, depending on the precursor used, is detected by their radioactivity in aliquots of the gradient using standard scintillation spectrophotometry counting techniques. Significant changes in the $^3H/^{14}C$ ratio indicate either differences in the position on the gradient of control and experimental membrane fractions or differences in the quantity or composition of the radiolabeled component(s) in the same membrane fraction. The double label method has the advantage of separating the control and experimental membrane fractions on the same gradient and thus avoiding small differences in the gradient position of the membrane fractions that may occur between control and experimental fractions separately owing to experimental variation between the continuous gradients (*see* Section 6.1.4. for comments on microscopic evaluation).

3.5. Specificity of Insult

An insult can act directly on the myelin (*see* Section 4.) or upon the oligodendrocyte or Schwann cell, resulting in abnormal myelin deposition or demyelination. During development, an insult could affect normal myelin deposition via the oligodendrocyte or Schwann cells by directly hindering their maturation and/or proliferation, resulting in fewer mature myelin-producing cells available for myelination. This could result in hypomyelination but not necessarily dysmyelination. An insult to myelin-producing cells could also act to prevent normal synthesis of myelin components by inhibition of: protein synthesis; enzyme(s) involved in the synthesis of a myelin lipid component; posttranslational modification of proteins; and organelles involved in myelin synthesis. These direct effects on the metabolism of myelin-producing cells could result in hypomyelination and/or dysmyelination, and, if dysmyelination resulted in an unstable form of myelin, demyelination. An insult to myelin-producing cells could also affect assembly of the proteins and lipids to form myelin membrane. These alterations could also lead to hypomyelination and/or dysmyelination. The synthesis and assembly of myelin components into myelin membrane are particularly vulnerable to an insult during the developmental period of rapid myelination.

In the adult, the components of myelin membrane, whether assembled early in development or as an adult, turn over and are replaced as is typical of other membranes (e.g., Sabri et al.,

1974; Miller and Morell, 1978; Singh and Jungalwala, 1979). The constituent moieties of the molecules that turn over are reutilized with varying efficiencies for the synthesis of the newly synthesized molecules that are destined to be inserted into the myelin (e.g., Miller et al., 1977).

Oligodendrocyte metabolism in the adult is also involved with the slow but continuous net accumulation of myelin that may be occurring in the adult. The sum of the metabolic commitment for turnover of myelin components in established myelin and the slow net accumulation of additional myelin in the adult is probably less than the metabolic expenditures of the oligodendrocyte rapidly synthesizing myelin during developmental myelination. Nevertheless, since a single oligodendrocyte can have from 10 to 40 separate myelin wrappings to support metabolically, considerable metabolic expenditures are probably invested by the oligodendrocyte for myelin-related functions even in the adult.

In addition to myelin disorders resulting from a direct insult, an insult-induced dysfunction of other tissues can lead to myelin abnormalities and/or demyelination. An insult affecting the thyroid can result in abnormal myelin deposition during development (Walters and Morell, 1981). In principle, any insult affecting the normal temporal expression of extracellular factors that play a role in maturation and expression of myelin components and the myelination process can cause abnormal myelin deposition. In the mature nervous system, myelin disorders can also result from an insult-induced dysfunction of other tissues, e.g., toxic diabetes can result in segmental demyelination in the PNS. Demyelination also occurs during injury or death of neurons (secondary demyelination). Distinguishing between direct and indirect effects of an insult is often a difficult step in elucidating the mechanism by which the insult affects myelin. Because of the heterogeneity of cell types and their development, special problems in gaining experimental access to the brain owing to the blood–brain barrier, and the complicated and diverse ramifications that can result from perturbing a single aspect of brain metabolism, the use of other techniques, such as tissue culture models, may be a necessary adjunct to the animal model.

4. Nonbiological Toxins

4.1. General Considerations

In the last few decades, there has been an increasing awareness and concern of exposure in the home, workplace, and outside environment to agents that can cause biological damage in relatively small concentrations. Whether the toxic effects of most of these agents are specific for myelin or the myelin-producing cells is doubtful; however, several factors may result in abnormal myelin deposition and/or demyelination being a prominent feature of these toxic agents. Myelin-related aspects of toxic injury in the CNS and PNS have been reviewed (Cammer, 1980; Blakemore, 1984; Thomas, 1984; Bondy and Prasad, 1988; Konat, 1984; Morell and Mailman, 1987; Tiffany-Castiglioni et al., 1989).

The exposure of the oligodendrocyte or the Schwann cell (or their precursors) during a critical period of development to toxic factors could result in a delayed or irreversible alteration in the normal deposition of myelin. In addition, as mentioned in Section 3., any alteration in development that affects the expression of factors controlling the developmental events leading to normal myelin deposition could potentially result in abnormal myelin deposition. Of note is the period of developmental myelination that appears to be a particularly susceptible period to insult (Wiggins, 1986) probably because of the immense metabolic requirements of the myelin-producing cells during this period. Even after the bulk of the myelin membrane has been deposited, because of the metabolic requirements of the oligodendrocyte and Schwann cell in support of the turnover of myelin components of such a vast membrane system, these cells could provide a vulnerable target for toxic agents that interfere with intermediary metabolism or energy processes, e.g., oxidative phosphorylation.

The high lipid-to-protein ratio of myelin, and thus its high hydrophobic character, may be of importance in the susceptibility of myelin to toxic agents. A general characteristic of most agents believed to act directly on myelin is that that they are hydrophobic or have a hydrophobic region in their molecular structure, which is separated from the polar or charged portion of the molecule.

It is difficult to establish a structure–pathology relationship for compounds associated with abnormal myelin deposition and/or demyelination because of the diversity of the structure of these compounds. For organic molecules the structural diversity includes: phenolic rings with polychloro substitution (e.g., hexachlorophene); substituted heterocyclic rings (e.g., isoniazid); and the amide of a short chain, unsaturated carboxylic acid (e.g., acrylamide). Despite the apparent importance of the hydrophobic portion of the molecule, in general there is no straightforward relationship between hydrophobicity and effectiveness of the compound or even its mode of action (e.g., organotin compounds, Walsh and DeHaven, 1988; organolead compounds, Bondy, 1988).

Many toxins have been found to have multiple effects, e.g., triethyltin (Cammer, 1980), which can be related to primary demyelination. However, many of these compounds are also known to be toxic to many types of cells, e.g., sodium cyanate and carbon monoxide, and thus are potentially capable of producing both primary and secondary demyelination (Thomas, 1984; Blakemore, 1984). The organometallic toxic agents not only act as more classical heavy metal inhibitors binding to porphyrins and reacting with –SH groups, but have been reported to compete with specific mono- and divalent cations, to interact with DNA, and to accumulate in large quantities in mitochondria. Methylmercury and triethyllead have been reported to inhibit the conversion of galactosylceramide to sulfatide (Grundt et al., 1974).

Abnormal myelin deposition and demyelination in tissue can be assessed by histological examination with light and electron microscopy (Raine, 1984) and by biochemical analysis. Histologic examination of tissue utilizing lipophilic staining can reveal extensive areas of demyelination and anatomical localization of these areas when compared to tissue from control animals. Other stains can be used to determine the relative preservation of other elements of tissue and the appearance of other cell types around the areas of demyelination, e.g., macrophages. The use of monospecific antisera and/or monoclonal antibodies directed against myelin-specific or myelin-enriched components has also been used histologically, especially when assessing the temporal appearance of these components during CNS and PNS development (e.g., Sternberger, 1984). The use of electron microscopy has revealed that myelin loss can proceed via several types of ultrastructural alterations of the myelin, e.g., separation of the myelin lamellae at the intraperiod lines and vesiculation of the myelin. The apparent phagocytic role of cellular infiltrates in the area of demyelination during some toxic exposures is more clearly revealed in electron micrographs.

In development of an animal model for myelin pathology owing to toxic agents, several components of the experimental paradigm should not be overlooked. The stage of development of the CNS or the PNS in the animal model should correspond to that of the human pathology being studied. For some compounds, such as hexachlorophene and 6-aminonicotinamide, the susceptibility to and possibly the mode of action of the insult may be a function of nervous system development (Cammer, 1980; Thomas, 1984; Konat, 1984). The route and duration (acute or chronic) of administration can be important (Suzuki, 1971). The kinetics of metabolism and/or excretion and the metabolic fate (i.e., its metabolites) may be a consideration in the selection of the animal model; in the latter case, the metabolic product may be involved in producing myelin pathology (Cremer, 1958; Spencer et al., 1980). The level of the toxic component in the tissue of the animal model should be measured. Although the level of toxic agent in a tissue may not be a good predictor of the clinical pathology (e.g., Morell and Mailman, 1987; Blaker et al., 1981), the measurement of toxic agent in a tissue becomes especially important when the route of administration is via the drinking water or food where variability in results within or between experiments can result from differences in the amount of toxic agent ingested by an individual animal. (For the quantitation of metals, atomic absorption has proven useful, e.g., Trachman et al., 1977.)

Comparison of microscopic and biochemical parameters of available human biopsy and/or autopsy tissue, if available, with those of animals treated with the toxic agent is probably the best way to establish the correspondence of an animal model to the effects of the toxic agent in humans. Parameters examined in the histologic comparison of human and an animal model will depend on the paradigm and could include such toxic effects as: hypomyelination; the presence of cellular infiltrates (Tellez-Nagel et al., 1977; Ludwin, 1978); abnormal appearance of Schwann cells and/or oligodendroglia (Rawlins and Uzman, 1970; Suzuki and Zagoren, 1974); and the microscopic appearance of demyelination. Demyelination, e.g., vesiculation or edema with separation of myelin lamellae at the intraperiod lines, should be comparable for the animal model and the human. Differences in the microscopic appearances of demyelination between the model and the human would imply a different mechanism in the human and the animal model, although similarity of microscopic details of demyelination does not ensure that the mechanism of the animal model corresponds to the mechanism of demyelination in the human.

4.2. Biochemical Evaluation

The progress of demyelination in whole brain can be assessed by following the decrease in whole brain of: a myelin-specific protein; a myelin-enriched lipid in brain relative to control; or 2',3'-cyclic nucleotide-3'-phosphohydrolase activity relative to control (e.g., Kung et al., 1989). Biochemical markers for other cell types are useful in determining the specificity of the toxic agent (Kung et al., 1989; Harry et al., 1985). The relative loss of specific or enriched myelin components may not be the same during demyelination (Wender et al., 1978). Cholesterol esters, not found in normal myelin, can often be detected during demyelination and this is usually attributed to the metabolism of cholesterol by phagocytic cells during demyelination when these cells are present in the demyelinating lesion (Wender et al., 1978).

Biochemical evaluation of the myelin in the animal model is important if dysmyelination is suspected to be a consequence of the toxic factor. As discussed in Section 3., isolation of the myelin should be on a continuous gradient. Attempts to characterize dysmyelination can be complicated by the occurrence of demyelination, since partially degraded myelin may not have the same relative lipid and protein composition as found in normal myelin. Thus, caution should be used when interpreting an abnormal membrane fraction as representing dysmyelination, since this fraction could in fact result from demyelination of normally constituted myelin.

Another potential problem occurs when the toxic agent is introduced before or during developmental myelination. Since the relative composition of myelin changes as a function of age, differences in myelin composition from those in age-matched controls could represent developmental delay, i.e., normally constituted myelin representing the relative composition of younger animals and not dysmyelination. The delay in developmental myelination during severe protein-calorie undernourishment results in the isolation of myelin that differs in relative composition from that in age-matched controls (Wiggins et al., 1976).

The use of radioactively-labeled precursors of the protein and lipid, e.g., leucine and glycerol, respectively, can be used to determine the effect of toxic factors on protein and/or lipid synthesis of myelin components. The precursor is injected in vivo, the animals killed after a short period of time, and the incorporation of the precursor into protein or lipid of brain homogenate, microsomes, and myelin is determined. Comparison of the synthesis of radiolabeled protein or lipid in brain homogenates from control and experimental animals will indicate whether there is an inhibition of protein or lipid synthesis in brain. A similar comparison of the incorporated radioactively-labeled proteins or lipids into myelin from control and experimental animals will indicate whether there is a toxic inhibition of myelin. By comparing the synthesis of radioactively-labeled proteins and lipids in microsomes in control animals to the same ratio derived from experimental animals, preferential inhibition of myelin synthesis can be determined relative to the synthesis of other membranes. Preferential inhibition of specific myelin components can be determined by comparing the radioactive labeling of individual protein or lipid components to the total radioactivity of myelin proteins or lipids, respectively, in experimental and control animals.

A variation of this approach is to use a "double label" protocol for determining the effect of a toxic agent on the synthesis of lipids and proteins. The protocol utilizes a radiolabeled precursor to lipid or protein to follow synthetic rates in control and toxin-treated animals; for one of the animal groups the precursor is labeled with ^{14}C, whereas the other group is injected with the same precursor labeled with ^{3}H. After sacrificing the animals, the brains from one control and one toxin-treated animal are homogenized together and this combined homogenate is used for all subsequent procedures, such as isolation of subcellular membranes, including myelin, and the isolation of lipids or proteins. The reason for using the "double label" protocol is that losses in the isolation of myelin, subcellular components, and proteins and lipids from control and toxin-treated animals in the mixed brain homogenate will be the same.

4.3. Variability of Biochemical Data in Determining Toxic Effects

Since biological variation among animals decreases the precision of data among animal replicates, consideration should be given to using littermates or age-matched animals of the same sex to reduce biologic variability within animal groups. In procedures involving the use of radiolabeled compounds, considerable variability encountered between animals can result from the variability in the injection of the radiolabeled precursor. Another approach to reduce the scatter owing to biological variation is measuring the relative changes of one component as compared to a second, both isolated from the same animal. This approach uses one of these components as a "floating baseline" and measures the change of the second component only in terms of the "baseline" value. An example, as discussed above, is the ratio of incorporation of radiolabeled amino acid into myelin protein relative to its incorporation into whole brain microsomes (baseline value). This approach assumes that the change observed between the baseline and the second parameter will be independent of the absolute value of the baseline value under the conditions used and, in the case of a toxic agent that preferentially affects myelin protein synthesis, will vary as a function of the magnitude of the insult. If for a particular experimental paradigm the assumptions are correct, then the approach does have the advantages of limiting interanimal variation, reducing variability involved in the radiolabeled precursor, and leaving open the option to determine changes with respect to other parameters such as changes relative to the amount of tissue or a tissue component, e.g., total protein or lipid. It has the disadvantage that it does not account for variable losses between control and experimental animals during the isolation of different fractions, such as a whole brain subcellular microsomal fraction and myelin.

The double isotope technique presented in this section and in Section 3. can minimize variation. This approach minimizes variation owing to interanimal differences, resulting from the injection of the radiolabeled precursor injected in vivo and variation resulting from differential losses during the isolation of subcellular fractions and their components. One disadvantage of this method is that it does not leave open the option to relate the experimental data to subcellular parameters, e.g., total protein or lipid in microsomes, of control or experimental animals. A second disadvantage is that this method introduces a new source of variation during the quantitation of radioactive isotopes which is owing to the overlap of the ^{3}H and ^{14}C energy spectra. This source of variation can be minimized by using appropriate techniques during scintillation counting. The use of the double label protocol requires a "reverse label" experiment such that the experiment be repeated by reversing the isotopically labeled precursor received by each animal group in the first experiment. This control insures that the differences in the $^{3}H/^{14}C$ ratios do not merely reflect differences in metabolic stability of the radiolabeled atom in the precursor. For example, in a double label paradigm, designed to compare the synthesis of glycerophospholipids in control and toxin-treated animals, the choice of [2-^{3}H]glycerol with [U-^{14}C]glycerol as precursors for phospholipids would not be judicious. The ^{3}H in [2-^{3}H]glycerol is much more metabolically labile than the ^{14}C of [U-^{14}C]glycerol (Benjamins and McKhann, 1973); a more appropriate choice for the ^{3}H precursor would be [1,3-^{3}H]glycerol.

4.4. Additional Experimental Considerations

The measurement of membrane protein and lipid synthesis in control and toxin-treated animals by quantitating a representative component has several other potential problems. Not all the components of the membrane may be inhibited to the same degree (e.g., Konat and Clausen, 1976; Konat and Offner, 1982). Results obtained from the measurement of a single lipid and/or protein component should be verified by measuring other membrane constituents. The quantitation of radiolabeled protein and lipid in the membrane component subsequent to in vivo injection of the radiolabeled precursor represents a number of different steps, including: transport from the site of injection to the cell; uptake by the cell; synthesis of the protein or lipid (including posttranslational modification of the protein); transport from the site of synthesis to the membrane; and incorporation into the membrane. Each of these steps represents a potential site for the action of a toxic agent.

When in vitro experiments are used to study the effects of toxic agents, confirmation of the proposed mechanisms should be verified by the use of an animal model (Ganser and Kirschner, 1985). Conversely, because of the complexity of the effects of toxic agents in vivo, it is important to verify the conclusions about the mechanism of toxic agents by in vitro experiments. In in vitro experiments that require the presence of myelin, organotypic cultures or tissue slices can be useful models. The tissue slices can be obtained from either toxin-treated or nontreated animals to approach a chronic or acute model, respectively. In some instances, where the interaction of a toxic agent with myelin is being investigated, freshly isolated myelin may suffice. Specific effects on oligodendroglial or Schwann cells may be assessed by using freshly prepared cells from animals or primary cultures of these cells. In some instances, subcellular fractionation may be required to obtain unambiguous results, e.g., as with mitochondria or golgi. One general caution in using the in vitro model to verify or extend in vivo results concerns the concentration of the toxic agent. Although the use of data derived from the quantitation of a toxic agent in nervous tissue of the animal model can be used to arrive at an in vitro concentration that is physiologically significant, the concentration of toxic agent in the in vitro media may not reflect intracellular levels since the cells can concentrate the toxic substance to several times its concentration in the surrounding media.

5. Nutritional Disorders of Myelin

5.1. *Effects of Protein-Calorie Undernutrition in Humans*

Myelin and the nervous system in general of the adult are spared the effects of protein-calorie undernutrition as compared to most other tissues of the body (Winick and Noble, 1966). However, *in utero* and in the infant where developmental myelination is an ongoing process, protein-calorie undernutrition can result in hypomyelination (e.g., Winick and Rosso, 1969; Martinez, 1982). *In utero,* despite the maternal reserve of nutrients to ameliorate the effects of maternal undernourishment of the fetus (Winick, 1976), maternal protein-calorie undernutrition can result in hypomyelination in the fetus (Winick, 1976; Martinez, 1982). In principle, CNS deficits could be a consequence of a nutritional insult on oligodendroglial maturation and/or developmental myelination. The initiation of developmental myelination in the PNS precedes that of the CNS, and, to the extent that PNS myelination occurs *in utero,* the effects of undernourishment of the fetus can be ameliorated by the mother. In studies of protein-calorie undernutrition in the human infant, it is difficult to attribute effects to the absence of a particular nutrient, e.g., protein, calorie, or essential fatty acids, since in severe undernutrition all of these nutrients are usually absent (Holman et al., 1981).

5.2. *Effects of Protein-Calorie Undernutrition in Animal Models*

In the rat, protein-calorie undernutrition established *in utero* and continued through developmental myelination results in decreased body weight, brain weight, and myelin deposition relative to age-matched controls and parallels the effects of protein-calorie undernutrition in the human. Additional effects of this protein-calorie undernutrition paradigm include decreases in: brain protein and lipid synthesis; whole brain DNA; and number of oligodendrocytes. In addition, fewer axons are myelinated and the ratio of number of myelin lamellae to axon diameter is decreased although the ultrastructure of the myelin lamellae appear normal (e.g., Krigman and Hogan, 1976; Wiggins et al., 1985; Lai et al., 1980). When dietary protein alone is restricted in the diet of rats during the first several weeks of life, a similar diminution of body weight, brain weight, and myelin deposition occurs with respect to well-nourished controls (e.g., Nakashi et al., 1975; Reddy et al., 1979; Reddy and Horrocks, 1982).

As judged by decreases in myelin recovered and whole brain determination of myelin-enriched lipids and myelin-specific proteins, the overall effect of decreased myelin deposition resulting from protein-calorie undernutrition in the animal models are comparable to the myelin deficit of the human infant subjected to severe undernutrition, although quantitative deficits depend on the protocol of the undernutrition model (Martinez, 1982) (*see* the discussion of experimental techniques in the latter part of this section). In the rat model, both the development of the mature oligodendrocyte and the process of myelination have been reported to be susceptible to an undernutrition insult (e.g., Wiggins, 1986; Robain and Ponsot, 1978; Lai et al., 1980), although the former has been questioned by Wiggins (1986). Undernutrition could effect oligodendrocyte maturation indirectly by interfering with the expression of regulatory factors (*see* Fig. 1), whereas an undernutrition paradigm probably directly affects developmental myelination by failing to supply sufficient energy source and metabolic precursors for the synthesis of myelin components that are synthesized at a very rapid rate during this period.

5.3. Effects of Essential Fatty Acid (EFA) Deficiency in Humans

Unsaturated fatty acids can be divided into three groups based on the number of carbon atoms from the terminal methyl group to the carbon of the first double bond. The fatty acids from two of these groups (*n*-3 and *n*-6) are essential fatty acids and they or a precursor, of the same group, must be supplied by diet. The *n*-9 series of fatty acids can be synthesized *de novo* but under normal dietary conditions are not as elongated or unsaturated as the *n*-3 and *n*-6 series. During EFA deficiency, longer chain, polyunsaturated fatty acids of the *n*-9 series are synthesized.

EFA deficiency can occur in premature infants, postsurgical patients, patients severely affected with Crohn's disease, and others receiving total parenteral nutrition that is missing EFA. Severe EFA deficiency can also occur in individuals with cystic fibrosis as a result of fat maladsorption caused by lipase insufficiency. Little is known about the consequences of EFA deficiency in humans.

5.4. Animal Models of EFA Deficiency

Most animal models of EFA deficiency concerned with the effects of altered fatty acid metabolism and composition on myelin have utilized rats or mice (there appear to be no significant differences between the rat and mouse models). An EFA deficient regimen is usually initiated with the pregnant dam and continued with the pups after weaning. Some general and myelin-associated deficits of EFA deficiency, many resembling protein-calorie undernutrition but less severe, include: decreased brain and body weight; decreased total brain DNA and protein; reduction in the amount of myelin isolated with a parallel decline in myelin enriched lipids and myelin specific proteins in brain; and change in the fatty acid composition of myelin phospholipids (e.g., Galli, 1973; Alling et al., 1973; Sun et al., 1974). The changes in the acyl composition of myelin phospholipids are consistent with the changes in fatty acid metabolism expected as a result of EFA deficiency, i.e., increased long chain, polyunsaturated, *n*-9 fatty acids and decreased polyunsaturates of the *n*-3 and *n*-6 series.

5.5. Effects of Vitamin Deficiency

Vitamin deficiency in the human during pregnancy and in infants can affect normal brain development. Animal models of vitamin deficiency *in utero* and during postnatal development demonstrate that myelin synthesis is decreased, and that effects appear in both neurons and glial cells as might be anticipated from the important role of vitamins in intermediary metabolism. Butterworth (1990) has recently reviewed effects of vitamin deficiency on developing brain in humans and in animal models. In addition to effects on myelin deposition, vitamin deficiency, including that of B_1, B_6, and B_{12} has been associated with demyelination.

5.6. Experimental Considerations in Animal Models

In rats, two different protocols have been used commonly to establish postnatal protein-calorie undernutrition. One model achieves undernutrition by allowing limited access to food. After birth, the litter is culled to 10 or fewer pups (fewer than this number of pups have been used in the control group to insure access to the dam). The control pups are allowed free access to the mother, but the undernourished pups are allowed access to the mother for a limited number of hours during a 24-h period; the rest of the pups are placed in a separate cage in which they can have access to water. The number of hours per day of access to the mother will determine the severity of undernourishment. By scheduling different hours of access of pups to the mother in several experimental litters, litters can be obtained that represent a gradation of undernourishment. Pups should be weighed each day. Normally, a litter is weaned when the pups are about 21 days old, and a restricted diet is maintained for undernourished rats. Since undernourished pups are smaller than controls, they may have difficulty in reaching the food and/or water if size of the rat determines its availability, e.g., if the food and water are normally placed near the top of the cage. Since experimental and control groups should be weaned at the same age, an alternate method must be devised for access to nourishment. Wiggins (1979) has found a lower mortality of undernourished pups by using a protocol of limited access to nourishment in which the newborn pups are separated from the mother for only a short period of time, but the access time to nourishment is decreased steadily as the pups get older. Consequences of this protocol may not be straightforward since: 1. Oligodendrocyte development and myelination temporally asynchronous in various brain regions are superimposed upon a schedule of nourishment that is constantly changing; and 2. As a model of human undernourishment, this assumes a constantly increasing severity of undernutrition.

In a second model of protein-calorie undernutrition, the control group is treated as in the first model, but undernourishment is achieved by overcrowding the litter with 12 or more pups so that, although the pups are constantly with the mother, free access to nourishment is not possible since the number of pups that can be fed at any given time is limited. This model assumes that over a period of time access to nourishment of the pups is uniformly reduced for the litter, which may not be valid based on the weight gain of individual pups. To the extent that pups are not equally deprived of food, the precision of measured parameters of the experiment may be lowered. Both of these models have advantages and disadvantages (Wiggins, 1979), and protocols have been reported that have used a combination of these two models (e.g., Krigman and Hogan, 1976).

Besides the use of data that are the average of several animals per data point, other approaches have been taken that attempt to reduce the variability in data reflecting variation among individual animals. One approach has been to use litters for experimental and control groups derived from dams who conceived within 12 h of each other and who gave birth within 12 h of each other, thus reducing variations that may arise from differences in gestational age. This approach requires mating several additional females to ensure that litters will be available that meet the temporal requirements. Another approach has been to take the control and experimental pups from the same litter. For example, in the first undernutrition model presented above, half of the litter would be allowed continual access to the mother (control), whereas the second half of the litter is removed from the cage for a specific period of time each day (undernourished). Since it is not unusual for the dam to give birth to more than 10 pups, utilizing the additional pups in the litter can be achieved by dividing the litter at birth with one group remaining with the mother and the second group fostered by a lactating dam.

Because of asynchrony of both the maturation of oligodendrocytes and developmental myelination in the CNS and the different stages of development that are potentially vulnerable to undernutrition, the age at which the undernutrition is initiated and terminated and its severity are important aspects in designing the undernutrition paradigm (e.g., Wiggins and Fuller, 1978; Miller, 1990).

As described above, an EFA-deficiency rat model can be established that is characterized by hypomyelination and dysmyelination (with respect to the lipid acyl composition). A standard purified diet that is fat-free can be used to establish EFA deficiency. The control diet is the same as the EFA-deficient diet except that a source of EFA is added, e.g., corn oil, 3% by weight. Coconut oil, hydrogenated to remove the small amount of linoleic acid, can be added to the EFA-deficient diet to match the caloric value between the control and EFA-deficient diet. EFA deficiency is usually initiated *in utero* during which a pregnant dam is fed an EFA-free diet and is maintained on the diet through the suckling period. We have found a reduction in the number of pups per litter when pregnant rats are placed on the EFA-deficient diet before the eighth day of pregnancy. After birth, females are removed from the litter since their requirement for EFA is ~1/3 that of the male (Pudelkewicz et al., 1968). Control and experimental litter size should not exceed 10. After weaning, pups are maintained on the EFA diet. Increasing EFA deficiency is reflected in the rise of the triene to tetraene ratio of acyl groups primarily owing to the rise in 20:3 (n-9) fatty acid (can be synthesized *de novo* but is virtually absent in controls) and a decrease in 20:4 (n-6), (arachidonate, an essential fatty acid).

5.7. Assessment of Deficits Resulting from Undernutrition

Estimation of the severity of protein-calorie undernutrition can be judged by the decrease in body weight relative to controls and parallels the severity of the undernutrition regimen. Relative decrease in body weight of experimental animals also parallels the decrease in whole brain myelin deposition. Brain weight decreases are also commonly used to estimate the effects of protein-calorie undernutrition; however, loss of brain weight in experimental animals may not be an accurate assessment of decreased myelin deposition especially when the weight loss is compared to myelin deposition in specific brain regions (Wiggins and Fuller, 1978).

In EFA deficiency, decreases of body and brain weight have also been used as general assessments of severity. However, a more commonly accepted and more direct assessment of severity of EFA deficiency in tissues is the increase in triene to tetraene ratio of acyl groups. The total trienes and tetraenes of tissue acyl groups can be determined by extraction of the lipid from the sample, alkaline methanolysis, and separation and quantitation using gas or thin-layer chromatography techniques (e.g., Miller et al., 1981). For animals fed an EFA-deficient diet, the rate of change of the triene to tetraene ratio differs among various tissues, necessitating that an estimation of the severity of EFA deficiency with respect to myelin be determined using lipid from isolated myelin.

Myelin can be isolated from undernourished rats and the lipid and protein composition can be determined using techniques described in Section 3. The relative composition of myelin lipids and proteins of protein-calorie undernourished rats differs from myelin isolated from well-nourished, age-matched controls. (As described previously, the lipid and protein composition of protein-calorie undernourished rats resembles myelin composition found in younger control rats, leading to the hypothesis that protein-calorie undernutrition results in a developmental lag in myelination (*see* review by Wiggins [1982]). Because of the altered myelin composition, assessment of hypomyelination should be determined by comparing the amount of myelin isolated from control and experimental rats in lieu of using a representative myelin component. Since the effect of protein-calorie undernutrition on myelin deposition can vary considerably between various brain regions (Wiggins and Fuller, 1978), estimates of hypomyelination using whole brain myelin represent a weighted average.

In EFA deficiency, the relative composition of myelin proteins and lipid classes is not altered with respect to controls, although an exception to this generalization may be proteolipid protein (compare McKenna and Campagnoni [1979] with Miller et al. [1984]). The methods used to determine metabolic effects of undernutrition are generally independent of the nutritional protocol. Using radiolabeled isotopes of lipid and protein precursors in experiments, the effect of protein-calorie undernutrition and EFA deficiency on myelin synthesis and turnover has been determined (e.g., Wiggins et al., 1976; Menon and Dhopeshwarkar, 1984; Miller et al., 1981,1984).

6. Infectious Agents

6.1. Viruses

6.1.1. Human CNS Demyelination and Viruses

A number of viruses have been suspected, and in some cases shown to correlate with the pathogenesis of a number of myelin disorders, particularly in diseases involving primary demyelination. The experimental investigations that involve putative viral-associated myelin pathology can be divided into those concerned with:

1. Establishment of a viral pathogenesis and identification of the virus;
2. Determining the acquisition and course of the viral infection (e.g., route of infection, cellular specificity or susceptibility of the oligodendrocyte, and investigation of the mechanism of latency exhibited by some neurotropic viruses); and
3. Determining the mechanism by which the virus produces myelin pathology.

Demonstration of a viral association with the human pathology is a necessary step in demonstrating a viral etiology in myelin pathogenesis, whereas the use of animal models is important in determining the acquisition, course, and mechanism of the virus infection (for overview, *see* Gonzalez-Scarano and Tyler, 1987). Some aspects of the viral infection, e.g., determination of the cell surface component that serves as the viral receptor, are more easily investigated using tissue culture techniques.

Evidence for a viral etiology can be obtained by direct or indirect methods (e.g., McFarlin and Koprowski, 1990; Dörries and Ter Meulen, 1984). Direct methods could involve identification of virions in oligodendrocytes or Schwann cells from biopsy or autopsy of disease tissue via electron microscopy; demonstration of viral DNA (or RNA) or m-RNA of viral proteins in pathological tissue by various molecular hybridization techniques; the identification of viral proteins in myelin-producing cells by immunological techniques; and/or the demonstration of viral isolates from pathological tissues using tissue culture techniques. Indirect methods generally use immunological techniques (e.g., radioimmunoassay) to demonstrate the presence of antibody to viral antigen(s) in cerebrospinal fluid or blood.

The demonstration of virus by the use of direct or indirect methods does not always establish with certainty a viral pathogenesis of the myelin disorder. For instance, although hybridization techniques can positively indicate the presence of virus DNA, the viral genome may be latent. In a similar vein, demonstration of antibody to virus does not preclude the possibility that the antibody detected was elicited during a previous infection that has been subsequently cleared. However, demonstration that the presence of a particular virus is significantly higher than in the general population and that a high correlation exists between the myelin disorder and the presence of the particular virus provides strong support for a causal relationship between the specific virus and the myelin disorder. Complicating the demonstration of a causal relationship between a particular virus and a myelin disorder is the prospect that a myelin disorder when defined by clinical and pathologic criteria may be produced by several different viruses. Such a circumstance could preclude the demonstration of a particular virus as the causal agent for a myelin disorder by the criterion of a high correlation between the disorder and consistent demonstration of the presence of a particular virus or its genome in multiple cases of the disorder.

The human CNS demyelinating disease most discussed in terms of a virus etiology is multiple sclerosis (MS), although the viral pathogenesis of MS has not been proven. MS is a primary demyelinating disease exhibiting multiple lesions that can involve the spinal cord, brain, and optic tract. CNS lesions appear as discrete areas of demyelination. In brain, the lesions often are found in, but not restricted to, the periventricular regions. Histologically, the initial involvement appears to be associated with myelin loss followed by oligodendrocyte death, although it is quite possible that normally appearing oligodendrocytes have a metabolic lesion and that this precedes the demyelination process. Various reports have indicated that there is a preferential loss of some myelin components during the early stages of demyelination (for a more detailed account of these changes in MS, *see* Norton and Cammer, 1984b). Early in the development of the lesion, inflammatory cells, including lymphocytes and macrophages, are seen in the lesion area (the role of these cells in the putative immune-related pathogenesis and related experimental models are presented in the following section, which concerns the role of the immune system in demyelination). With the loss of myelin and oligodendrocytes, astroglia migrate into the lesion area to form scar tissue.

MS is generally diagnosed between the second and fourth decade of life and, in most cases, has a chronic relapsing-remitting clinical course with incomplete recovery from an episode. Prevalence is approximately 50% higher in women. The risk of MS shows a correlation with race, histocompatibility haplotypes, and within families (Traugott and Raine, 1984). Epidemiological data indicate a higher incidence with increasing latitude (Kurtzke, 1980). Some epidemiologic studies based on migration data have suggested that the risk of MS as a function of latitude is established below the age of 15 years and, after this age, residence at a different latitude does not change the risk of contracting MS (Kurtzke, 1980).

Many investigators have suggested that MS pathogenesis involves a latent virus. This is consistent with the incidence and risk factors that correlate with MS. Attempts to demonstrate by direct and indirect methods a correlation between the presence of a portion or all of a specific viral genome or the virion itself in MS tissue have had only partial success. Various reports that have propounded a specific virus in the pathogenesis of MS have been criticized, generally based on the lack of high correlation between tissue samples obtained from a large number of MS patients and demonstration of the virus (or a portion of the viral genome) in the tissue.

Another category of acute demyelinating disorders exists in which demyelination, usually fatal, occurs subsequent to a viral infection (reviewed by Raine, 1984b; Traugott and Raine, 1984). The virus associated with the infection varies, but the disorder is acute, demyelination is disseminated, inflammation of the white matter occurs, and involvement of other structures associated with the brain, e.g., meninges and vascular system, can occur to varying degrees. Although direct evidence of virus in the brain has only been shown infrequently, the correlation of previous virus infection shortly before the onset of the demyelinating disorder and the intense inflammatory response has led to a strong belief that this group of demyelinating disorders is virus-induced and mediated through the immune response; however, the details of the mechanism are unclear.

Progressive multifocal leukoencephalopathy, an acute demyelinating disorder, resembles MS with respect to the presence of multifocal lesions, but differs from MS and the acute disseminated demyelinating disorders mentioned in the previous paragraph in that no inflammatory response occurs. In addition, some axonal loss occurs and abnormal-appearing astrocytes are found in the lesion area. Of note is the demonstration of papovavirus inclusions in oligodendrocytes in progressive multifocal leukoencephalopathy (Padgett et al., 1971). Because papovavirus infections can occur in association with AIDS, the frequency of progressive multifocal leukoencephalopathy has been increasing.

6.1.2. Animal Models of Viral-Induced CNS Myelin Disorders

A wide range of viruses (e.g., mouse hepatitis virus, Theiler's murine encephalomyelitis virus, herpes simplex virus, rubella, border disease virus) and animal hosts have been used to develop animal models of viral-induced demyelination. Common problems in developing an animal model of viral-induced demyelination are that the pathology produced by the virus often involves other cell types, e.g., neurons and astrocytes (i.e., not all of the demyelination is primary) (Dal Canto and Rabinowitz, 1982) and the high mortality of animals receiving the virus inoculum. For a model using a particular animal and virus, enhanced specificity for the oligodendrocyte and reduced animal mortality may be achieved by judiciously selecting a specific strain of the virus, the dose and route of administration of the virus, and the age of the animal (e.g., Lavi et al., 1984, 1985).

If the virus is known or suspected, the ideal case would be to select an animal model that matches the clinical course and pathology of the human disorder. In the case of MS this would mean that not only would the disease in the animal be restricted to the CNS, exhibit a chronic relapsing-remitting course, and be accompanied by an inflammatory response, but the demyelinating lesions should also be multifocal, exhibit a propensity for the vascular regions, involve astrogliosis of demyelinated areas, and be consonant with the risk factors of MS. Such an ideal situation may not be achievable in the animals commonly used for models because of the host specificity of the virus and/or the animal's response (e.g., immune response) to the viral infection may differ from the human. Thus, it is not uncommon to use an animal model that focuses on a particular aspect of the pathology characteristic of the human disorder and may involve a virus that may be related to a human pathogen, but itself is not known to be a human pathogen (Dal Canto and Rabinowitz, 1982). This piecemeal approach has been particularly useful in understanding aspects of viral latency and reactivation in the nervous system and the possible role of the inflammatory response in primary demyelination.

6.1.3. Sampling Tissue for Biochemical Evaluation

General considerations concerning subfractionation of tissue components, biochemical analysis of these subfractions, and interpretation of the data were presented earlier in this chapter. Demyelination occurs in a nonuniform manner in the

CNS, resulting in areas of demyelination in close proximity to normal appearing white matter (an extreme case of this situation is encountered with the multifocal lesions of MS). This close proximity of a focal lesion to normal appearing tissue results in sampling difficulties. Even when the tissue sample appears to represent a demyelinating lesion, there is probably heterogeneity in the sample since the demyelinating disease will probably not be at the same stage in the entire sample. Thus, biochemical data derived from the analysis of myelin or other tissue components will represent a weighted average that is dependent on sampling technique and the extent that the biochemical composition is a function of the disease progress. Tissue samples, including the myelin isolated from this tissue, excised from grossly normal appearing white matter for analysis as a "normal control" can, in fact, have an altered composition resulting from demyelinating lesions that may become apparent under microscopic examination and histological tests (Allen and McKeown, 1979). In extensive, disseminated demyelination, substantial amounts of brain tissue can be used for comparison with normal brain to provide useful information on the alteration of specific components (Kamoshita et al., 1968), although these samples must also represent a weighted average reflecting the uniformity of the disease process.

6.1.4. Microscopic Evaluation

With the use of lipophilic stains, light microscopy, at low magnification, can be used to obtain a general assessment of significant myelin lesions in tissue sections of brain or spinal cord. *In situ* detection of specific compounds in tissue sections can be achieved utilizing monospecific antisera or monoclonal antibodies used in conjunction with an enzyme covalently bound to the primary antibody or to a secondary antibody which is specific for the immunoglobulin class of the primary antibody (Sternberger et al., 1970; Hsu et al., 1981). The use of commercially available soluble substrates that are converted to an insoluble product at the binding site of the enzyme-linked antibodies has proven useful in detecting, and under carefully controlled conditions, providing a semiquantitative estimate of viral antigens and specific or highly enriched antigens present in specific cells and myelin. Cell-specific antibodies can also be useful in an inflammatory response to characterize the lymphocytes found in the lesion area (Traugott et al., 1982). An excellent reference indicating sources of commercially available primary monospecific antisera and monoclonal antibodies, enzyme-linked secondary antibodies, and appropriate enzyme substrates is *Linscott's Directory of Immunological and Biological Reagents.*

Although the use of immunohistochemistry is an important tool in understanding the events involved in demyelination in animal models, special care must be taken in selection of the immunological reagents and their use to prevent ambiguous results. Some of the potential problems involve:

1. Determining the antibody specificity under the assay conditions used with the experimental animal tissue that includes crossreaction with structurally related molecules (e.g., antibodies made against galactosylceramide often bind to sulfatide) and nonspecific binding owing to the charge or hydrophobicity of cell components (e.g., the nonspecific binding often encountered with antibody prepared against myelin basic protein);
2. Variability in the temporal appearance of the chromophoric product in the visualization of the bound enzyme–antibody complex that can be misleading in the determination of the presence and abundance of a tissue component; and
3. Improper selection or use of immunological reagents (e.g., use of the whole antibody molecule in lieu of $F(ab')_2$ fragments or without blocking tissue receptors for the F_c portion of the antibody can result in binding of the antibody to F_c receptors of oligodendrocytes that could be misinterpreted as specific antigen-antibody binding)(Aarli et al., 1975; Ma et al., 1981).

For a review of the use of immunological reagents in detecting myelin components *see* Sternberger (1984).

Electron microscopic comparison of brain tissue from an animal model can be used in visualization of the abnormal appearance of myelin and the oligodendrocyte, and, if applicable to the disease model, in attempting to provide a clearer understanding of the role of cellular infiltrates in the demyelination process (e.g., Lavi et al., 1984). Ultrastructural data are also useful in localizing virions in and damage to cells other than the oligodendrocyte.

Even though histologic procedures are less useful in identifying and quantitating individual molecular species than biochemical approaches, they provide a greater opportunity for trying to determine the temporal sequence of events. Examination of lesions that exhibit various degrees of myelin and oligodendroglial disorganization provide a basis for speculating on the sequence of events that lead to demyelination. However, subtle changes in cellular events such as alteration in oligodendroglial metabolism, may not be detectable by cytochemical methods. Furthermore, knowledge of the temporal sequence of events leading to demyelination does not insure that it is possible to distinguish events that directly contribute to demyelination from those events that are a consequence of the demyelination process.

6.1.5. Proteases and Primary Demyelination

The role of proteases in the pathogenesis of demyelinating diseases, e.g., MS, and related animal models, e.g., experimental allergic encephalomyelitis (EAE), has been investigated (for a review of the earlier literature, *see* Smith, 1977). Although initial studies focused on myelin basic protein (Einstein et al., 1968), which is readily degraded by several different proteases, subsequent experiments have shown other myelin proteins may be hydrolyzed by the action of proteases during demyelination (Lees and Chan, 1975; Fishman et al., 1977; Sato et al., 1982; Maruthi Mohan and Sastry, 1987). Potential sources of proteases found in demyelinating lesions could include the oligodendrocyte, endogenous proteases found in myelin, and cells such as macrophages that are present at the lesion site as part of an inflammatory response (e.g., Cammer et al., 1978,1986; Vanguri and Shin, 1988).

One method of quantitating and identifying proteases involves microdissection of tissue followed by microchemical assay of proteases. Using endogenous substrates or model peptides, protease activity can be determined with fluorescamine to determine the increase in primary amine resulting from protease activity. Alternatively, tissue proteases can be assayed using fluorescent peptides as substrates (Hirsch and Parks, 1979; Hirsch, 1981). Although both fluorescent assays are very sensitive, each has its shortcoming. Fluorescamine as the detecting reagent gives high backgrounds owing to endogenous primary amines. On the other hand, a specific protease derived from either different tissues of the same species or the same tissue from different species can vary in its ability to hydrolyze a particular fluorescent substrate, complicating the identification of a tissue protease based on substrate specificity (Hirsch, 1981). Despite the problems, this approach to investigating tissue proteases in demyelinating tissue is useful. The use of small tissue samples, increasing the likelihood of homogeneous tissue samples, and the small aliquots required for the protease microassay, which allow a portion of the sample to be used for other chemical determinations, are advantages of this technique.

6.1.6. Phagocytosis and Demyelination

In addition to the possibility that cells recruited during inflammation to the lesion site participate in the pathogenesis of demyelination by the secretion of proteases as described above, they may also participate in demyelination by acting to phagocytize pieces of myelin. The use of electron microscopy has revealed single and multilamellar membrane fragments (presumably myelin) within phagocytic cells located in demyelinating lesions in both experimental animal models and humans (Lampert, 1978; Raine, 1984b; Sergott et al., 1988). Various reports have presented electron micrographs that appear to show cells participating in the disruption of the myelin sheath. In some cases these cells have insinuated pseudopodia between myelin lamellae that otherwise appear fairly compact (e.g., Sergott et al., 1988). Magnification of 40,000 to 60,000× appears to be adequate in visualizing these events.

Phagocytic cells are associated with: viral-related inflammatory and noninflammatory demyelinating diseases in humans; viral and nonviral primary demyelinating disorders in animal models; and Wallerian degeneration (secondary demyelination) in animal models (Lampert, 1978; Raine, 1984b; Perry et al., 1987; Sergott et al., 1988). The ubiquitous finding of myelin phagocytic cells occurring during myelin loss may indicate that phagocytosis is probably not part of the pathogenesis of myelin disorders, but rather represents a response to the presence of membrane debris.

6.1.7. Perturbation of Oligodendrocyte Metabolism and Demyelination

As described earlier, there is an immense metabolic requirement of the oligodendrocyte during the developmental period of rapid myelination. Subsequently, a large continuing metabolic commitment of the oligodendrocyte must also be required for the replacement of lipids and proteins that turn over in the large number of myelin sheaths that may be supported by a single oligodendrocyte, as well as for any continuing slow accumulation of myelin that may occur. Viral infection of oligodendrocytes during developmental myelination could result in hypomyelination. A viral infection, subsequent to this developmental period, could result in hypomyelination by interfering with the post-developmental slow accumulation of myelin. In addition, a viral infection during the post developmental period could have a profound effect on the steady-state of myelin metabolism, i.e., the turnover rate of myelin components relative to their replacement rate, by hindering the replacement of myelin components that turn over (Pleasure et al., 1973). Because individual myelin components turn over at various rates (Sabri et al., 1974; Singh and Jungalwala, 1979; Miller et al., 1977), a possible consequence of hindering the replacement of these components could be an alteration in myelin composition, which if unstable, could lead to demyelination. One way in which viruses interfere with host-cell metabolism is by inhibiting host protein synthesis (Kääriäinen, 1984). Mice infected with the JHM strain of mouse hepatitis virus comprise an animal model in which any immune involvement appears to accompany viral effects on oligodendrocytes, as exemplified by unusual cellular morphology particularly with respect to the plasma membrane (e.g., Powell and Lampert, 1975; Fleury et al., 1980).

The effect of viral infection on oligodendrocyte metabolism in animal models has not been extensively investigated and can be problematic. Injection of isotope-enriched or radiolabeled molecules into viral-infected and control animals followed by temporal comparison of the metabolic products isolated from brain homogenates may yield results that are difficult to attribute directly to viral-induced changes in oligodendrocytes, since results will represent a weighted average of the metabolism of all brain cells, infected and noninfected. A greater specificity could be obtained by using labeled precursors to follow the rate of synthesis of myelin-enriched or myelin-specific molecules such as galactosylceramide or proteolipid protein. Although this approach represents a weighted average between infected and noninfected oligodendrocytes, if sufficient numbers of oligodendrocytes are infected and/or the viral-induced metabolic alteration is of sufficient magnitude, this approach could yield useful information concerning alterations of oligodendrocyte metabolism. The use of antibodies might also be useful in studying viral effects on oligodendrocyte metabolism (Oldstone et al., 1984). Although considerable information about viral effects on oligodendrocyte metabolism can be obtained from animal models, some aspects of virus perturbation of cell metabolism, such as membrane transport, are more amenable to the use of tissue culture techniques (Kohn, 1979; Pasternak et al., 1988).

6.1.8. Border Disease and Abnormal Myelin Deposition

Border disease is a viral infection in which an infection of pregnant sheep can result in fetal death or malformation of organs in lambs that survive the infection *in utero.* The characteristic aberration in the CNS is a combination of hypomyelination and dysmyelination. Hypomyelination in the newborn lamb appears to be the result of a slowing of the rate of myelin deposition and thinner myelin sheaths; these effects occur without inflammation (Barlow and Storey, 1977; Potts et al., 1985). The extent of decrease of myelin content in newborn lambs that results from congenital virus infection varied with respect to brain region (reduction in myelin content: cerebellum > cerebral cortex = spinal cord > brain stem) and similar reductions were found in adult sheep that had been congenitally infected by virus (Potts et al., 1985). The most prominent features of the myelin sheath in electron micrographs were the smaller number of lamellae/sheath, fewer myelinated axons, and a lade of compaction of some myelin sheaths (Barlow and Storey, 1977). Biochemical analysis revealed decreased total lipid, phospholipid, decreased hexose lipid (presumably cerebroside and sulfatide), changes in the fatty acid pattern, and the appearance of cholesterol esters (Patterson et al., 1975). Although the chemical composition does not reflect that of normal myelin, the presence of cholesterol esters, usually indicating myelin breakdown, and the extent of loss of aberrant myelin that would not sediment with normal myelin in the step gradient procedure make it likely that the reported myelin composition may not accurately reflect the altered myelin composition produced in this model (*see* previous comments on use of continuous gradients in the isolation of myelin resulting from dysmyelination). Nevertheless, the lack of myelin compaction, similar to that seen in quaking mice, a genetic model, may indicate that this is a model of viral-induced dysmyelination, and analysis of the myelin lipid and protein based on an alternative approach to myelin isolation would be of interest. Unfortunately, the use of sheep for the animal model limits the availability of this to most research laboratories. Thus, establishment of this myelin disease in a laboratory animal would be important.

6.1.9. Human PNS Demyelination and Viruses

Viral-related PNS demyelination is associated with an acute inflammatory demyelinating syndrome commonly referred to as Guillain-Barré syndrome (GBS), which is usually diagnosed based on clinical symptoms, and, when available, tissue pathology, which is characterized by inflammation and segmental demyelination (for a detailed description of clinical criteria and pathology, *see* Arnason, 1984). Although GBS has been associated with a significant number of other initiating factors including nonviral infections, surgical trauma, and vaccination, 60–70% of GBS patients have had a very recent viral infection. The most common viruses associated with GBS are cytomegalovirus, human immunodeficiency virus, Epstein-Barr, vaccinia, and variola.

The pathology of GBS is consistent irrespective of the putative initial etiology. The distribution and age of myelin lesions in the PNS does not appear to form any pattern, but sites of inflammation correlate with demyelination. At the microscopic level, one type of demyelinating lesion is characterized at the early stages by myelin sheath changes occurring at the node of Ranvier and Schmidt-Lanterman incisures. Subsequently, macrophages, but not lymphocytes, are found in intimate contact with the myelin in the developing lesion. A second type of lesion, in which macrophages are near but not in direct contact with the myelin, proceeds via separation of myelin lamellae along the interperiod lines or splitting of the major dense lines followed by vesicle formation and loss of the disrupted myelin. The pathology of GBS demyelination resembles aspects of CNS demyelination associated with inflammation. The possible role of macrophages and proteases in the demyelinating process was described above with respect to the CNS. In GBS, virus has not been detected in Schwann cells whose myelin process is undergoing demyelination.

6.1.10. Animal Models of Virus-Related Human PNS Demyelination

Experimental allergic neuritis (EAN) is an extensively studied animal model in which the clinical symptoms and microscopic pathology closely resemble GBS. Induction of EAN is produced in several animal species, including rabbit and rat, by the injection of either PNS myelin protein or lipid (Brostoff et al., 1977; Saida et al., 1979). Onset occurs a few weeks after injection, and in the commonly used acute model, animals may recover or die depending on the severity of the disease. PNS demyelination resembling many of the features of EAN has also been produced by transfer of EAN serum (Saida et al., 1978). EAN, and its relationship to GBS is discussed in more detail in the following Section, which considers the role of the immune response in demyelinating diseases.

6.1.11. Marek's Disease

Marek's disease virus, an oncogenic herpes virus of chickens, can instigate demyelination in the PNS of chickens. The pathology of this demyelination resembles that of GBS and EAN (Pepose et al., 1981). Virions and virus antigen are not detectable in Schwann cells that have myelin processes, nor can virus be reactivated from these cells. However, the virus can be demonstrated in other PNS cells including nonmyelinating Schwann cells and in lymphocytes. This animal model and others in which humoral and/or cellular immunity appear to play a role in demyelination will be discussed in the following section.

6.2. Nonviral Infections

Infection by nonviral organisms can cause demyelination generally in concert with effects on other tissues. The effect of the organism can be indirect via the secretion of toxic compound(s).

Diphtheria toxin. The effects of diphtheria toxin result from an infection by Corynebacterium diphtheriae, which secretes a toxin responsible for demyelination in the PNS (the toxin does not cross the blood-brain barrier). Since laboratory animals are not a natural host for the organism, the toxic effect can be studied by injection. Introduction of the toxin by injection has resulted in a segmental demyelination resembling that found in human disease. The injection of diphtheria toxin into animal models has been useful in establishing the basis of the anatomical localization of demyelination, i.e., the effect of myelinated nerves in the PNS where the blood-nerve barrier is not well established (Waksman, 1961). Inhibition of protein synthesis in the Schwann cell has been proposed as a putative mechanism by which the toxin causes demyelination (Pleasure et al., 1973). In addition to the use of this model to investigate the bacterial disease, it can also be used as a model of other human demyelinating neuropathies (Kaplan, 1980).

7. Immune Mediated Demyelination

As discussed in the previous Section, in many demyelinating disorders of the CNS and PNS, a virus is associated or suspected as the primary factor in the disorder. Most viral-related, demyelinating disorders are also accompanied by inflammation in which lymphocytes are a prominent feature (e.g., MS and GBS). Efforts to develop animal models of autoimmune-mediated demyelination, especially models that show a clinical and pathological resemblance to either MS or GBS, have focused on the related questions of: how a virus can initiate an autoimmune response; and the mechanism by which the autoimmune response brings about demyelination (e.g., Wucherpfennig and Weiner, 1990; Huppert and Wild, 1986).

7.1. General Mechanisms of Viral-Initiated Autoimmunity

Several mechanisms have been proposed for viral-induced demyelination.

1. Molecular mimicry, in which an immune response to an epitope on one molecule is crossreactive with the epitope on a second unrelated molecule, could involve a shared or structurally similar epitope between a virus and myelin (or myelin-producing cell). By this mechanism, a variety of viruses could elicit an immune response crossreacting to the same myelin or oligodendrocyte component and result in identical pathology. For example, monoclonal antibody produced against measles virus and monoclonal antibody produced against herpes simplex virus both crossreact with the same cellular protein (Fujinami et al., 1983).
2. A virus infection could modify the membrane surface of an oligodendrocyte or Schwann cell, resulting in a different presentation of the antigen and failure of the immune system to recognize the antigen as self, e.g., membrane modification resulting from virus budding or expression of intracellular components on the cell surface (Huppert and Wild, 1986; Wild et al., 1981).
3. Demyelination could result from a nonspecific, general activation of B and T lymphocytes (Huppert and Wild, 1986). Subsequent to a previous injection of a non-nervous system antigen, antigen placed in CNS or PNS tissue resulted in inflammation and activation of polyclonal T and B lymphocytes that was accompanied by primary demyelination

(Wisniewski and Bloom, 1975). Suggestions have been made that an early step in the demyelinating process could be the release of proteases during the early events of the inflammatory response (*see* comments on "bystander demyelination" by Norton and Cammer, 1984b). This model would have to account for the localization of the demyelination as well as other characteristics of the observed pathology that are discussed at the end of this section.

4. Dysfunction of the immune system's ability to distinguish self from nonself can result in an autoimmune disorder. T-cell dysfunction resulting in an autoimmune response has been associated with demyelinating disorders, e.g., canine distemper and experimental allergic encephalomyelitis (*see below*).

The four putative mechanisms presented are not mutually exclusive, and autoimmunity may involve a combination, e.g., numbers 1 and 4.

7.2. Models Focusing on the Role of the Immune Response in CNS Demyelination

Difficulties in developing animal models that replicate exactly all aspects of human myelin disorders, including initiation and clinical, pathological, and biochemical changes, have necessitated a piecemeal approach in which animal models emphasize a particular aspect of the disorder. This has been a very productive approach to the study of the role of the immune response in inflammatory demyelinating disorders.

MS (described in Section 6.) is the most commonly used prototype of a human inflammatory demyelinating disorder in the CNS (possibly initiated by a virus or several different viruses), and experimental allergic encephalomyelitis (EAE) is the most common model that has been used in several animal species to understand the role of an autoimmune response in demyelination (Brostoff, 1984; Kumar et al., 1989). Guinea pigs, mice, rats, and rabbits are among the many species used for an animal model of EAE. In some animal species, susceptibility varies with the different strains, and susceptible strains can vary in their response to various protocols used to establish EAE.

EAE can be produced by injection of white matter homogenate, with or without adjuvant, or purified components (e.g., MBP or PLP) with adjuvant. Differences in the EAE model have been reported to result from the age of the animal; source, quantity, and purity of the CNS antigens, site of injection, and type of bacterium used with the adjuvant. Variations in the protocol of EAE have produced models with varied clinical courses including acute, chronic, and chronic relapsing.

In acute EAE, which develops within a few weeks after injection, pathological examination of the brain shows perivascular infiltration of mononuclear cells with subsequent appearence of T cells followed by B cells. Acute EAE exhibits significant inflammation but, by analogy with MS, relatively small amounts of demyelination; at the microscopic level, demyelination appears to result from the action of macrophages. The pathology of acute EAE appears to be more closely related to acute disseminated encephalomyelitis, which can result from rabies immunization, or to acute MS (a fatal and rare form of MS) than to chronic MS (Raine, 1984b). The loss of motor function followed by either recovery or death is the typical course of acute MS.

Chronic EAE can be produced by using immature guinea pig, relatively resistant to acute EAE (Stone and Lerner, 1965). Onset of the first symptoms varies from 2 to 10 weeks and differences in the time of average onset vary between strains. Lymphocytes, monocytes, and lesser numbers of plasma cells are found in the lesion area. In some strains of guinea pig, extensive inflammation can occur with loss of both glia and neurons, whereas in other strains and in larger animals with chronic EAE demyelinating lesions generally occur with preservation of axons. Animals develop permanent paralysis.

Chronic relapsing EAE, a clinical course more reminiscent of MS, can be produced with a single injection by modifying the injection protocol used to produce chronic, progressive EAE with high frequency in strain 13 of guinea pig (Wisniewski and Keith, 1977). Distribution of focal, demyelinating lesions is strain-specific in the guinea pig, with either a predominance of lesions in the spinal cord or a distribution of lesions between brain, optic nerve, and spinal cord. Perivascular inflammation occurs with infiltration into brain and spinal cord of lymphocytes, mononuclear cells, and some plasma cells. In older plaque areas, gliosis occurs with some oligodendrocyte loss. Remyelination in the older plaque areas is also found, but astroglial scarring is not common (Lassmann and Wisniewski, 1979). Some neuronal loss is also noted in the older plaques. The pathology of chronic relapsing EAE shows a greater frequency of remyelination with less frequent astroglial scarring in older plaques, and a lack of sudanophilic staining of myelin degradation products during demyelination as compared to MS.

The clinical presentation of chronic relapsing EAE usually manifests as a single attack consisting of motor dysfunction of varying severity that occurs within 2–3 mo after a single inoculation with guinea pig spinal cord. Subsequent recovery follows with several cycles of exacerbation and remission occurring over the next 19 mo. To date, the chronic relapsing EAE model provides the best in vivo model with respect to the clinical and pathological characteristics of MS, despite the discrepancies noted above.

The cellular and humoral immune responses in EAE have been studied extensively. T-cell requirements in development of EAE have been strongly supported by several different experimental findings.

1. It has been shown that transfer of T cells from EAE animals to previously untreated animals results in transfer of EAE (Mokhtarian et al., 1984).
2. Surgical depletion of thymus-derived cells (T cell) followed by irradiation in rats has been shown to prevent MBP-induced EAE and antibody against MBP, whereas thymus-depleted rats reconstituted with thymocytes were susceptible to MBP-induced EAE and produced antibody to MBP (Gonatas and Howard, 1974).
3. EAE can be abolished by injection of antibody to helper T cells (Kumar et al., 1989).

4. Immunization with a synthetic peptide corresponding to a portion of the T-cell receptor prevented MBP-induced EAE (Vandenbark et al., 1989).

Despite the strong evidence supporting an obligatory role for T cells in MBP-induced EAE, the exact manner in which a T-cell mediated autoimmune response results in demyelination and the requirement for other elements of the immune system remains unclear. Any proposed immune-mediated mechanism for MBP-induced autoimmune demyelination would need to explain how MBP, believed to be sequestered within the myelin structure, can serve as a target in the immune response.

Whether a humoral-mediated immune response is important in demyelinating disorders in the human or in animal models such as EAE remains uncertain. Lack of a primary role of antibodies in demyelination occurring in EAE is based on experimental evidence including: Unlike the passive transfer of T cells, EAE cannot be transferred to naive animals by serum from EAE animals; a lack of correlation of antibody titer with severity of EAE; and arrival of plasma cells, which secrete antibody, at the lesion after the onset of demyelination in EAE.

Other experiments indicate that antibodies that bind to myelin components may participate in the demyelination process. Macrophages, which appear to play an early role in demyelination by disruption and phagocytosis of myelin, increase phagocytosis and metabolism of myelin in vitro when the myelin is preincubated with antisera containing antibodies to the major myelin proteins and galactosylceramide (Trotter et al., 1986). Monoclonal antibody, prepared against a surface glycoprotein of oligodendrocytes and myelin, injected into MPB-treated mice before the onset of acute EAE accelerated the onset and increased the severity of the disease. This antibody, when injected into mice with MBP-associated, chronic relapsing EAE, enhanced the size of demyelinating lesions and increased the severity and duration of clinical symptoms, and in the recovery phase of the disease induced a relapse (Schluesener et al., 1987; Linington, 1988). Polyclonal monospecific antisera to galactosylceramide (myelin- and oligodendrocyte-enriched) and monoclonal antibody to myelin-associate glycoprotein (myelin-specific) have been shown to cause primary demyelination (Sergott et al., 1984,1988).

7.3. Models Focusing on the Role of the Immune Response in PNS Demyelination

GBS is the stereotypical human demyelinating disorder in the PNS, and the animal model used to study the role of the immune response in GBS is experimental allergic neuritis (EAN). The clinical correspondence between GBS and EAN has been found to be extremely good (Brostoff, 1984). EAN is often cited as the PNS counterpart of EAE because of similar experimental protocols and the associated clinical and pathological findings. The major clinical difference is that EAN represents the most common form of GBS, an acute monophasic disease, whereas the EAE model, which represents the most prevalent form of MS, is the chronic relapsing EAE model. Thus, EAN and EAE will differ to the extent that the acute monophasic and chronic relapsing disease course reflect differences in the mechanism of the disease.

EAN was initially produced by injection into rabbits of PNS homogenates in complete Freund's adjuvant (Waksman and Adams, 1955), which resulted in a PNS-restricted inflammatory response. In mice, chickens, and monkeys the same inoculum also produces a PNS-restricted inflammatory response, whereas in the guinea pig, rat, and sheep inflammation also occurs in the CNS. Subsequently, this protocol, in guinea pig, was shown to result in lymphocytes sensitized to P_0, the most abundant protein in PNS myelin and P_2, a low molecular weight basic protein, both unique to the PNS myelin. However, this procedure did not result in sensitized lymphocytes to P_1, identical to the 18.5 kDa MBP found in PNS and CNS myelin and which elicits EAE when injected (Carlo et al., 1975). A peptide derived from the cleavage of bovine P_2 (amino terminal 21 amino acids) was shown to produce EAN in rabbit similar to that produced by the injection of whole spinal cord homogenate or purified PNS myelin (Brostoff et al., 1977). Over a prolonged injection schedule, galactosylceramide, common to CNS and PNS myelin, produced

EAN when injected intramuscularly in complete Freund's adjuvant with bovine serum albumin initially and in incomplete adjuvant in subsequent injections. Clinical symptoms were typical of EAN and histologically resembled EAN with cellular infiltrates of phagocytic mononuclear cells, primarily macrophages; however, small lymphocyte infiltrates seen in EAN were absent. Antibodies that bound to galactosylceramide, did not crossreact with other sphingolipids although sulfatide was not tested (Saida et al., 1979). EAN, like EAE, can be transferred by cells and is considered to be a cell-mediated immune disorder. However, EAN can also be transferred by an injection into nerve parenchyma of EAN serum with a source of complement added to the inoculum (Saida et al., 1978). Systemic injection and injection in the absence of complement did not produce EAN.

A subset of peripheral neuropathy patients, not classified as GBS, has been shown to have an immunoglobulin paraproteinemia in which demyelination occurs (Latov et al., 1980). An IgM monoclonal antibody has been shown to bind to myelin-associated glycoprotein, present in PNS and CNS myelin (Latov et al.,1981; Ilyas et al., 1984); to a glycolipid found in the PNS but not the CNS (Ilyas et al., 1984); and to HSB-2, a human T-cell line, and human peripheral blood lymphocytes (Dobersen et al., 1985; Miller et al., 1987). The glycolipid has been shown to be a sphingoglycolipid that does not contain sialic acid, but does contain an unusual sulfated glucuronic acid as a terminal sugar that is also found as part of the carbohydrate portion of myelin-associated glycoprotein (Chou et al., 1985,1986). The paraprotein binds to the carbohydrate portion of the antigens. The role of the antibody in demyelination is not known; however, other unrelated peripheral neuropathies have been associated with antilipid antibody (Salazar-Grueso et al., 1990).

7.4. Models Focusing on Virus-Induced and Immune-Mediated Demyelination

Although the etiology of MS and most cases of GBS are putatively virus-associated, the animal models most associated with these human diseases, EAE and EAN, use host antigen rather than virus to induce immune-mediated demyelination. However, a considerable number of animal models of virus-induced demyelination have been developed. Although demyelination occurs in these animal models, in many of these models demyelination is owing to lytic infections and the infectious agents do not preferentially infect oligodendrocytes. Animal models of virus-induced demyelination should meet criteria that reflect the human disorder, although in reality an exact replication of the human disorder in animals is not likely to be achieved and is often not necessary to investigate a particular aspect of the disorder. Some of the criteria that might be considered in establishing the model are:

1. Specificity for infecting oligodendrocytes or Schwann cells in lieu of neurons or other glial cells;
2. Cell lysis should not occur since it is not a hallmark of most primary demyelinating diseases;
3. Whether the host response is inflammatory and the type of cells recruited by this response;
4. Temporal course of demyelination, i.e., acute, chronic progressive (indicating possibly a persistent infection?), chronic relapsing (indicating possibly latency with sporadic activation?);
5. Whether there is preservation of the myelin-producing cell associated with the lesion and, if not, the temporal sequence of loss, i.e., does the myelin loss occur before or after the loss of the myelin-producing cell; and
6. Anatomical distribution of the lesion in brain and whether it is disseminated or focal.

Two examples of virus infections with possible involvement of the immune system in demyelination are mouse hepatitis virus (MHV) and Theiler's virus. MHV infection in mice and rats has been used frequently as a model of viral-induced demyelination (*see* Lavi and Weiss, 1989). Strain JHM of MHV is primarily neurotropic and has been the most frequently used demyelinating strain of MHV, but causes high mortality at low doses, especially in young mice. Strain A59 is neurotropic and hepatotropic, also causes demyelination, and is less virulent than JHM. After an acute encephalitis during the first 2 weeks of infection, a persistent infection develops accompanied by primary demyelination during which virus is very difficult to recover but viral RNA can be detected by *in situ* hybridization. The role of the immune system in demyelination associated with this infection is not clear; however, Watanabe et al. (1983) have shown that lymphocytes from MHV (JHM)-infected rats can be stimulated by MBP in vitro and injected into virus-free rats to produce demyelination resembling EAE. MHV (A59) has been shown to induce the major histocompatability complex antigen on the surface of oligodendrocytes that do not normally express this antigen (Suzumura et al., 1986). Persistent presence of antibody occurs during the chronic demyelinating phase of MHV infection (Lavi and Weiss, 1989). Whether any of these immune responses are involved in chronic demyelination of MHV is not known.

Theiler's murine encephalomyelitis virus has also been shown to cause demyelination in the CNS of mice (Lipton, 1975). The course of the disease goes through remission followed by exacerbation. Immunosuppression in the early phases of the disease results in a decrease in demyelination, implying that the immune system may play a role in the demyelinating process (Lipton and Dal Canto, 1976).

8. Summary

The development and utilization of animal models to understand the etiology and mechanism of human myelin disorders requires an input of diverse information. This scope of information includes: an understanding of known aspects of the human myelin disorder; developmental and biochemical aspects of the normal events that result in myelin deposition; properties of the myelin membrane; the way in which the animal host responds to an insult; and properties of the chemical or biological insult. Some of this information can be obtained from the technical literature. However, in order to refine old animal models and develop new ones, a broad view of the research problem is required as well as the collaborative efforts of a broad spectrum of research investigators including biochemists, cell and developmental biologists, histologists, immunologists, pathologists, neurologists, and virologists.

Acknowledgments

I wish to thank Francisco Gonzalez-Scarano, David Klurfeld, Gregory Konat, Susan Weiss, and Richard Wiggins for helpful comments. Supported in part by US Public Health Grants HD 19661 and NS26884 and a grant from the Lannan Foundation.

References

Alling C., Bruce A., Karlsson I., and Svennerholm L. (1973) The effect of dietary lipids on the central nervous system, in *Dietary Lipids and Postnatal Development* (Galli C., Facini G., Pecile A., eds.), pp. 203–215. Raven, New York.

Aarli J. A., Aparicio S. R., Lumsden C. E., and Tönder O. (1975) Binding of normal human IgG to myelin sheaths, glia and neurons. *Immunology.* **28,** 171–185.

Agrawal H. C., Banik N. L., Bone A. H., Davison A. N., Mitchell R. F., and Spohn M. (1970) The identity of a myelin-like fraction isolated from developing brains. *Biochem. J.* **120,** 635–642.

Agrawal H. C., Trotter J. L., Burton R. M., and Mitchell R. F. (1974) Metabolic studies on myelin. Evidence for a precursor role of a myelin subfraction. *Biochem. J.* **140,** 99–109.

Agrawal H. C. and Hartman B. K. (1980) Proteolipid protein and other proteins of myelin, in *Proteins of the Nervous System,* 2nd ed. (Bradshaw R. A. and Schneider D. M., eds.), 145–169. Raven, New York.

Allen I. V. and McKeown S. R. (1979) A histological, histochemical and biochemical study of the macroscopically normal white matter in multiple sclerosis. *J. Neurol. Sci.* **41,** 81–91.

Arnason B. G. W. (1984) Acute inflammatory demyelinating polyradiculoneuropathies, in *Peripheral Neuropathy,* Vol. 2 (Dyck P. J., Thomas P. K., Lambert E. H., and Bunge R., eds.), pp. 2050–2100, Saunders, Philadelphia.

Banik N. L and Smith M. E. (1977) Protein determinants of myelination in different regions of developing rat central nervous system. *Biochem. J.* **162,** 247–255.

Barlow R. M. and Storey I. J. (1977) Myelination of the ovine CNS with special reference to Border disease. I. Qualitative aspects. *Neuropathol. Appl. Neurobiol.* **3,** 237–253.

Benjamins J. A. and McKhann G. M. (1973) [2-^{3}H]Glycerol as a precursor for phospholipids in rat brain: Evidence for lack of recycling. *J. Neurochem.* **20,** 1111–1120.

Benjamins J. A., Miller S. L., and Morell P. (1976) Metabolic relationships between myelin subfractions: Entry of galactolipids and phospholipids. *J. Neurochem.* **27,** 565–570.

Blaker W. D., Krigman M. R., Thomas D. J., Mushak P., and Morell P. (1981) Effect of triethyltin on myelination in the developing rat. *J. Neurochem.* **36,** 44–52.

Blakemore W. F. (1984) The response of oligodendrocytes to chemical injury. *Acta Neurol. Scand.* **70 (Suppl. 100),** 33–38.

Bondy S. C. (1988) The neurotoxicity of organic and inorganic lead, in *Metal Neurotoxicity* (Bondy S. C. and Prasad K. N., eds.), pp. 1–17, CRC Press, Boca Raton, FL.

Bondy S. C. and Prasad K. N., eds. (1988) *Metal Neurotoxicity*, pp. 1–189, CRC Press, Boca Raton, FL.

Bourre J. M., Jacque C., Delassalle A., Nguyen-Legros J., Dumont O., Lachapelle F., Raoul M., Alvarez A., and Baumann N. (1980) Density profile and basic protein measurements in the myelin range of particulate material from normal developing mouse brain and from neurological mutants (Jimpy, Quaking, Trembler, Shiverer, and its mld allele) obtained by zonal centrifugation. *J. Neurochem.* **35,** 458–464.

Brody B. A., Kinney H. C., Kloman A. S., and Gilles F. H. (1987) Sequence of central nervous system myelination in human infancy. 1. An autopsy study of myelination. *J. Neuropathol. Exp. Neurol.* **46,** 283–301.

Brostoff S. (1984) Immunological responses to myelin and myelin components, in *Myelin* (Morell P., ed.), pp. 405–439. Plenum, New York.

Brostoff S., Levit S., and Powers J. M. (1977) Induction of experimental allergic neuritis with a peptide from myelin P_2 basic protein. *Nature* **268,** 752,753.

Butterworth R. F. (1990) Vitamin deficiencies and brain development, in *(Mal)nutrition and the Infant Brain* (van Gelder N. F., Butterworth R. F., and Drujan B. D., eds.), pp. 207–224, Wiley-Liss, New York.

Cammer W. (1980) Toxic demyelination: Biochemical studies and hypothetical mechanisms, in *Experimental and Clinical Neurotoxicology* (Spencer P. S. and Schaumburg H. H., eds.), pp. 239–256, Williams & Wilkins, Baltimore.

Cammer W., Bloom B. R., Norton W. T., and Gordon S. (1978) Degradation of basic protein in myelin by neutral proteases secreted by stimulated macrophages: A possible mechanism of inflammatory demyelination. *Proc. Natl. Acad. Sci. USA* **75,** 1554–1558.

Cammer W., Brosnan C. F., Basile C., Bloom B. R., and Norton W. T. (1986) Complement potentiates the degradation of myelin proteins by plasmin: Implications for a mechanism of inflammatory demyelination. *Brain Res.* **364,** 91–101.

Campagnoni A. T., Roth H. J., Hunkeler M., Pretorius P. J., and Campagnoni C. W. (1989) Expression of myelin basic protein genes in the developing mouse brain, in *Developmental Neurobiology* (Evrard P. and Minkowski A., eds.), pp. 95–109, Raven, New York.

Carlo D. J., Karkhanis Y. D., Bailey P. J., Wisniewski H. M., and Brostoff S. W. (1975) Experimental allergic neuritis: evidence for the involvement of P_0 and P_2 proteins. *Brain Res.* **88,** 580–584.

Chou K. H., Ilyas A. A., Evans J. E., Quarles R. H., and Jungalwala F. B. (1985) Sructure of a lipid reacting with monoclonal IgM in neuropathy and with HNK-1. *Biochem. Biophys. Res. Commun.* **128,** 383–388.

Chou K. H., Ilyas A. A., Evans J. E., Costello C., Quarles R. H., and Jungalwala F. B. (1986) Structure of sulfated glucuronyl glycolipids in the nervous system reacting with HNK-1 antibody and some IgM paraproteins in neuropathy. *J. Biol. Chem.* **261,** 11717–11725.

Cochran F. B. Jr., Yu R. K., and Ledeen R. W. (1982) Myelin gangliosides in vertebrates. *J. Neurochem.* **39,** 773–779.

Cremer J. E. (1958) The biochemistry of organotin compounds. The conversion of tetraethyltin into triethyltin in mammals. *Biochem. J.* **68,** 685–692.

Dal Canto M. C. and Rabinowitz S. G. (1982) Experimental models of virus-induced demyelination of the central nervous system. *Ann. Neurol.* **11,** 109–127.

Doberson M. J., Gascon P., Trost S., Hammer J. A., Goodman S., Noronha A. B., O'Shannesessy D. J., Brady R. O., and Quarles R. H. (1985) Murine monoclonal antibodies to the myelin-associated glycoprotein react with large granular lymphocytes of human blood. *Proc. Natl. Acad. Sci. USA* **82,** 552–555.

Dobbing J. and Smart J. L. (1973) Early undernutrition, brain development and behavior, in *Ethology and Development* (Barnett S. A., ed.), pp. 16–36. Lippincott, Philadelphia.

Dörries R. and Ter Meulen V. (1984) Detection and identification of virus-specific, oligoclonal IgG in unconcentrated cerebrospinal fluid by immunoblot technique. *J. Neuroimmunol.* **7,** 77–89.

Einstein E. R., Csejtey J., and Marks N. (1968) Degradation of encephalitogen by purified acid proteinase. *FEBS Lett.* **1,** 191–195.

Eng L. F. and Noble B. P. (1968) The maturation of rat brain myelin. *Lipids* **3,** 157–161.

Eng L. F., Chao F. -C., Gerstl B., Pratt D., and Tavaststjerna M. G. (1968) The maturation of human white matter myelin. Fractionation of the myelin membrane proteins. *Biochem.* **7,** 4455–4465.

Fishman M. A., Agrawal H. C., Alexander A., Golterman J., Martenson R. E., and Mitchell R. F. (1975) Biochemical maturation of human central nervous system myelin. *J. Neurochem.* **24,** 689–694.

Fishman M. A., Trotter J. L., and Agrawal H. C. (1977) Selective loss of myelin proteins during autolysis. *Neurochem. Res.* **2,** 247–257.

Fleury H. J. A., Sheppard R. D., Bornstein M. B., and Raine C. S. (1980) Further ultrastructural observations of virus morphogenesis and myelin pathology in JHM virus encephalomyelitis. *Neuropathol. Appl. Neurobiol.* **6,** 165–179.

Folch J., Lees M., and Sloane-Stanley G. H. (1957) A simple method for the isolation and purification of total lipids from animal tissue. *J. Biol. Chem.* **226,** 497–509.

Fraher J. P. (1976) The growth and myelination of central and peripheral segments of ventral motoneurone axons. A quantitative ultrastructural study. *Brain Res.* **105,** 193–211.

Fraher J. P., Kaar G. F., Bristol D. C., and Rossiter J. P. (1988) Development of ventral spinal motoneurone fibres: A correlative study of the growth and maturation of central and peripheral segments of large and small fibre classes. *Prog. Neurobiol.* **31,** 199–239.

Fujinami R. S., Oldstone M. B. A., Wroblewska Z., Frankel M. E., and Koprowski H. (1983) Molecular mimicry in virus infection: Crossreaction of measles virus phosphoprotein or of herpes simplex virus protein with human intermediate filaments. *Proc. Natl. Acad. Sci. USA* **80,** 2346–2350.

Galli C. (1973) Dietary lipids and brain development, in *Dietary Lipids and Postnatal Development* (Galli C., Facini G., Pecile A., eds.), pp. 191–202. Raven, New York.

Ganser A. L. and Kirschner D. A. (1985) The interaction of mercurials with myelin: Comparison of in vitro and in vivo effects. *Neurotoxicol.* **6,** 63–78.

Gilles F. H., Shankle W., and Dooling E. C. (1983) Myelinated tracts: Growth patterns, in *The Developing Human Brain* (Gilles F. H., Leviton A., and Dooling E. C., eds.), pp. 117–183. Wright/PSG, Boston.

Gonatas N. K. and Howard J. C. (1974) Inhibition of experimental allergic encepthalomyelitis in rats severely depleted of T cells. *Science* **186,** 839–841.

Gonzales-Scarano F. and Tyler K. L. (1987) Molecular pathogenesis of neurotropic viral infections. *Ann. Neurol.* **22,** 565–574.

Greenfield S., Norton W. T., and Morell P. (1971) Quaking mouse: Isolation and characterization of myelin protein. *J. Neurochem.* **18,** 2119–2128.

Grundt I., Offner H., Konat G., and Clausen J. (1974) The effect of methylmercury chloride and triethyllead chloride on sulphate incorporation into sulphatides of rat cerebellum slices during myelination. *Environ. Physiol. Biochem.* **4,** 166–171.

Harry G. J., Goodrum J. F., Krigman M. R., and Morell P. (1985) The use of synapsin I as a biochemical marker for neuronal damage by trimethyltin. *Brain Res.* **326,** 9–18.

Hirsch H. E. (1981) Proteinases and demyelination. *J. Histochem. Cytochem.* **29,** 425–430.

Hirsch H. E. and Parks M. E. (1979) A thiol proteinase highly elevated in and around the plaques of multiple sclerosis. Some biochemical parameters of plaque activity and progression. *J. Neurochem.* **32,** 505–513.

Holman R. T., Johnson S. B., Mercuri O., Itarte H. J., Ridrigo M. A., and DeThomas M. E. (1981) Essential fatty acid deficiency in malnourished children. *Am. J. Clin. Nutr.* **34,** 1534–1539.

Horrocks L. A. (1973) Composition and metabolism of myelin phosphoglycerides during maturation and aging. *Prog. Brain Res.* **40,** 383–395.

Hsu S. -M., Raine L., and Fanger S. H. (1981) Use of avidin-biotin-peroxidase complex (ABC) in immunoperoxidase techniques: A comparison between ABC and unlabeled antibody (PAP) procedures. *J. Histochem. Cytochem.* **29,** 577–580.

Huppert J. and Wild T. F. (1986) Virus-related pathology; is the continued presence of the virus necessary? *Adv. Virus Res.* **31,** 357–385.

Ilyas A. A., Quarles R. H., MacIntosh T. D., Dobersen M. J., Trapp B. D., Dalakas M. C., and Brady R. O. (1984) IgM in a human neuropathy related to paraproteinemia binds to a carbohydrate determinant in the myelin-associated glycoprotein and to a ganglioside. *Proc. Natl. Acad. Sci. USA* **81,** 1225–1229.

Kääriäinen L. (1984) Inhibition of cell functions by RNA-virus infections. *Ann. Rev. Microbiol.* **38,** 91–109.

Kamoshita S., Rapin I., Suzuki S., and Suzuki S. (1968) Spongy degeneration of the brain. A chemical study of two cases including isolation and characterization of myelin. *Neurology* **18,** 975–985.

Kaplan J. G. (1980) Neurotoxicity of selected biological toxins, in *Experimental and Clinical Neurotoxicology* (Spencer P. S. and Schaumburg H. H., eds.), pp. 631–648. Williams & Wilkins, Baltimore.

Kohn A. (1979) Early interactions of viruses with cellular membranes. *Adv. Virus Res.* **24,** 223–276 .

Konat G. and Clausen J. (1976) Triethyllead-induced hypomyelination in the developing rat forebrain. *Exp. Neurol.* **50,** 124–133.

Konat G. and Offner H. (1982) Effects of triethyllead on post-translational processing of myelin proteins. *Exp. Neurol.* **75,** 89–94.

Konat G. (1984) Triethyllead and cerebral development: An overview. *Neurotoxicol.* **5,** 87–96.

Krigman M. R. and Hogan E. L. (1976) Undernutrition in the developing rat: effect on myelination. *Brain Res.* **107,** 239–255.

Kumar K., Kono D. H., Urban J. L., and Hood L. (1989) The T-cell receptor repertoire and auto immune diseases. *Ann. Rev. Immunol.* **7,** 657–682.

Kung M. -E., Nickerson P. A., Sansone F. M., Olson J. R., Kostyniak P. J., Adolf M. A., and Roth J. A. (1989) Effect of chronic exposure to hexachlorophene on rat brain cell specific marker enzymes. *Neurotoxicol.* **10,** 201–210.

Kurtzke J. F. (1980) Multiple sclerosis: An overview, in *Clinical Neuroepidemiology* (Rose F. C., ed.), pp. 170–195, Pitman Press, Bath, UK.

Laatsch R. H., Kies M. W., Gordon S., and Alvord E. C., Jr. (1962) The encephalomyelitic activity of myelin isolated by ultracentrifugation. *J. Exp. Med.* **115,** 777–788.

Lai M., Lewis P. D., and Patel A. J. (1980) Effects of undernutrition on gliogenesis and glial maturation in rat corpus callosum. *J. Comp. Neurol.* **193,** 965–972.

Lampert P. W. (1978) Autoimmune and virus-induced demyelinating diseases. A review. *Am. J. Pathol.* **91,** 176–208.

Lassman H. and Wisniewski H. M. (1979) Chronic relapsing experimental allergic encephalomyelitis. Clinicopathological comparison with multiple sclerosis. *Arch. Neurol.* **36,** 490–497.

Latov N., Sherman W. H., Nemni R., Galassi G., Shyong J. S., Penn A. S., Chess L., Olarte M. R., Rowland L. P., and Osserman E. F. (1980) Plasma-cell dyscrasia and peripheral neuropathy with a monoconal antibody to peripheral nerve myelin. *N. Engl. J. Med.* **303,** 618–621.

Latov N., Braun P. E., Gross R. B., Sherman W. H., Penn A. S., and Chess L. (1981) Plasma cell dyscrasia and peripheral neuropathy: Identification of the myelin antigens that react with human paraproteins. *Proc. Natl. Acad. Sci. USA* **78,** 7139–7142.

Lavi E., Gilden D. H., Wroblewska Z., Rorke L. B., and Weiss S. R. (1984) Experimental demyelination produced by the A59 strain of mouse hepatitis virus. *Neurology* **34,** 597–603.

Lavi E., Gilden D. H., Highkin M. K., and Weiss S. R. (1986) The organ tropism of mouse hepatitis virus A59 in mice is dependent on dose and route of inoculation. *Lab. Anim. Sci.* **36,** 130–135.

Lavi E. and Weiss S. R. (1989) Coronaviruses in *Clinical and Molecular Aspects of Neurotropic Virus Infection* (Gilden D. H. and Lipton H. L., eds.), pp. 101–139. Kluwer Academic, Boston.

Lees M. B. and Chan D. S. (1975) Proteolytic digestion of bovine brain white matter proteolipid. *J. Neurochem.* **25,** 595–600.

Linington C., Bradl L., Lassmann H., Brunner C., and Vass K. (1988) Augmentation of demyelination in rat acute allergic encephalomyelitis by circulating mouse monoclonal antibodies directed against a myelin/oligodendrocyte glycoprotein. *Am. J. Pathol.* **130,** 443–454.

Lipton H. L. (1975) Thieler's virus infection in mice: An unusual biphasic disease process leading to demyelination. *Infect. Immun.* **11,** 1147–1155.

Lipton H. L. and Dal Canto M. C. (1976) Thieler's virus-induced demyelination: Prevention by immunosuppression. *Science* **192,** 62–64.

Lowry O. H., Rosebrough N. J., Farr A. L., and Randall R. J. (1951) Protein measurement with the Folin phenol reagent. *J. Biol. Chem.* **193,** 265–275.

Ludwin S. K. (1978) Central nervous system demyelination and remyelination in the mouse. An ultrastructural study of cuprizone toxicity. *Lab. Investig.* **39,** 597–612.

Ma B. I., Joseph B. S., Walsh M. J., Potvin A. R., and Tourtellotte W. W. (1981) Multiple sclerosis serum and cerebralspinal fluid immunoglobulin binding to F_c receptors of oligodendrocytes. *Ann. Neurol.* **9,** 371–377.

Martinez M. (1982) Myelin lipids in the developing cerebrum, cerebellum, and brainstem of normal and undernourished children. *J. Neurochem.* **39,** 1684–1692.

Maruthi Mohan P. and Sastry P. S. (1987) Susceptibility of the Wolfgram proteins and stability of 2',3'-cyclic nucleotide 3'-phosphodiesterase of rat brain myelin to limited proteolytic digestion. *J. Neurochem.* **48,** 1083–1089.

McFarlin D. E. and Koprowski H. (1990) Neurological disorders associated with HTLV-1. *Curr. Topics Microbiol. Immunol.* **160,** 99–119.

McKenna M. C. and Campagnoni A. T. (1979) Effect of pre- and postnatal essential fatty acid deficiency on brain development and myelination. *J. Nutr.* **109,** 1195-1204.

McMorris F. A. and Dubois-Dalcq M. (1988) Insulin-like growth factor I promotes cell proliferation and oligodendroglial commitment in rat glial progenitor cells developing in vitro. *J. Neurosci. Res.* **21,** 199–209.

Menon N. K. and Dhopeshwarkar G. A. (1984) Incorporation of [U-^{14}C] isoleucine into myelin in essential fatty acid (EFA) deficiency in the rat. *Nutr. Rep. Intern.* **29,** 783–789.

Miller S. L. (1990) Effects of Undernutrition on Myelin Deposition, in *(Mal)nutrition and the Infant Brain* (van Gelder N. F., Butterworth R. F., and Drujan B. D., eds.), pp. 175–190. Wiley-Liss, New York.

Miller S. L., Klurfeld D. M., Weinsweig D., and Kritchevsky D. (1981) Effect of essential fatty acid deficiency on synthesis and turnover of myelin lipid. *J. Neurosci. Res.* **6,** 203–210.

Miller S. L., Klurfeld D. M., Loftus B., and Kritchevsky D. (1984) Effect of essential fatty acid deficiency on myelin proteins. *Lipids* **19,** 478–480.

Miller S. L., Benjamins J. A., and Morell P. (1977) Metabolism of glycerophospholipids of myelin and microsomes in rat brain. *J. Biol. Chem.* **252,** 4025–4037.

Miller S. L. and Morell P. (1978) Turnover of phosphatidylcholine in microsomes and myelin in brains of young and adult rats. *J. Neurochem.* **31,** 771–777.

Miller S. L., Kahn S. N., Perussia B., and Trinchieri G. (1987) Comparative binding of murine and human monoclonal antibodies reacting with myelin-associated glycoprotein to myelin and human lymphocytes. *J. Neuroimmunol.* **15,** 229–242.

Miller S. L. Ötvös-Papp E., Prichett W., and Meyer W. D. (1989) Detection and partial biochemical characterization of a novel 57,500 dalton protein in rat brain myelin. *J. Neurosci. Res.* **22,** 262–268.

Mokhtarian F., McFarlin D. E., and Raine C. S. (1984) Adoptive transfer of myelin basic protein-sensitized T cells produces chronic relapsing disease in mice. *Nature* **309,** 356–358.

Morell P. (ed.) (1984) *Myelin,* Plenum, New York.

Morell P. and Mailman R. B. (1987) Selective and nonselective effects of organometals on brain neurochemistry, in *Neurotoxicants and Neurobiological function: Effects of Organoheavy Metals* (Tilson H. A. and Sparber S. B., eds.), pp. 202–229. John Wiley & Sons, New York.

Nakhasi H. L., Toews A. D., and Horrocks L. A. (1975) Effect of postnatal protein deficiency on the content and composition of myelin from brains of weanling rats. *Brain Res.* **83,** 176–179.

Nishizawa Y., Kurihara T., and Takahashi Y. (1981) Immunochemical localization of 2',3'-cyclic nucleotide 3'-phosphodiesterase in the central nervous system. *Brain Res.* **212,** 219–222.

Norton W. T. and Cammer W. (1984a) Isolation and characterization of myelin, in *Myelin* (Morell P., ed.), pp. 147–195. Plenum, New York.

Norton W. T. and Cammer W. (1984b) Chemical pathology of diseases involving myelin, in *Myelin* (Morell P., ed.), pp. 369–403. Plenum, New York.

Norton W. T. and Poduslo S. E. (1973a) Myelination in rat brain: Changes in myelin composition during brain maturation. *J. Neurochem.* **21,** 759–773.

Norton W. T. and Poduslo S. E. (1973b) Myelination in rat brain: Method of myelin isolation. *J. Neurochem.* **21,** 749–757.

Oldstone M. B. A., Rodriguez M., Daughaday W. H., and Lampert P. W. (1984) Viral perturbation of endocrine function: disordered cell function lead to disturbed homeostasis and disease. *Nature* **307,** 278–281.

Padgett B. L., Walker D. L., ZuRhein G. M., and EckRoade R. J. (1971) Cultivation of a papova-like virus from human brain wih progressive multifocal leucoencephalopathy. *Lancet* **1,** 1257–1260.

Pasternak C. A. Whitaker-Dowling P. A., and Widnell C. C. (1988) Stress-induced increase of hexose transport as a novel index of cytopathic effects in virus infected cells: Role of the L protein in the action of vesicular stomatitis virus. *Virology* **166,** 379–386.

Patterson D. S. P., Terlecki S., Foulkes J. A., Sweasey D., and Glancy E. M. (1975) Spinal cord lipids and myelin composition in Border disease (hypomyelinogenesis congentia) of lambs. *J. Neurochem.* **24,** 513–522.

Pepose J. S., Stevens J. G., Cook. M. L., and Lampert P. W. (1981) Marek's disease as a model for the Landry-Guillain-Barré syndrome. Latent viral infection in nonneuronal cells accompanied by specific immune responses to peripheral nerve and myelin. *Am. J. Pathol.* **103,** 309–320.

Perry V. H., Brown M. C., and Gordon S. (1987) The macrophage response to central and peripheral nerve injury. A possible role for macrophages in regeneration. *J. Exp. Med.* **165,** 1218–1223 .

Pleasure D. E., Feldmann B., and Prockop D. J. (1973) Diphtheria toxin inhibits the synthesis of myelin proteolipid and basic proteins by peripherial nerve in vitro. *J. Neurochem.* **20,** 81–90.

Potts B. J., Berry L. J., Osburn B. I., and Johnson K. P. (1985) Viral persistence and abnormalities of the central nervous system after congenital infection of sheep with Border disease virus. *J. Infect. Dis.* **151,** 337–343.

Powell H. C. and Lampert P. W. (1975) Oligodendrocytes and their myelin-plasma membrane connections in JHM mouse hepatitis virus encephalomyelitis. *Lab. Investig.* **33,** 440–445.

Pudelkewicz C., Seufert J., and Holman R. T. (1968) Requirements of the female rat for linoleic and linolenic acids. *J. Nutr.* **94,** 138–146.

Raff M. C. (1989) Glial cell diversification in rat optic nerve. *Science* **243,** 1450–1455.

Raine C. S. (1984) Morphology of myelin and myelination, in *Myelin* (Morell P., ed.), pp. 1–50. Plenum, New York.

Raine C. S. (1984b) The neuropathology of myelin diseases, in *Myelin* (Morell P., ed.), pp. 259–310. Plenum, New York.

Rawlins F. A. and Uzman B. G. (1970) Effect of AY-9944, a cholesterol biosynthesis inhibitor, on peripheral nerve myelination. *Lab. Investig.* **23,** 184–189.

Reddy P. V., Das A., and Sastry P. S. (1979) Quantitative and compositional changes in myelin of undernourished and protein malnourished rat brains. *Brain Res.* **161,** 227–235.

Reddy T. S. and Horrocks L. A. (1982) Effects of neonatal undernutrition on the lipid composition gray matter and white matter in rat brain. *J. Neurochem.* **38,** 601–605.

Robain O. and Ponsot G. (1978) Effects of undernutrition on glial maturation. *Brain Res.* **149,** 379–397.

Rorke L. B. and Riggs H. E. (1969) *Myelination of the Brain in the Newborn,* pp. 1–91. J. B. Lippincott Company, Philadelphia.

Roussel G. and Nussbaum J. L. (1981) Comparative localization of Wolfgram W1 and myelin basic proteins in the rat brain during ontogenesis. *Histochem. J.* **13,** 1029–1047.

Sabri M. I., Bone A. H., and Davison A. N. (1974) Turnover of myelin and other structural proteins in the developing rat brain. *Biochem. J.* **142,** 499–507.

Saida T., Saida K., Dorfman S. H. Silberberg D. H., Sumner A. J., Manning M. C., Lisak R. P., and Brown M. J. (1979) Experimental allergic neuritis induced by sensitization with galactocerebroside. *Science* **204,** 1103–1106.

Saida T., Saida. K., Silberberg D. H., and Brown M. J. (1978) Transfer of demyelination by intraneural injection of experimental allergic neuritis serum. *Nature* **272,** 639–641.

Salazar-Grueso E. F., Routbort M. J., Martin J., Dawson G., and Roos R. P. (1990) Polyclonal IgM Anti-GM_1 ganglioside antibody in patients with motor neuron disease and variants. *Ann. Neurol.* **27,** 558–563.

Sapirstein V., Trachtenberg M., Lees M. B., and Koul O. (1978) Regional developmental studies on myelin and other carbonic anhydrases in rat CNS, in *Myelination and Demyelination* (Palo J., ed.), pp. 117–133. Plenum, New York.

Sato S., Quarles R. H., and Brady R. O. (1982) Susceptability of the myelin-associated glycoprotein and basic protein to a neutral protease in highly purified myelin from human and rat brain. *J. Neurochem.* **39,** 97–105.

Schluesener H. J., Sobel R. A., Linington C., and Weiner H. L. (1987) A monoclonal antibody against a myelin oligodendrocyte glycoprotein induces relapses and demyelination in central nervous system autoimmune disease. *J. Immunol.* **139,** 4016–4021.

Sergott R. C., Brown M. J., Silberberg D. H., and Lisak R. P. (1984) Antigalactocerebroside serum demyelinates optic nerve in vivo. *J. Neurol. Sci.* **64,** 297–303.

Sergott R. C., Brown M. J., Lisak R. P., and Miller S. L. (1988) Antibody to myelin-associated glycoprotein produces central nervous system demyelination. *Neurol.* **38,** 422–426.

Singh H. and Jungalwala F. B. (1979) The turnover of myelin proteins in adult rat brain. *Intern. J. Neurosci.* **9,** 123–131.

Smith M. E. (1973) A regional survey of myelin development: some compositional and metabolic aspects. *J. Lipid Res.* **14,** 541–551.

Smith M. E. (1977) The role of proteolytic enzymes in demyelination in experimental allergic encephalomyelitis. *Neurochem. Res.* **2,** 233–246.

Smith M. E. and Sedgewick L. M. (1975) Studies of the mechanism of demyelination: Regional differences in myelin stability *in vitro*. *J. Neurochem.* **24,** 763–770.

Smith P. K., Krohn R. I., Hermanson G. T., Mallia A. K., Gartner F. H., Porvenzano M. D., Fujimoto E. K., Goeke N. M., Olson B. J., and Klenk D. C. (1985) Measurement of protein using Bicinchoninic acid. *Anal. Biochem.* **150,** 76–85.

Söderberg M., Edlund C., Kristensson K., and Dallner G. (1990) Lipid composition of different regions of the human brain during aging. *J. Neurochem.* **54,** 415–423.

Spencer P. S., Couri D., and Schaumburg, H. H. (1980) n-Hexane and methyl n-butyl ketone, in *Experimental and Clinical Neurotoxicology* (Spencer P. S. and Schaumburg H. H., eds.), pp. 456–475. Williams & Wilkins, Baltimore.

Sternberger L. A., Hardy P. H., Jr., Circulis J. J., and Meyer H. G. (1970) The unlabeled antibody enzyme method of immunohistochemistry: preparation and properties of soluble antigen-antibody complex (horseradish peroxidase-antihorseradish peroxidase) and its use in identification of spirochetes. *J. Histochem. Cytochem.* **18,** 315–333.

Sternberger N. H. (1984) Patterns of oligodendrocyte function seen by immunocytochemistry, in *Oligodendroglia* (Norton W. T., ed.), pp. 125–173. Plenum, New York.

Sternberger N. H., Itoyama Y., Kies M. W., and Webster H. de F. (1978) Myelin basic protein demonstrated immunocytochemically in oligodendroglia prior to myelin sheath formation. *Proc. Natl. Acad. Sci. USA* **75,** 2521–2524.

Sternberger N. H., Quarles R. H., Itoyama Y., and Webster H. de F. (1979) Myelin-associated glycoprotein demonstrated immunocytochemically in myelin and in myelin-forming cells of developing rat. *Proc Natl. Acad. Sci. USA* **76,** 1510–1514.

Stone S. H. and Lerner E. M. (1965) Chronic disseminated allergic encephalomyelitis in guinea pigs. *Ann. NY Acad. Sci.* **122,** 227–241.

Sun G. Y., Go J., and Sun A. Y. (1974) Induction of fatty acid deficiency in mouse brain: Effects of fat deficient diet upon acyl group composition of myelin and synaptosome-rich fractions during development and maturation. *Lipids* **9,** 450–454.

Suzuki K. (1971) Some new observations in triethyl-tin intoxication of rats. *Exp. Neurol.* **31,** 207-213.

Suzuki K. and Zagoren J. C. (1974) Degeneration of oligodendroglia in the central nervous system of rats treated with AY9944 or triparanol. *Lab. Invest.* **31,** 503–515.

Suzumura A., Lavi E., Weiss S. R., and Silberberg D. H. (1986) Coronavirus infection induces H-2 antigen expression on oligodendrocytes and astrocytes. *Science* **232,** 991–993.

Tellez-Nagel I., Korthals J. K., Vlassara H. V., and Cerami A. (1977) An ultrastructural study of chronic sodium cyanate-induced neuropathy. *J. Neuropathol. Exp. Neurol.* **36,** 351–363.

Thomas P. K. (1984) The peripheral nervous system as a target for toxic substances. *Acta Neurol. Scand.* **70 (Suppl. 100),** 21–26.
Tiffany-Castiglioni E., Sierra E., Wu J. -N., and Rowles T. K. (1989) Lead toxicity in neuroglia. *Neurotoxicol.* **10,** 417–444.
Trachman H. L., Tyberg A. J., and Branigan P. D. (1977) Atomic absorption spectrometric determination of sub-part-per-million quantities of tin in extracts and biological materials with a graphite furnace. *Anal. Chem.* **49,** 1090–1093.
Traugott U., Reinherz E. L., and Raine C. S. (1982) Monoclonal anti-T cell antibodies are applicable to the study of inflammatory infiltrates in the central nervous system. *J. Neuroimmunol.* **3,** 365–373.
Traugott U. and Raine C. S. (1984) The neurology of myelin diseases, in *Myelin* (Morell P., ed.), pp. 311–335. Plenum, New York.
Trotter J., DeJong L. J., and Smith M. E. (1986) Opsonization with antimyelin antibody increases the uptake and intracellular metabolism of myelin in inflammatory macrophages. *J. Neurochem.* **47,** 779–789.
Vandenbark A. A., Hashim G., and Offner H. (1989) Immunization with a synthetic T-cell receptor V-region peptide protects against experimental autoimmune encephalomyelitis. *Nature* **341,** 541–544.
Vanguri P. and Shin M. L. (1988) Hyrolysis of myelin basic protein in human myelin by terminal complement complexes. *J. Biol. Chem.* **263,** 7228–7234.
Waehneldt T. V. (1978) Protein heterogeneity in rat CNS myelin subfractions, in *Myelination and Demyelination* (Palo J., ed.), pp. 117–133. Plenum, New York.
Waksman B. H. (1961) Experimental study of diphtheritic polyneuritis in the rabbit and guinea pig III. The blood-nerve barrier in the rabbit. *J. Neuropathol. Exp. Neurol.* **20,** 35–77.
Waksman B. H. and Adams R. D. (1955) Allergic neuritis: An experimental disease in rabbits induced by the injection of peripheral nervous tissue and adjuvants. *J. Exper. Med.* **102,** 213–235.
Walters S. N. and Morell P. (1981) Effects of altered thyroid states on myelinogenesis. *J. Neurochem.* **36,** 1792–1801.
Walsh T. J. and DeHaven D. L. (1988) Neurotoxicity of the alkyltins, in *Metal Neurotoxicity* (Bondy S. C. and Prasad K. N., eds.), pp. 87–107. CRC Press, Boca Raton, FL.
Watanabe R., Wege H., and ter Meulen V. (1983) Adoptive transfer of EAE-like lesions from rats with coronavirus-induced demyelinating encephalomyeliis. *Nature* **305,** 150–153.
Webster H. deF. and Favilla J. (1984) Development of peripheral nerve fibers, in *Peripheral Neuropathy, Volume I* (Dyck P. J., Thomas P. K., Lambert E. H., and Bunge R., eds.), pp. 329–359. Saunders, Philadelphia
Wender M., Adamczewska-Goncerzewicz Z., Mularek O., and Zgorzalewicz B. (1978) The effect of intoxication with alkylnitrosourea derivatives on cerebral myelin, in *Myelination and Demyelination* (Palo J., ed.), pp. 487–498. Plenum, New York.

Wiggins R. C. (1979) A comparison of starvation models in studies of brain myelination. *Neurochem. Res.* **4,** 827–830.

Wiggins R. C. (1982) Myelin development and nutritional insufficiency. *Brain Res. Rev.* **4,** 151–175.

Wiggins R. C. (1986) Myelination: A critical stage of development. *Neurotoxicology* **7,** 103–120.

Wiggins R. C., Miller S. L., Benjamins J. A., Krigman M. R., and Morell, P. (1976) Myelin synthesis during postnatal nutritional deprivation and subsequent rehabilitation. *Brain Res.* **107,** 257–273.

Wiggins R. C. and Fuller G. N. (1978) Early postnatal starvation causes lasting brain hypomyelination. *J. Neurochem.* **30,** 1231–1237.

Wiggins R. C. and Fuller G. N. (1979) Relative synthesis of myelin in different brain regions of postnatally undernourished rats. *Brain Res.* **162,** 103–112.

Wiggins R. C., Bissel A. C., Durham L., and Samorajski. T. (1985) The corpus callosum during postnatal undernourishment and recovery: A morphometric analysis of myelin and axon relationships. *Brain Res.* **328,** 51–57.

Wild T. F., Bernard A., and Greenland T. (1981) Measles Virus: Evolution of a persistent infection BGM cells. *Arch. Virol.* **67,** 297–308.

Winick M. (1976) Clinical Nutrition, in *Malnutrition and Brain Development,* pp. 3–34. Oxford University Press, New York.

Winick M. and Noble A. (1966) Cellular response in rats during malnutrition at various ages. *J. Nutr.* **89,** 300–306.

Winick M. and Rosso P. (1969) The effect of severe early malnutrition on cellular growth of human brain. *Pediat. Res.* **3,** 181–184.

Wisniewski H. M. and Bloom B. R. (1975) Primary demyelination as a nonspecific consequence of a cell-mediated immune reaction. *J. Exper Med.* **141,** 346–359.

Wisniewski H. M. and Keith A. B. (1977) Chronic relapsing experimental allergic encephalomyelitis: An experimental model of multiple sclerosis. *Ann. Neurol.* **1,** 144–148.

Wood P. and Bunge R. P. (1984) The biology of the oligodendrocyte, in *Oligodendroglia* (Norton W. T., ed.), pp. 1–46. Plenum, New York.

Wucherpfennig K. W. and Weiner H. L. (1990) Immunologic mechanisms in chronic demyelinating diseases of the central and peripheral nervous system, in *Immunologic Mechanisms in Neurologic and Psychiatric Disease* (Waksman B. K., ed.), pp. 105–116. Raven, New York.

Animal Models of the Cerebellar Ataxias

Roger F. Butterworth

1. The Cerebellum

1.1. Introduction

The cerebellum is composed of the cerebellar cortex, internal white matter, and deep cerebellar nuclei. These nuclei are the fastigial, interpositus, and dentate nuclei; they mediate most of the output of the cerebellum. This output is directed primarily to motor regions of the brain stem and cerebral cortex.

The cerebellum receives afferent projections from the periphery via the spinocerebellar tracts, as well as from brain stem and cerebral cortex.

1.2. Cellular Organization of the Cerebellar Cortex

Cerebellar cortex contains five distinct types of neurons, namely Purkinje, stellate, granule, Golgi, and basket cells. These cells are arranged in three layers as shown in a simplified schematic form in Fig. 1.

The molecular layer is composed of granule cell axons (also known as parallel fibers) as well as stellate cells and basket cells. These interneurons are excited by the parallel fibers. The Purkinje layer is composed of a single layer of Purkinje cells that provide the sole output of the cerebellar cortex. Purkinje cells have a large dendritic tree extending into the molecular layer. The inner layer, the granular layer, consists of Golgi neurons and granule cells. The number of neurons in the cerebellar granular

From: *Neuromethods: Vol. 21 Animal Models of Neurological Disease, I*
Eds: A. Boulton, G. Baker, and R. Butterworth

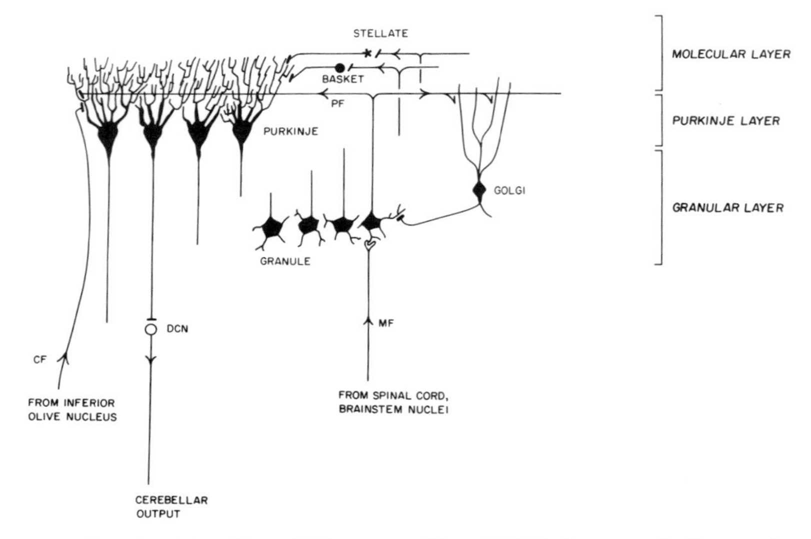

Fig. 1. CF: climbing fiber; MF: mossy fiber; DCN: deep cerebellar nucleus.

layer (approx 10^{11} cells) exceeds the total number of cells in the remaining regions of mammalian brain.

1.3. Cerebellar Afferent Systems

Input to the cerebellar cortex is mediated by two excitatory systems, namely the climbing fibers originating in the inferior olive and the mossy fibers whose cell bodies are localized in brainstem and spinal cord. Each Purkinje cell receives a single climbing fiber input. However, a single climbing fiber may make connections with up to 10 Purkinje cells. Mossy fiber afferents terminate on granule cell dendrites. The granule cells in turn send axons into the molecular layer where they give rise to the parallel fibers, which form synaptic contacts with the dendrites of Purkinje cells. Each Purkinje cell receives excitatory inputs from up to 2×10^5 parallel fibers. Whereas the afferent climbing fibers and mossy fibers are excitatory, the interneurons in the cerebellar cortex (stellate cells, basket cells, Golgi neurons) are inhibitory. Stellate and basket cells form inhibitory synaptic contacts with Purkinje neurons in the molecular layer. The Golgi neurons receive parallel fiber input but then go on to form synapses with granule cells as part of a feedback loop.

1.4. Role of Cerebellum in the Control of Movement

The cerebellum plays a key role in the initiation and coordination of voluntary movements. Asynergia or limb ataxia is the lack of synergy, which manifests itself as a combination of abnormally coordinated movements. Holmes, in the 1920s, studied patients from the First World War who had cerebellar lesions owing to gunshot wounds. He was able to demonstrate that each side of the cerebellum is implicated in the smooth control of the arm on that side. Patients with cerebellar damage could not carry out movements involving several joints of the arm. This disability is known as decomposition of movement, which may include hypermetria or excessive extent of movement (overshoot) or hypometria in which the limb stops short. Lesions in the lateral part of cerebellum produce limb ataxia, whereas lesions in the vermis produce gait ataxia. Disorders in the articulation of speech (dysarthria) are also seen in cerebellar disease.

1.5. Cerebellar Neurotransmitters: Key Role of the Amino Acids

Neurotransmitters such as acetylcholine and the biogenic amines (dopamine, noradrenaline, and serotonin) occur in very low concentrations in cerebellum. On the other hand, excitatory amino acids such as glutamate and aspartate, and the inhibitory amino acid γ-aminobutyric acid (GABA) are present in relatively high concentrations. Stimulation of Purkinje cells leads to the release of GABA, and the GABA synthetic enzyme glutamic acid decarboxylase (GAD) has been detected by immunohistochemical techniques in Purkinje cell terminals in deep cerebellar nuclei. Furthermore, iontophoretic application of GABA onto single neurons in lateral vestibular nucleus mimics the effects of Purkinje cell stimulation. Taken together, these findings strongly support GABA as the Purkinje cell neurotransmitter (*see*, e.g., McGeer et al., 1978).

Other cerebellar neurons also appear to use GABA as a neurotransmitter. Basket cells show GAD immunostaining and uptake studies in cerebellar preparations using radiolabeled GABA show heavy labeling in basket, stellate, and Golgi neurons.

Based on selective destruction of cerebellar granule cells by neurotoxins, X-irradiation, or in mouse mutants (discussed more fully in Sections 3.4. and 3.6., this chapter), there is now a strong consensus of opinion that the granule cell neurontransmitter is glutamate. The other cerebellar afferent system, namely the climbing fibers, on the other hand, appear to be aspartatergic and/or glutamatergic (*see* Section 3.3.).

2. The Cerebellar Ataxias

2.1. Classification of the Cerebellar Ataxias

The hereditary ataxias or cerebellar ataxias are a poorly defined heterogeneous group of hereditary disorders. The prevalence of this group of disorders has been estimated to be between 6 and 23 per 100,000 in various world communities. Classification depends on clinical symptomatology, neuropathological evaluation, and, where available, biochemical findings. There is substantial disagreement concerning classification of these disorders and the classification described in Table 1 must therefore be regarded as tentative.

Friedreich's Ataxia is perhaps the best known of the hereditary ataxias, being first described by Friedreich in 1863. This disorder is characterized by chronic progressive loss of coordinated movements and decreasing muscle stretch reflexes. Neuropathological evaluation reveals degeneration of the posterior columns of spinal cord as well as lesions in cerebellar cortex and nuclei. The disease is inherited in an autosomal recessive manner.

Roussy-Levy Syndrome is a gait ataxia of somewhat later onset and milder course than Friedreich's Ataxia.

Charcot-Marie-Tooth Disease resembles both Friedreich's Ataxia and Roussy-Levy Syndrome. Three different patterns of inheritance of Charcot-Marie-Tooth Disease (autosomal dominant, autosomal recessive, and X-linked) have been described. Ataxia, scoliosis, and upper motor neuron disease characterize the autosomal recessive type.

Table 1
Classification of the Cerebellar Ataxias

Spinocerebellar forms
Friedreich's ataxia
Roussy-Levy syndrome
Charcot-Marie-Tooth disease
Refsum's disease
Bassen-Kornsweig syndrome (Abetalipoproteinemia)
Cerebellar forms
Pierre Marie's "hereditary cerebellar ataxia"
Olivocerebellar atrophy (Holmes)
Olivopontocerebellar atrophies (OPCA)
Intermittent cerebellar ataxia of childhood

Refsum's Disease is an apparently distinct clinical entity characterized by spino-cerebellar ataxia in which patients have an abnormal oxidative degradation of phytanic acid.

In the Bassen-Kornzweig Syndrome, also known as acanthocytosis, degeneration of posterior columns of spinal cord as well as cerebellar lesions are the classic neuropathological findings. The disorder is characterizied by the absence of circulating low density lipoproteins, acanthocytosis, and retinitis pigmentosa.

In 1893, Pierre Marie described the clinical condition that he called "hérédoataxie cérébelleuse" or hereditary cerebellar ataxia. Cerebellar signs characterize this disorder. Muscle stretch reflexes are increased in contrast to the decreased reflexes reported in Friedreich's Ataxia. Spinal cord signs as well as lesions of posterior and lateral columns have also been reported. Onset of the disease is in adulthood and the mode of inheritance appears to be autosomal dominant.

The Olivopontocerebellar Atrophies (OPCA) are a group of hereditary ataxic disorders. Based on genetic, clinical, and pathological findings, five types of OPCA have been described:

1. OPCA Type I, dominant type ("Menzel" Ataxia)
2. OPCA Type II, recessive type
3. OPCA Type III, dominant type with retinal degeneration

4. OPCA Type IV, dominant type ("Schut-Haymaker" type)
5. OPCA Type V, dominant type with dementia and extrapyramidal signs

2.2. Biochemical Abnormalities in the Cerebellar Ataxias

Various biochemical abnormalities have been reported in patients with cerebellar ataxias (Table 2). In Friedreich's Ataxia, there is evidence to suggest that mitochondrial malic enzyme is reduced in cultured fibroblasts from these patients (Stumpf et al., 1982). Furthermore, obligate heterozygotes have reduced malic enzyme activities, suggesting that the enzyme defect in Friedreich's Ataxia patients is genetically determined (Stumpf et al., 1983). Biochemical pathological studies on autopsied spinal cord from Friédreich's Ataxia patients reveal selective reductions of glutamate in lumbar gray matter and posterior columns (Butterworth and Giguère, 1984). It was suggested that this loss may be the result of the loss of primary sensory afferent fibers and of descending corticospinal tracts in Friedreich's Ataxia. Taurine levels in spinal cord were increased in Friedreich's Ataxia patients. Losses of glutamate with concomitantly increased taurine were also reported in autopsied cerebellum from patients with Friedreich's Ataxia (Huxtable et al., 1979).

Abnormalities of lipid metabolism have also been reported in the Cerebellar Ataxias. Patients with Refsum's Disease have an abnormal excess of lipids containing phytanic acid (Richterich et al., 1965). Low fasting cholesterol and betalipoprotein levels have been described in 13 members of a kindred with hypobetalipoproteinemia and the clinical symptoms of a spino-cerebellar degeneration similar to Friedreich's Ataxia (Aggerbeck et al., 1974).

In Olivopontocerebellar Atrophy (POCA), both dominant and recessive forms, reduced cerebellar concentrations of glutamate and aspartate have been reported (Perry et al., 1981). In type IV OPCA, marked reductions of activities of glutamate dehydrogenase were found (Plaitakis et al., 1980), and it was suggested that genetic defects of this enzyme might underlie some forms of spinocerebellar ataxias.

Table 2
Biochemical Abnormalities in the Cerebellar Ataxias

Disorder	Biochemical findings	Reference
Friedreich's ataxia	Malic Enzyme Deficiency	Stumpf et al., 1982
	Reduced glutamate, aspartate in spinal cord	Robinson, 1968
	Reduced glutamate in spinal cord	Butterworth and Giguère, 1984
	Increased taurine in spinal cord	Butterworth and Giguère, 1984
	Reduced glutamate in cerebellum	Huxtable et al., 1979
	Increased taurine in cerebellum	Huxtable et al., 1979
Refsum's Disease	Phytanic acid oxidation defect	Richterich et al., 1965
Bassen-Kornsweig Disease	Low density lipoprotein deficiency	Aggerbeck et al., 1974
Olivopontocerebellar Atrophy (Type I)	Reduced glutamate, aspartate in cerebellum	Perry et al., 1981
Olivopontocerebellar Atrophy (Type II)	Glutamate dehydrogenase deficiency	Plaitakis et al., 1980
	Reduced glutamate, aspartate in cerebellum	Perry et al., 1981
Intermittent Ataxia of Childhood	Pyruvate Dehydrogenase Deficiency	Blass et al., 1970
	Carnitine Acetyltransferase Deficiency	DiDonato et al., 1979

In young children with intermittent ataxic syndromes, deficiencies of both pyruvate dehydrogenase (Blass et al., 1970; Butterworth, 1985) and carnitine acetyltransferase (DiDonato et al., 1979) have been reported.

3. Animal Models of the Cerebellar Ataxias

3.1. Introduction

A great deal of useful information has been acquired by the study of animal models of cerebellar ataxia. Not only have such studies provided evidence for the selective vulnerability of certain cerebellar neurons to toxic, genetic, and viral insults, but also, the use of neurotoxic substances, X-rays, and genetic mutants has assisted in the elucidation of ontogenic aspects of cerebellar function and synaptic connectivity as well as in the assignment of cerebellar neurotransmitters. Results of studies of animal models of viral-induced cerebellar ataxia, in particular, could have important implications for the pathogenesis of certain human cerebellar ataxias.

3.2. Measurement of Ataxia in the Laboratory Rat

Reports of ataxia or loss of coordination in experimental animals are frequently based on visual, generally highly subjective, assessment. In a more systematic approach to the measurement of ataxia in the laboratory rat, Jolicoeur et al. (1979) reported on a standardized battery of neurological tests designed to analyze quantitatively ataxia and related neurological signs. Analyses of gait by measurement of length, width, and angle of steps as well as forelimb extension, hindlimb extension, and righting reflexes were reported in several experimental models of cerebellar ataxia. This battery of tests constitutes a sensitive and reliable technique for the detection, quantitation, and differentiation of ataxic syndromes in experimental animals.

3.3. Neurotoxin Models of Cerebellar Ataxia

3.3.1. 3-Acetylpyridine-Induced Cerebellar Deafferentation

3-Acetylpyridine (3 AP) is an antimetabolite of nicotinamide. A single ip injection of 3 AP (65 mg/kg) to male rats results in partial degeneration of facial and hypoglossal nuclei and total destruction of the inferior olivary nucleus (Desclin and Escubi, 1974). During the acute stage of intoxication, animals experience difficulties in breathing and swallowing, symptoms that subside after a few days. Such symptoms are consistent with lesions of ambiguous and hypoglossal nuclei. Following 3 AP administration, animals are ataxic and easily lose their balance. Such symptoms are generally interpreted as being cerebellar in origin (Desclin and Escubi, 1974). Since no lesions of cerebellum (cortex or nuclei) *per se* are observed following 3 AP treatment, impairment of cerebellar function is attributed to the destruction of the inferior olivary nucleus with resulting loss of afferent cerebellar climbing fibers. Degenerating climbing fibers are observed in both molecular and granular layers of cerebellum as early as 12 h after 3 AP treatment.

The specific destruction of the climbing fibers, which constitute a major excitatory input to cerebellum (Fig. 1) without any observable lesions of other cerebellar elements, has been used as a tool to identify the neurotransmitter from the olivocerebellar projection. Both glutamate and aspartate have been proposed as possible candidates for the climbing fiber neurotransmitter. Twenty-one days following a single ip injection of 3 AP to rats, levels of aspartate were found to be significantly decreased in three regions of cerebellum, as well as in the synaptosomal fraction of cerebella homogenates from 3-AP treated animals (Rea et al., 1980, Table 3). The lower levels of aspartate in the synaptosomal fraction and in cerebellar cortex of 3 AP- treated rats are consistent with the notion that climbing fibers may use this amino

Table 3
Cerebellar Aspartate and Glutamate Content Following 3 AP Treatment

Brain region	Treatment	n	Amino acid concentration, nmol/mg prot	
			Aspartate	Glutamate
Cerebellar cortex	Saline	14	22.2 ± 0.4	166 ± 4
	3 AP	11	19.6 ± 0.6*	162 ± 5
Cerebellar synaptosomes	Saline	11	9.5 ± 0.5	26.5 ± 1.4
	3 AP	14	8.1 ± 0.2*	22.2 ± 0.7*

* $P < 0.5$ compared to saline-treated group (data from Rea et al., 1980).

acid as neurotransmitter. However, decreases of glutamate in cerebellum of 3 AP-treated rats have also been reported (Butterworth et al., 1978; Rea et al., 1980), suggesting that glutamate may also (or additionally) be a climbing fiber neurotransmitter. Two independent studies have measured the effect of 3 AP on the size of the releasable (neurotransmitter) pool of aspartate or glutamate in cerebellum. K^+-stimulated, Ca^{2+}-dependent release of these amino acids was assessed in cerebellar slices in vitro (Flint et al., 1981; Wiklund et al., 1982). 3 AP treatment did not result in any significant changes of K^+-stimulated aspartate or glutamate release from cerebellar slices in the study by Flint et al. (1981), but caused a selective 26% decrease of aspartate release from cerebellar slices in the study reported by Wiklund et al. (1982).

3.3.2. 3-Acetylpyridine-Induced Cerebellar Deafferentation: Example Protocol

It has been suggested that the neurotoxicity of 3 AP may be species- and strain-dependent (Guidotti et al., 1975; Woodhams et al., 1978). Therefore, care should be taken in the choice of experimental animal to be used. Adult male Sprague-Dawley or Wistar rats weighing 200–250 g are administered a single ip injection of 3 AP (Sigma Chemical Co.) dissolved in physiological saline at a dose of 65 mg/kg body wt. Control rats are administered equal vols of physiological saline and are fed identical diets to the 3 AP treatment group. Within a few hours of 3

AP administration, rats show tremors as well as difficulties in swallowing and breathing. Because of such difficulties, it has been suggested that rats be fed powdered rat chow mixed with sweetened condensed milk (Rea et al., 1980). Using this feeding regimen, fewer than 20% of 3 AP-treated rats died. After a period (up to 4 d) of initial weight loss, surviving animals gain weight and go on to show permanent signs of cerebellar dysfunction that include intention tremors, loss of coordination, and impaired locomotion (Jolicoeur et al., 1979).

Histopathological examination 14–21 d following 3 AP administration reveals almost complete loss of inferior olivary neurons (Desclin and Escubi, 1974; Wiklund et al., 1982).

3.3.3. Other Neurotoxins

Microinjections of *kainic acid* into rat cerebellum elicit a profound destruction of Purkinje, stellate, basket, and Golgi II cells while leaving granule cells unaltered (Herndon and Coyle, 1977). The neurotoxicity of kainic acid involves its interaction with glutamate receptors. In the cerebellum, all neuronal cells (except the granule cells themselves) receive input from the parallel fibers of the granule cells. Therefore, if glutamate is the granule cell neurotransmitter, all cells except the granule cells should possess these glutamate receptors. Kainic acid lesions produced a 50% loss of cerebellar GABA but no change in glutamate, consistent with the possibility that glutamate is the granule cell neurotransmitter and GABA is the neurotransmitter of other cerebellar cells (Tran and Snyder, 1979).

Methylazoxymethanol acetate (MAM) is a potent, short-acting nucleic acid alkylating agent that kills dividing cells. Morphological and neurochemical evaluation of the brains of rats treated with MAM during early fetal life indicates that groups of neurons in the process of developing are eliminated; neurons dividing before or after that date are spared (Johnston and Coyle, 1979). MAM injection into newborn mice leads to a reduction of cerebellar size, decrease in granule cell number, and disorganization of Purkinje cells. It has been suggested that this type of dysgenesis is morphologically similar to certain hereditary cerebellar ataxias (Slevin et al., 1982). Mice whose cerebella were rendered granuloprival by neonatal MAM injections had decreased adult cerebellar concentrations of glutamate consistent with the assignment of this amino acid as the neurotransmitter of cerebellar granule cells.

3.4. Rodent Mutant Models

The last 10 years have seen the appearance of reports describing mutant rodents, particularly mice, with various genetically inherited deficiencies of cerebellar cellular organization. Such models have been particularly useful in studies aimed at understanding cerebellar neuronal interactions and function as well as in the identification of the neurotransmitter of the various types of cerebellar neurons. Mutant rodent models can be conveniently divided into two types: those affecting primarily granule cells and those resulting in deficiencies of Purkinje cells.

3.4.1. Granule Cell-Deficient Mutants

3.4.1.1. Weaver (wv) Mouse

Weaver (wv) is an autosomal recessive mutation that results in an almost complete loss of cerebellar granule cells (Sidman et al., 1965). In normal development, granule cell proliferation in the external granule cell layer of cerebellar cortex in the mouse begins on postnatal days 3–4. Granule cells then migrate deeper into cerebellum past the Purkinje cells to form the internal granular layer. In wv/wv mice, granule cells are lost in the external granule layer. Heterozygote animals suffer a modest loss of granule cells (Hirano and Dembitzer, 1975). Affected wv/wv animals have severe ataxia, hypotonia, and a fine tremor (Sidman, 1968). In the wv/wv cerebellar cortex, glutamate concentrations remain the same during at least the first 10 d postnatally, whereas levels in normal mice increase twofold (Roffler-Tarlov and Turey, 1982) (Table 4).

3.4.1.2. Staggerer (sg) Mouse

Staggerer (sg) is an autosomal recessive mutation that, like weaver, results in severe loss of cerebellar granule cells. However, wv and sg genes are located on different chromosomes and the two diseases follow different pathological courses. In

Table 4
Cerebellar Amino Acid Changes in Ataxic Mutants

Mutant	Granule cell loss	Purkinje cell loss	Amino acid changes			
			Glutamate	Aspartate	GABA	Taurine
Weaver mouse	+++	+	↓	•	↑	↑
Staggerer mouse	+++	+++	↓	NC	NC	NC
Reeler mouse	+++	•	↓	NC	↑	↑
Rolling mouse Nagoya	+++	•	↓	NC	↑	↑
Purkinje cell degeneration mouse	+	+++	NC*	NC	NC	↓
Nervous mouse	•	+++	NC	NC	NC	NC
Gunn rat	+	+++	↓	NC	NC	•

*NC: no change.

sg/sg cerebellum, granule cell proliferation is reduced both in the external granule layer and inner granule layer (cells that have migrated). The death of the internal granule cells may be secondary to failure of synapse formation with Purkinje cells in this mutant (Sotelo and Changeux, 1974). Sg/sg mice have abnormal righting reflexes after postnatal day 10 (Sidman et al., 1962). As was the case with weaver, deficits of glutamate and aspartate appear at postnatal day 10 (Table 4) (Roffler-Tarlov and Turey, 1982). However, in contrast to weaver, sg/sg aspartate values do not recover. Furthermore, concentrations of GABA in deep cerebellar nuclei are significantly reduced in sg/sg at postnatal day 4, indicating lack of full innervation of deep cerebellar nuclei by Purkinje cells in this mutant.

3.4.1.3. Reeler (rl) Mouse

Reeler (rl) is an autosomal recessive mutation that results in a pronounced laminar misalignment of cerebellar cortical neurons (Goffinet et al., 1984). The predominant abnormality in the rl/rl cerebellum is the malpositioning of Purkinje cells. The majority of granule cells are not in their normal positions.

3.4.1.4. Rolling Mouse Nagoya (rol)

Rolling mouse Nagoya (rol) is an ataxic mutant mouse discovered in Nagoya, Japan, in 1973. Inherited as an autosomal recessive trait, the mutant is characterized by severe ataxia 10–14 d after birth. Neurological symptoms and neuropathological findings are less severe than weaver or staggerer. Decreases of granule, basket, and stellate cells have been reported. Purkinje cells are unaffected. Cerebellar concentrations of glutamate are reportedly decreased in rol/rol (Muramoto et al., 1981) consistent with the loss of granule cells in this mutant.

3.4.2. Purkinje Cell-Deficient Mutants

3.4.2.1. Purkinje Cell Degeneration (pcd) Mouse

This mutant mouse loses selectively almost all of its Purkinje cells between postnatal days 20–40. Affected animals have an ataxic gait that is surprisingly mild considering the degree of cell loss in these mutants (Roffler-Tarlov et al., 1979). Amino acids in cerebellar cortex are unchanged, with the exception of taurine, in these animals but GABA content of deep cerebellar nuclei is decreased by 50%, consistent with loss of GABA nerve terminals. Densities of GABA/benzodiazepine receptors, as labeled by ^{3}H-flunitrazepam, are increased in deep cerebellar nuclei of pcd mice, demonstrating that reductions in Purkinje cell number are associated with receptor upregulation in these mutants (Rotter and Frostholm, 1986).

3.4.2.2. Stumbler (stu) Mouse

The stu mouse mutation arose spontaneously in a stock of C3H/HeJ mice at the Jackson Laboratory, Bar Harbor, ME. In the stu mutant, both Purkinje cells and granule cells are decreased in number from postnatal day 9 onward (Caddy and Sidman, 1981).

3.4.2.3. Lurcher (lc) Mouse

Heterozygous mice that carry this autosomal dominant mutation have a rocking gait and impaired balance. Almost all Purkinje cells degenerate between postnatal days 9–30. The major Purkinje cell afferents are also affected; most of the granule cells and 75% of inferior olivary neurons are also lost postnatally (Caddy and Biscoe, 1976).

3.4.2.4. Nervous (nr) Mouse

Nervous mutant mice have a selective degeneration of Purkinje cells with relative sparing of granule cells (Roffler-Tarlov and Sidman, 1978). Measurement of cerebellar amino acids from nr/nr mice revealed no significant reductions of glutamate but reduced GABA levels in deep cerebellar nuclei, consistent with the loss of Purkinje cell nerve terminals in this mutation. In addition to ataxia, the nervous mutant exhibits deficits of exploration and habituation (Lalonde and Botez, 1985).

3.4.2.5. The Gunn Rat

The Gunn rat is a mutant that shows autosomal recessive hereditary unconjugated hyperbilirubinemia (Gunn, 1938). Homozygous j/j rats develop jaundice shortly after birth. Histological evaluation reveals selective loss of Purkinje cells in this mutant. Granule cells are also reduced in number. Measurement of amino acid content of cerebellum from 8-mo-old Gunn mutant rats revealed a selective reduction of glutamate consistent with the loss of granule cells (Mikoshiba et al., 1980).

3.5. Virus-Induced Animal Models of Ataxia

Cerebellum is selectively vulnerable to several viruses. For example, in parvovirus infection, granule cells are lysed, whereas in arenavirus infections, cerebellar lesions result from an immunopathologic process. The parvovirus feline ataxia (FAV) causes a disease in kittens characterized by unsteady, ataxic gait. The ataxic state persists as the animal matures (Kilham and Margolis, 1966). Neuropathological studies reveal loss of the external granule cells. There is severe cerebellar hypoplasia, depletion of granule cells and disorganization of cerebellum with Purkinje cells scattered through all layers (Herndon et al., 1971).

Infection of the neonatal hamster with a parvorirus, rat virus strain PRE 308, leads to destruction of the rapidly-dividing external germinal cells to produce a hypoplastic granuloprival cerebellum. More than 95% of cerebellar granule cells are depleted using this virus and election microscopic evaluation of infected hamster cerebella showed a significant reduction of parallel fiber synapses and granule cell dendrites (Young et al., 1974). Other cell types appeared to be normal. Analysis of free amino acids in cerebella of infected animals revealed a selective decrease of glutamate. Partial granule cell depletions correlated with the decrease of glutamate in this model. It was therefore suggested that these findings support the candidacy of glutamate as granule cell neurotransmitter.

It has been hypothesized that certain hereditary ataxias such as ataxia talengiectasia may have a viral etiology (Weiner et al., 1978).

3.6. X-Irradiation-Induced Loss of Cerebellar Granule Cells

Exposure of rats to low-level X-irradiation during the first two weeks postnatally results in prevention of the acquisition of cerebellar stellate cells and of late-forming granule cells (Altman,

Table 5
Amino Acid Content of Crude Synaptosomal Fraction from Cerebella of Rats: Effects of X-Irradiation

Amino acid	Amino acid concentration, nmol/mg prot.	
	Control	X-irradiation
Glutamate	39.1 ± 2.0	29.5 ± 1.2**
Aspartate	13.6 ± 0.8	11.6 ± 0.4*
GABA	5.7 ± 0.2	6.4 ± 0.3
Taurine	18.2 ± 0.9	17.1 ± 1.0
Glycine	3.5 ± 0.4	4.2 ± 0.3
Alanine	2.1 ± 0.2	1.9 ± 0.1

*$p < 0.05$, **$p < 0.005$ compared to control values by Student's *t*-test (data from Rhode et al., 1979).

1975). Measurement of amino acids in synaptosomal fractions from cerebella of X-irradiated rats revealed a selective loss of glutamate and aspartate (Table 5) (Rohde et al., 1979).

References

Aggerbeck L. P., McMahon J. P., and Scanu A. M. (1974) Hypobetalipoproteinemia: clinical and biochemical description of a new kindred with "Friedreich's Ataxia." *Neurology* **24,** 1051–1063.

Altman J. (1975) Experimental reorganization of the cerebellar cortex. V. Effects of early X-irradiation schedules that allow or prevent the acquisition of basket cells. *J. Com. Neurol.* **165,** 31–48.

Blass J. P., Avignan J., and Uhlendorf B. W. (1970) A defect in pyruvate decarboxylase in a child with an intermittent movement disorder. *J. Clin. Invest.* **49,** 423–432.

Butterworth R. F. (1985) Pyruvate dehydrogenase deficiency disorders, in *Cerebral Energy Metabolism and Metabolic Encephalopathy* (D. W. McCandless, ed.), Plenum, New York, pp. 121–141.

Butterworth R. F. and Giguére J. F. (1984) Amino acids in autopsied human spinal cord: selective changes in Friedreich's Ataxia. *Neurochem. Pathol.* **2,** 7–17.

Butterworth R. F., Hamel E., Landreville F., and Barbeau A. (1978) Cerebellar ataxia produced by 3-acetylpyridine in rat. *Can. J. Neurol. Sci.* **5,** 131–133.

Caddy K. W. T. and Biscoe T. J. (1976) The number of Purkinje cells and olive neurones in the normal and Lurcher mutant mouse. *Brain Res.* **111,** 396–398.

Caddy K. W. T. and Sidman R. L. (1981) Purkinje cells and granule cells in the cerebellum of the Stumbler mutant mouse. *Dev. Brain Res.* **1,** 221–236.

Desclin J. C. and Escubi J. (1974) Effects of 3-acetylpyridine on the central nervous system of the rat as demonstrated by silver methods. *Brain Res.* **77,** 349–364.

DiDonato S., Rimoldi M., Moise A., Bertagnoglio B., and Uziel G. (1979) Fatal ataxic encephalopathy and carnitine acetyltransferase deficiency: a functional defect of pyruvate oxidation? *Neurology* **29,** 1578–1583.

Flint R. S., Rea M. A., and McBride W. J. (1981) In vitro release of endogenous amino acids from granule cell-, stellate cell- and climbing fibre-deficient cerebella. *J. Neurochem.* **37,** 1425–1430.

Goffinet A. M., So K. F., Yamamoto M., Edwards M., and Caviness V. S. (1984) Architectonic and hodological organization of the cerebellum in reeler mutant mice. *Dev. Brain Res.* **16,** 263–276.

Guidotti A., Biggio G., and Costa E. (1975). 3-acetylpyridine: a tool to inhibit the tremor and the increase of cGMP content in cerebellar cortex elicited by harmaline. *Brain Res.* **96,** 201–205.

Gunn C. K. (1938) Hereditary acholuric jaundice in a new mutant strain of rats. *J. Hered.* **29,** 137–139.

Herndon R. M. and Coyle J. T. (1977) Selective destruction of neurons by a transmitter agonist. *Science* **198,** 71,72.

Herndon R. N., Margolis G., and Kilham L. (1971) The synaptic organization of the malformed cerebellum induced by perinatal infection with feline pauleukopenia virus (PLV). I. Elements forming the cerebellar glomeruli. *J. Neuropathol. Exp. Neurol.* **30,** 196–205.

Hirano A. and Dembitzer H. M. (1975) The fine structure of staggerer cerebellum. *J. Neuropathol. Exp. Neurol.* **34,** 1–11.

Huxtable R., Azari J., Reisine T., Johnson P., Yamamura H., and Barbeau A. (1979). Regional distribution of amino acids in Friedreich's Ataxia brains. *Can. J. Neurol. Sci.* **6,** 255–258.

Johnston M. and Coyle J. (1979) Histological and neurochemical effects of fetal treatment with methylazoxymethanol on rat neocortex in adulthood. *Brain Res.* **170,** 135–155.

Jolicoeur F. B., Rondeau D. B., Hamel E., Butterworth R. F., and Barbeau A. (1979) Measurement of ataxia and related neurological signs in the laboratory rat. *Can. J. Neurol. Sci.* **6,** 209–215.

Kilham L. and Margolis G. (1966) Viral etiology of spontaneous ataxia of cats. *Am. J. Pathol.* **48,** 991–1011.

Lalonde R. and Botez M. I. (1985) Exploration and habituation in nervous mutant mice. *Behav. Brain Res.* **17,** 83–86.

McGeer P. L., Eccles J. C., and McGeer E. G. (1978) *Molecular Neurobiology of the Mammalian Brain*. Plenum, New York, pp. 202–206.

Mikoshiba K., Kohsaka S., Takamatsu K., and Tsukada Y. (1980) Cerebellar hypoplasia in the Gunn rat with hereditary hyperbilirubinemia: immunohistochemical and neurochemical studies. *J. Neurochem.* **35,** 1309–1318.

Muramoto O., Kanazawa I., and Ando K. (1981) Neurotransmitter abnormality in rolling mouse Nagoya, an ataxic mutant mouse. *Brain Res.* **215,** 295–304.

Perry T. L., Kish S. J., Hansen S., and Currier R. D. (1981) Neurotransmitter amino acids in dominantly inherited cerebellar disorders. *Neurology* **31,** 237–242.

Plaitakis A., Nicklas W. J., and Desnick R. J. (1980) Glutamate dehydrogenase deficiency in three patients with Spinocerebellar Syndrome. *Ann. Neurol.* **7,** 297–303.

Rea M. A., McBride W. J., and Rohde B. H. (1980) Regional and synaptosomal levels of amino acid neurotransmitters in the 3-acetyl pyridine deafferented rat cerebellum. *J. Neurochem.* **34,** 1106–1108.

Richterich R., Van Mechelen P., and Rossi E. (1965) Refsum's Disease (heredopathia atactica polyneuritiformis): An inborn error of lipid metabolism with storage of 3,7,11,15-tetra-methylhexadecanoic acid. I. Report of a case. *Am. J. Med.* **39,** 230–236.

Robinson N. (1968) Chemical changes in the spinal cord in Friedreich's Ataxia and motor neurone disease. *J. Neurol. Neurosurg. Psychiatr.* **31,** 330–333.

Roffler-Tarlov S. and Sidman R. L. (1978) Concentrations of glutamic acid in cerebellar cortex and deep nuclei of normal mice and weaver, staggerer and nervous mutants. *Brain Res.* **142,** 269–283.

Roffler-Tarlov S., Beart P. M., O'Gorman S., and Sidman R. L. (1979) Neurological and morphological consequences of axon terminal degeneration in cerebellar deep nuclei of mice with inherited Purkinje cell degeneration. *Brain Res.* **168,** 75–95 .

Roffler-Tarlov S. and Turey M. (1982) The content of amino acids in the developing cerebellar cortex and deep cerebellar nuclei of granule cell-deficient mutant mice. *Brain Res.* **247,** 65–73.

Rohde B. H., Rea M. A., Simon J. R., and McBride W. J. (1979) Effects of X-irradiation induced loss of cerebellar granule cells on the synaptosomal levels and the high affinity uptake of amino acids. *J. Neurochem.* **32,** 1431–1435.

Rotter A. and Frostholm A. (1986) Cerebellar benzodiazepine receptor distribution: an autoradiographic study of the normal C57BL/6J and Purkinje cell degeneration mutant mouse. *Neurosci. Lett.* **1,** 66–71.

Schoenberg B. S. (1979) Epidemiology of the inherited ataxias, in *Advances in Neurology, vol. 21* (R. A. P. Kark, R. N. Rosenberg, and L. J. Schut, eds.), Raven, New York, pp. 15–32.

Sidman R. L. (1968) Development of interneuronal corrections in brain of mutant mice, in *Physiological and Biochemical Aspects of Nervous Integration* (F. D. Carlson, ed.), Prentice Hall, Engelwood Cliffs, New Jersey, p. 163.

Sidman R. L., Green M. C., and Appel S. H. (1965) *Catalog of the Neurological Mutants of the Mouse,* Harvard University Press, Cambridge, UK, pp. 66,67.

Sidman R. L., Lane P. W., and Dickie M. M. (1962) Staggerer, a new mutation in the mouse affecting the cerebellum. *Science* **137,** 610–612.

Slevin J. T., Johnston M. V., Biziere K., and Coyle J. T. (1982) Methylazoxymethanol acetate ablation of mouse cerebellar granule cells: effects on synaptic neurochemistry. *Dev. Neurosci.* **5,** 3–12.

Sotelo C. and Changeux J. P. (1974) Transsynaptic degeneration "en cascade" in the cerebellar cortex of staggerer mutant mice. *Brain Res.* **67,** 519–526.

Stumpf D. A., Parks J. K., Eguren L. A., et al. (1982) Friedreich Ataxia: III Mitochondrial malic enzyme deficiency. *Neurology* **32,** 221–227.

Stumpf D. A., Parks J. K., and Parker W. D. (1983). Friedreich's Disease: IV. Reduced mitochondrial malic enzyme activity in heterozygotes. *Neurology* **33,** 780–783.

Tran V. T. and Snyder S. H. (1979) Amino acid neurotransmitter candidates in rat cerebellum: selective effects of kainic acid lesions. *Brain Res.* **167,** 345–353.

Weiner L. P., Herndon R. M., and Johnson R. T. (1978) Animal models of viral-induced ataxia: Implications for human disease, in *Advances in Neurology, vol. 21,* (R. A. P. Kark, R. N. Rosenberg, and L. J. Schut, eds.), Raven, New York, pp. 373–379.

Wiklund L., Toggenburger G., and Cuenod M. (1982). Aspartate: possible neurotransmitter in cerebellar climbing fibres. *Science* **216,** 78–80.

Woodhams P., Rodd R., and Balazs R. (1978) Age-dependent susceptibility of inferior olive neurones to 3-acetylpyridine in the rat. *Brain Res.* **158,** 194–198.

Young A. B., Oster-Granite M. L., Herndon R. M., and Snyder S. H. (1974) Glutamic acid: selective depletion by viral induced granule cell loss in hamster cerebellum. *Brain Res.* **73,** 1–13.

Animal Models for Lesch-Nyhan Disease

Roberta M. Palmour

1. Introduction

In 1964, in the Johns Hopkins clinics, Michael Lesch and William Nyhan saw two brothers with cerebral palsy, movement disorder, and an extremely high plasma uric acid (UA) level. Most striking in these patients was compulsive self-mutilatory behavior (SMB) involving intense biting of the digits and lips. Careful metabolic study showed profound hyperuricemia and hyperuricosuria in this syndrome, now known as Lesch-Nyhan disease (LND). When adjusted for body wt, total daily UA excretion was often 100 times normal (Nyhan et al., 1965). Because of the unusual association of hyperuricemia and a behavioral abnormality, other cases were quickly identified, and an X-linked recessive pattern of inheritance was deduced. In 1967, J. E. Seegmiller and his colleagues reported that erythrocytes from patients with LND possessed less than 0.001% of normal activity for the purine salvage enzyme hypoxanthine-guanine phosphoribosyltransferase (HPRT).

Although there are many defined single-gene human disorders with neurological symptoms or mental retardation, LND is one of the few examples of a specific behavioral phenotype secondary to a known enzymatic deficiency. This may account, at least in part, for the disproportionate research interest in this very rare disorder (prevalence between 1 in 500,000 and 1 in 200,000 live births). LND may also be a prototype for understanding certain types of progressive degenerative disease

From: *Neuromethods, Vol. 21: Animal Models of Neurological Disease, I*
Eds: A. Boulton, G. Baker, and R. Butterworth

(especially those with motor and subcortical impairment), certain forms of cerebral palsy, and some aspects of pathological aggression.

Several animal models of LND have been proffered, all presumably devised with the intent of understanding the behavioral or neurobiological correlates of the disorder. None of these animal models is adequate, but each has allowed some degree of hypothesis testing to proceed. Because any putative animal model for a human disease must be derived from what is known about the disorder, a brief review of some aspects of the biochemistry and genetics of LND is in order. Several detailed reviews of the pathophysiology and molecular genetics of the disorder are also available (Seegmiller, 1980; Kelley and Wyngaarden, 1983; Wilson et al., 1983; Stout and Caskey, 1989).

2. Lesch-Nyhan Disease

2.1. Basic Genetics and Biochemistry

The gene for HPRT, in humans and other mammals, is located on the long arm of the X chromosome (Xq26-27), so that for all practical purposes, only males are affected with LND. HPRT is expressed in all tissues and in both static and rapidly dividing cells; activity is particularly high in brain and in testis (Kelley et al., 1969). HPRT is constitutively expressed and is generally considered to be a "housekeeping enzyme." Complete sequences are known for both HPRT protein and the HPRT gene; there is no evidence for common structural polymorphism (Wilson et al., 1982; Jolly et al., 1983). The gene was cloned independently in two laboratories, and the feasibility of gene therapy is being assessed in several model systems (Brennand et al., 1982; Jolly et al., 1982; Willis et al., 1984; Chang et al., 1987).

Children affected with LND show a virtually complete deficiency of HPRT (Seegmiller et al., 1967). In about one-third of cases, residual immunoreactive HPRT-like protein can be detected, whereas the remainder are completely devoid of the enzyme protein (Wilson et al., 1986). In about 3% of patients with severe gouty arthritis and no neurological abnormality, partial

HPRT deficiency (10–30% of normal activity) has been documented (Kelley et al., 1967). There are also a small number of individuals with HPRT activity between 0.001 and 10% of normal; these persons may display mild to moderate neurological alteration (Kelley et al., 1969; Seegmiller, 1976,1980), as do some patients with other purine enzyme deficiencies (Hirschhorn et al., 1980; Simmonds et al., 1986; Barshop et al., 1988).

Recent reviews document the extensive biochemical and allelic heterogeneity of HPRT variants (Seegmiller, 1981; Kelley and Wyngaarden, 1983; Wilson et al., 1986). From the mass of data concerning molecular analysis of specific HPRT mutations, two points are germane to this chapter:

1. Detailed analysis of protein and DNA from 24 patients showed separate and distinct molecular alterations in at lease 14 cases, indicating that many mutations leading to HPRT deficiency must have arisen independently (Yang et al., 1984; Wilson et al., 1986); and
2. In only two of 17 patients with full LND was there a gene deletion (Wilson et al., 1986; Gordon et al., 1987). Thus, in most cases, a gene or group of genes has not been deleted from the chromosome. Rather, there has been mutation of one or a few nucleotides within the HPRT gene.

In normal cells, HPRT and a similar enzyme known as adenine phosphoribosyltransferase (APRT) salvage preformed purines by converting them to monophosphate nucleosides (Fig. 1). HPRT, therefore, catalyzes the formation of inosinic and guanylic acids from hypoxanthine (Hx) and guanine, respectively. Purines may also be synthesized *de novo* via a 10-step anabolic process that adds single carbon units to a ribose base (Watts, 1983; Wyngaarden and Kelley, 1983). In the first and rate-limiting step, phosphoribosylamine is synthesized from 5'-phosphoribosyl pyrophosphate (PRPP) and glutamine. PRPP also provides the phosphorylribose group for purine reutilization. Purine salvage enzymes have a slightly higher affinity for PRPP than does PRPP-amidotransferase, the first enzyme of *de novo* synthesis. In normal cells, the effective total rate of purine syn-

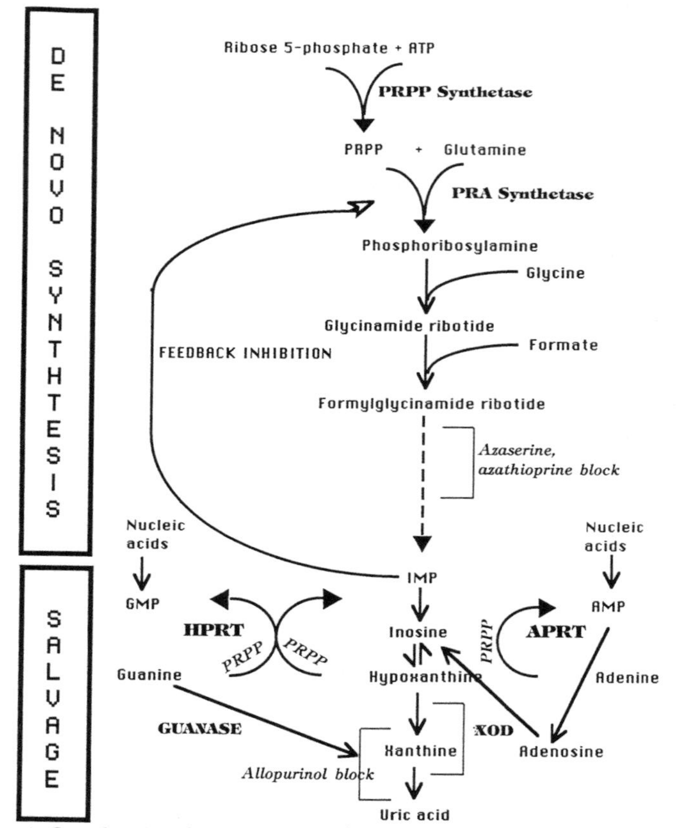

Fig. 1. Synthesis of purines via *de novo* and salvage pathways.

thesis reflects the cellular concentration of PRPP, which is poised to favor salvage synthesis (Fox and Kelley, 1971; Watts, 1983).

In addition, purine nucleotides regulate purine *de novo* synthesis by feedback inhibition of PRPP-amidotransferase (Wyngaarden and Ashton, 1959; Wyngaarden and Kelley, 1987). At least in cultured lymphocytes, this regulation appears to be compartmentalized; inosine monophosphate (IMP) derived from salvage conversion effectively inhibits PRPP-amidotransferase activity, whereas IMP derived from deamination of adenosine monophosphate (AMP) does not (Hershfield and Seegmiller, 1977). Thus, in LND patients and in cells cultured from those patients, despite relatively normal intracellular quantities of IMP, purine *de novo* synthesis is unregulated (Nyhan et al., 1965; Rosenbloom et al., 1968; Wood et al., 1973). This overproduction is attributable partly to elevated intracellular levels of PRPP and partly to failure of salvage synthesis of IMP (Greene at al., 1970; Fox and Kelley, 1971; Hershfield and Seegmiller, 1977). Thus, purines are both overproduced and unsalvaged, and the total body load of inosine, Hx, and UA is elevated by two separate mechanisms.

2.2. Neurological and Behavioral Characteristics

The development of neurological abnormality in patients with LND has been reviewed in detail by several authors (Hoefnagel et al., 1965; Nyhan, 1976; Seegmiller, 1976; Watts et al., 1982; Wilson et al., 1983; Stout and Caskey, 1989). Most importantly, affected males appear to be neurologically normal at birth and in the early postnatal period. By 6 mo of age, however, developmental milestones begin to be delayed, with nearly all cases showing motor impairment upon careful examination. The initial presentation of motor abnormality varies from case to case. One typical pattern is gradual loss of head control, followed by increasing hyperkinesis of arms and legs. In other cases, there may initially be tremors in the extremities that later progresses to chorea. Eventually, all patients develop hypertonia, spasticity, and contractures. Children with LND do not walk and typically must be restrained in bed or in a wheelchair. Death in LND is not a consequence of brain damage. LND patients typically die of kidney failure, pneumonia, or septic infection, most often before the age of 20.

Neuromuscular impairments, including dysarthria and feeding difficulties, pose particular problems for children with HPRT deficiency. For example, the growth retardation that has been noted by many authors may be the consequence of difficult feeding and frequent regurgitation. Calm, consistent nursing care and pharmacological control of hyperuricemia (Nyhan, 1976; Seegmiller, 1976; Mehta et al., 1990) have both been reported to improve feeding considerably. LND patients perform poorly on verbal IQ tests partly because they have problems in making themselves understood. However, at least as infants and young children, their alertness and responsiveness suggest that they are not grossly mentally retarded. Intellectual functioning unquestionably declines as patients become older. This is undoubtedly caused, in part, by poverty of stimulation but may also reflect increasing neuronal deterioration.

SMB occurs in at least 85% of patients with complete HPRT deficiency. Biting may begin as early as 1 yr of age or may be delayed until age 6–7 (Seegmiller, 1976; Watts et al., 1982; Mizuno, 1986). Many patients begin signaling some months in advance of the onset of biting by putting the hand in the mouth but not actually biting. In most cases, SMB intensifies at least through the first decade of active biting.

SMB is relatively specific in these children, in that they will always prefer to bite themselves is possible. However, if biting is made impossible by restraints or extraction of teeth, they may bang their heads, tangle their fingers in the spokes of their wheelchairs, or find other ingenious ways to inflict self-injury. A few patients learn to control their compulsive behavior so that restraints can be removed for periods of time; some also develop enough insight into their behavioral patterns to warn the nursing staff if restraints need to be applied (Seegmiller, 1976; Watts et al., 1982). Stress and psychological disturbance increase the frequency and severity of SMB episodes. At times, LND patients may also inflict their compulsive aggressive behavior on others. Behavioral symptoms typically worsen with age for at least 10 yr.

SMB in LND is not the consequence of congenital insensitivity to pain (Partington and Hennen, 1967). The children cry when they hurt themselves and are obviously frightened by their behavior. If restraints are removed, even for a short time, they will typically begin to cry and even ask to be restrained, but their hands nonetheless go immediately to the mouth. There is also no postmortem evidence of peripheral neuropathy (Sags et al., 1965). The possibility that SMB is a reflection of denervation supersensitivity has been suggested by several investigators (Lloyd et al., 1981; Goldstein et al., 1985; Mizuno, 1986). Knowledge of the specific neurotransmitter profiles in LND and of basic neurophysiological principles support this inference, but it is ethically difficult to explore at the clinical level.

2.3. Metabolic Basis of the LND Phenotype

2.3.1. Neuropathology

The pathogenesis of neurological and behavioral alterations in LND is poorly understood. Electromyograms (EMG), electroencephalograms (EEG), and CAT scans are all unremarkable (Berman et al., 1969; Watts et al., 1982). There have been isolated reports of seizure activity in affected children; these are attributable to hypoglycemia, anoxia, and the like. Although cortical thinning and loss of cerebellar granule cells have been reported in occasional patients, no diagnostic gross or ultrastructural brain alterations have been found at postmortem analysis (Sass et al., 1965; Rosenbloom et al., 1967; Crussi et al., 1969). Brain weight is often low for age. There is no evidence of gliosis.

The pattern of motor dysfunction (Hoefnagel et al., 1968) and the high specific activity of HPRT in normal human caudate and putamen are consistent with the inference that basal ganglionic function is impaired (Rosenbloom et al., 1967; Marsden, 1982). Other clinical features, such as impulsive aggressive behavior toward others, compulsive self-injury, and progressive hydrophobia suggest damage to the amygdala and other parts of the limbic system (Mark and Ervin, 1970). Unfortunately, these regions of the brain have been virtually ignored in studies of LND patients.

2.3.2. Purines and Purine Derivatives

Biochemical understanding of the behavioral phenotype in LND is far from complete. The idea that elevated central nervous system (CNS) UA subserves neurological dysfunction has been negated, as the concentration of UA in the cerebrospinal fluid (CSF) of LND patients consistently falls within normal limits (Rosenbloom et al., 1967; Sweetman, 1968; Sweetman and Nyhan, 1970). In addition, allopurinol, which diminishes serum UA, has no effect on neurological damage or SMB (Nyhan, 1976; Seegmiller, 1976). By contrast, CSF xanthine and Hx concentrations are elevated fourfold or greater in LND patients (Rosenbloom et al., 1967; Harkness et al., 1989), and extracellular levels of Hx in cultured $HPRT^-$ lymphocytes, fibroblasts, and neuroblastoma cells are also much higher than those found for HPRT+ cells (Rosenbloom et al., 1968; Wood et al., 1973a,b; Snyder et al., 1978). CSF guanine levels are normal, presumably as a consequence of abundant brain guanase activity (Seegmiller, 1976). It has been suggested that Hx, which supports oxygen radical formation, might have direct toxic effects in brain (Palmour et al., 1989a). Sweetman and Nyhan (1970) found reduced levels of adenine and adenosine (ADO) in serum and urine from LND patients. Although no known toxic metabolite has been found in CSF from LND patients, adequate contemporary chromatographic analyses of purine derivatives in these samples have not been reported.

Several authors have suggested that neurobehavioral abnormalities in LND may result from inadequate production of an essential metabolite rather than production of a neurotoxin (McKeran and Watts, 1976; Simmonds et al., 1987; Page and Nyhan, 1989). For example, purine *de novo* synthesis in brain might be inadequate to support the demand for guanine nucleotides, which are important regulators of chemical signal transduction. In fact, guanonsine 5'-triphosphate (GTP) levels in erythrocytes from LND patients are reduced about 75%, but intracellular concentrations of adenine and guanine nucleotides are, if anything, elevated in rapidly dividing HPRT-deficient cultured cells (Brenton et al., 1977; Snyder et al., 1978; Simmonds et al., 1987).

Allsop and Watts (1980) have shown that PRPP-amidotransferase is present in rat brain at various stages of development, and Edwards et al. (1989) found that CSF levels of cyclic adenosine monophosphate (cAMP) and cyclic guanosine 5'-monophosphate (cGMP) were elevated, not reduced, in CSF from LND patients. Nonetheless, *de novo* synthesis may not adequately supply the demand for purine nucleotides suggested by high levels of HPRT activity in normal human and primate basal ganglia and limbic brain (Kelley et al., 1969; Rosenbloom et al., 1967; Palmour, unpublished).

2.3.3. Catecholamines and Indoleamines

The presence of dystonia and other extrapyramidal signs occasioned the examination of catecholamine neurotransmitter systems in LND (Lloyd et al., 1981). Indeed, many markers of catecholamine content and function are altered in LND, as sum-

Table 1
Abnormalities of Biogenic Amine Biochemistry in LND

Enzyme or metabolite	Direction of change	References
Dopamine β-hydroxylase	Reduced	Rockson et al., 1974
Monoamine oxidase	Reduced	Breakefield et al., 1976 Skaper and Seegmiller, 1976
Sympathetic response to stress	Reduced	Lake and Zeigler, 1977
Brain DA, HVA, DOPAC	Reduced	Lloyd et al., 1981
CSF DA, HVA	Reduced	Silverstein et al., 1985
NE	Unchanged	Lloyd et al., 1981
Urinary MHPG	Elevated	Jankovic et al., 1988
5HT, 5HIAA in brain	Elevated	Lloyd et al., 1981
CSF 5HIAA	Elevated	Sweetman et al., 1977
CSF 5HIAA	Reduced	Castells et al., 1979

marized in Table 1. Most important was the report of selective loss of dopamine (DA), its metabolites—homovanillic acid (HVA) and 3,4-dihydroxyphenylacetic acid—and the enzymes of DA synthesis from striatum and nucleus accumbens in three brains examined postmortem. Although DA in caudate and putamen was severely reduced (70–90% compared to normal), substantia nigra (SN) cell bodies had normal levels of DA (Lloyd et al., 1981). Norepinephrine (NE) levels were essentially normal, whereas serotonin (5HT) was elevated, perhaps compensatorily, in some regions of the brain. Whether DA loss in LND results from failure of terminal arborization at a critical stage in development or from direct neurotoxicity remains to be established (Stout and Coskey, 1989; Palmour et al., 1990).

There is considerable indirect evidence and preliminary direct evidence to support postsynaptic DA receptor supersensitivity in LND patients (Casas-Bruge et al., 1985; Goldstein et al., 1985; Jankovic et al., 1988; Palmour et al., 1989a). CSF levels of monoamines and their metabolites did not appear to be strongly correlated with levels of UA or Hx (Silverstein et al., 1985). In this series, there was sequential decline of (HVA) and 5-hydroxyindole-3-acetic acid (5HIAA) with repeated sampling, consistent with the progressive neurotoxicity inferred from clinical evidence.

Amino acid levels have been infrequently documented in LND; this is undoubtedly a mistake, considering the cytotoxic effects of excitatory amino acids. In three postmortem brain samples, glutamine was elevated, and glutamate and γ-aminobutyric acid (GABA) were normal; but many other amino acids were reduced (Rassin et al., 1982). In CSF, amino acids were also generally reduced, but again, GABA and glutamate were normal (Harkness, 1989). Benzodiazepine receptor affinity was enhanced in postmortem brain, but GABA stimulation of binding was reduced (Kish et al., 1985).

2.4. What Is the Relationship Between HPRT Deficiency and the Behavioral Phenotype in LND Patients?

2.4.1. Primary and Secondary Changes in Inborn Errors of Metabolism

According to principles originally enunciated by Archibald Garrod (1902), metabolic changes in an inborn error of metabolism should be sufficient to explain the clinical phenotype. Because metabolic processes are both integrated and regulated, changes in the level of critical metabolites may reverberate, causing alterations that are secondary to the primary lesion, as well as those that attempt to compensate for failures of homeostasis (Fig. 2). Although many details of this cascade remain ambiguous for LND, considerable progress has been made nonetheless. For example, we understand quite thoroughly the metabolic basis of gouty arthritis in LND, and rationally treat hyperuricemia in these children with allopurinol rather than with azathioprine, which inhibits *de novo* synthesis, but does not affect unsalvaged purines (Seegmiller et al., 1967; Seegmiller, 1980).

Attempts to identify the metabolic triggers of neurological and behavioral abnormality are still preliminary. An important advance in understanding the neurological profile in phenylketonuria was recognition that elevated levels of phenylalanine secondarily alter the activity of other proteins—amino acid transport proteins, enzymes of neurotransmitter synthesis, and struc-

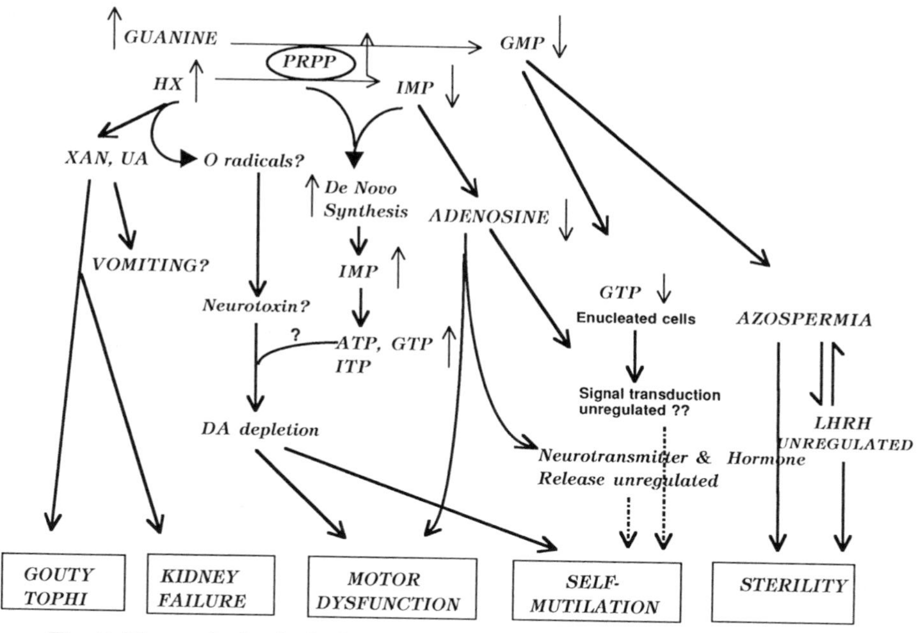

Fig. 2. The pathological phenotype in LND as a function of the cascade of metabolic alterations.

tural proteins. These changes are not the result of associated mutations at additional loci but rather from unbalanced metabolite pools, aberrant feedback regulation, and atypical allosteric regulation. Similarly, in LND, several enzymes of purine or pyrimidine metabolism display altered activities (Seegmiller, 1980). In part, elevated activity derives from an expanded pool of PRPP or, in the case of PRPP-amidotransferase, from failure of compartmentalized IMP feedback inhibition (Hershfield and Seegmiller, 1976). Monoamine oxidase and dopamine-β-hydroxylase are also altered in activity; but for these enzymes, we cannot link the changes to the presence or absence of a specific metabolite. Alterations of this type occur in metabolic diseases of many types and are said to be secondary to the primary genetic lesion.

Secondary alterations of receptors and other membrane proteins may be integral to an understanding of the phenotype in inborn errors of metabolism that affect the brain. As part of their normal function, these proteins respond to changes in the

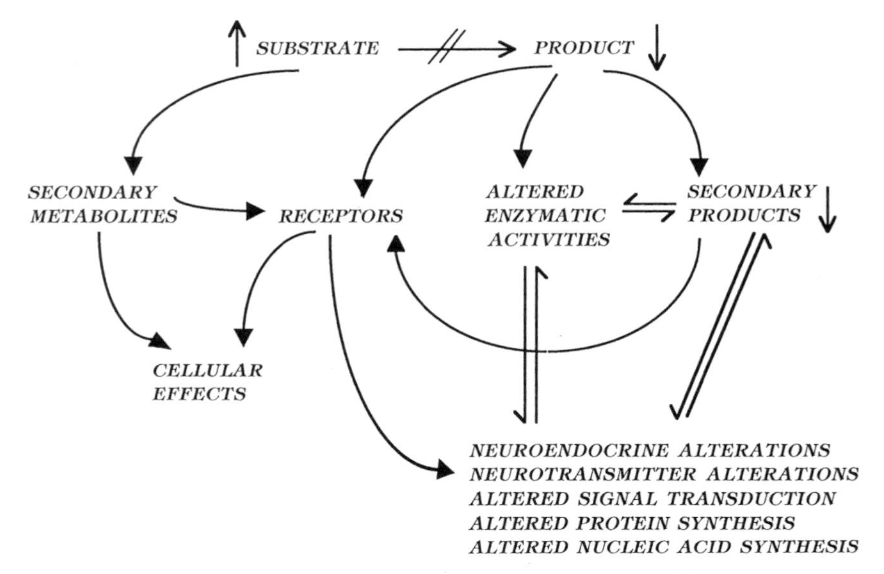

Fig. 3. Alterations of macromolecules and physiological function secondary to a defined inborn error of metabolism.

extracellular environment, either increasing or decreasing responsiveness to maximize inadequate transmission or minimize excessive stimulation. Secondary alterations of receptors or other signal transduction components might be expected in those inborn errors of metabolism characterized by increased or decreased concentrations of hormones, neuromodulators, or neurotransmitters. A simplified scheme of the complex nature of secondary and tertiary modifications is presented in Fig. 3.

2.4.2. HPRT Deficiency and SMB

If one takes the literature at surface value, it is clear that not all investigators are persuaded that SMB is necessarily the consequence of HPRT deficiency. The alternative, that the neurobehavioral phenotype is either a coincidence or the result of an associated but different mutation, is not supported by either clinical or genetic evidence. First, virtually all children with complete HPRT deficiency self-mutilate if they live long enough. The phenotype does not display much variability of expression, in contrast to many other (especially dominant) inherited diseases. Second, of 24 carefully studied cases, there was evidence for at least 14 unique mutational events (Wilson et al., 1986). The probability of chance association between HPRT deficiency and SMB in 14 separate mutations would be astronomically small. Only two patients had identifiable gene deletions (Wilson et al., 1986; Gordon et al., 1987). Consequently, it is not possible to argue that adjacent deleted genes account for DA depletion or SMB. Third, the degree of neurological impairment in complete or partial deficiencies of HPRT (Table 2) directly parallels the level of residual enzymatic activity (Page et al., 1981; Page and Nyhan, 1989).

Cases of partial HPRT activity may be particularly instructive with regard to the issue of a causal relationship between HPRT deficiency and the full LND phenotype. Most patients with complete LND have less than 1% activity, whereas most individuals with greater than 5% HPRT activity have no behavioral symptoms (Table 2). Gout or gouty arthritis is usually the only symptom if enzymatic activity exceeds 10% of normal (Seegmiller, 1978,1980). In some kindreds with HPRT activity of from 1 to 5% of normal, both gouty symptoms and neurological (motor) alterations are seen, but there is no SMB (Kelley et al., 1969; Seegmiller, 1980). In a single kindred, there is evidence for aggressive and self-injurious behavior but not the classical compulsive biting associated with about 25–30% normal HPRT activity.

Thus, in general, there is very good correlation between quantitative enzymatic activity and the clinical phenotypes (Seegmiller, 1980; Page et al., 1981; Page and Nyhan, 1989). However, both types of exceptions to this rule have been reported. Several kindreds with extremely low erythrocytic activity (using standard assay conditions) but a mild phenotype have an unstable enzyme that functions adequately in nucleated cells (Page et al., 1981; Page and Nyhan, 1989) or have mutations at cofactor binding sites (Seegmiller et al., 1980; Kelley and Wyngaarden, 1983). A few cases with higher residual enzymatic activity and a severe phenotype have low in vivo enzymatic function.

Table 2
Metabolic Abormalities in LND Patients

Symptoms	References
Purines elevated: Hx, xanthine, inosine, UA:	Many authors
Phosphoribosylpyrophosphate elevated	Wood et al., 1976
Purines overproduced: IMP→ATP, GTP, ITP	Many authors
Pyrimidines overproduced: UTP, TTP, CTP	Many authors
Activities of several enzymes increased	Seegmiller, 1976,1980
Adenine, ADO down in urine, CSF	Sweetman and Nyhan, 1970
Second messengers: GTP reduced in erythrocytes:	Simmonds et al., 1987
ATP, GTP elevated in lymphocytes	Snyder et al., 1978
cAMP, cGMP elevated in CSF	Edwards et al., 1989
Dopamine, DOPAC, HVA	
Down 70–90% postmortem	Lloyd et al., 1981
Down over time in CSF	Silverstein et al., 1985
Acetylcholine, Choline acetyltransferase	
Reduced 15–25% postmortem	Lloyd et al., 1981
Amino acid transmittters glutamine elevated; GABA, glutamate normal; other amino acids down:	Lloyd et al., 1981; Kish et al., 1985; Harkness et al., 1989
Urinary MHPG down:	Jankovic et al., 1989
Brain 5HT, 5HIAA elevated:	Lloyd et al., 1981
Urinary 5HIAA both down or up:	Sweetman et al., 1977 Castells et al., 1979
CSF 5HIAA down in longitudinal series:	Silverstein et al., 1985
DA receptors: functional supersensitivity; locus uncertain but probably not binding site	Casas-Bruge et al., 1985 Goldstein et al., 1985
ADO receptors: Functional supersensitivity A1	Palmour et al., 1989a
Benzodiazepine receptors: Increased receptor affinity	Kish et al., 1985

2.5. Diagnosis and Treatment

In the absence of a family history of LND, diagnosis before the age of 1 year is most often made by accident. In sophisticated medical settings, patients may come to biochemical attention through a workup for either developmental delay or a severe

Table 3
Degree of Residual HPRT Activity
and Associated Symptomatology in LND and Gout*

Residual HPRT activity	Gout	Motor impairment	Mental retardation	Self-mutilation
10–60% of normal	Yes	No	1 case	1 case
5–10% of normal	Yes	Sometimes	Few cases	No
2–5% of normal	Yes	Most yes, 1 no	20%	No
1.6–2% of normal	Yes	Yes	3 yes, 1 no	50%
0–1.6% of normal	Yes	Yes	Yes	Yes

*Sources: Page et al., 1981; Page and Nyhan, 1989; and Seegmiller, 1980.

urinary tract infection. Otherwise, the initial diagnosis is typically cerebral palsy (Mizuno, 1986; Seegmiller, 1976). However, experienced mothers or those alert to the possibility of LND by virtue of a previous family history will often report UA crystals in the diaper, as well as increased irritability and progressively decreased coordination after about 3 mo of age. Prevention of LND through prenatal diagnosis and selective abortion has been a reality for some time (Van Heeswijk et al., 1972). HPRT is expressed both in amniocytes and in choronic villi, so that prenatal diagnosis can be achieved in the first trimester (Stout and Caskey, 1989). Determination of enzymatic activity is far more reliable an indicator of HPRT deficiency than RFLP analysis, as no single haplotype is reliably associated with enzymatic deficiency (Wilson et al., 1986; Stout and Caskey, 1989).

Understanding the molecular nature of LND has led to excellent therapies for hyperuricemia and gout. Unfortunately, there is not yet an effective treatment for the neurological and behavioral manifestations of the disease. The beneficial effect of diazepam in reducing spasticity and contractures was recognized early in clinical experience (Seegmiller, 1976; Mizuno, 1986), but the current approach to SMB is physical restraint. Attempts to increase the levels of GMP, IMP, or cGMP (with adenine) have had little therapeutic value and toxic side effects (Stout and Caskey, 1989). Treatment with major antipsychotic medications (chlorpromazine, haloperidol) had no specific beneficial effect (Partington and Hennen, 1967; Watts et al., 1974), but Goldstein et al. (1986) have recently reported that fluphenazine, in a placebo-controlled trial, reduced SMB and irritability. In one patient, L-dopa + carbidopa was therapeutic, but in a second patient, it exacerbated dystonia (Watts et al., 1982; Jankovic et al., 1988). Oral 5-hydroxytryptophan has been reported to ameliorate or to have no effect on SMB, but it did improve sleep and reduce irritability (Mizuno and Yugari 1974; Frith et al., 1976; Anders et al., 1978; Nyhan et al., 1980). Propanolol likewise reduces irritability but does not block SMB (Rosenblatt and Palmour, unpublished observations). Bone marrow transplantation was attempted under less-than-ideal circumstances and with very equivocal results (Nyhan et al., 1986). We recently demonstrated a reduction of urinary and CSF purine levels after an experimental trial of microencapsulated xanthine oxidase given parenterally (Palmour et al., 1989).

3. Criteria for an Animal Model of LND

Animal models for human disease are generally useful in basic research because they either facilitate testing of hypotheses concerning the pathophysiological processes or they provide a substratum for trials of therapeutic agents. In developing relevant disease models, criteria for success will vary both with the specific disorder and with the question(s) under investigation. For an inborn error of metabolism with behavioral manifestations, the ideal model would probably be a nonhuman primate with the same enzymatic deficiency as patients with the disease, as this would preserve both biochemical specificity, evolutionary similarity, and the richness of the behavioral repertoire. Useful models to date are far short of this ideal. Despite many false starts, it is clear in retrospect that those models that have advanced our understanding of SMB in LND are marked by either behavioral or biochemical specificity. Those few models that continue to be pursued are specific at both levels.

In evaluating putative animal models of LND, the following questions may be useful guides:

1. Is SMB produced selectively, or does it occur primarily as the end point of a continuum of agitated motor behaviors;

2. Is HPRT activity reduced;
3. Is *de novo* purine synthesis increased; and
4. Are the secondary metabolic consequences of HPRT deficiency, such as, neuronal DA depletion, present?

4. Pharmacological Models of LND

Pharmacological models for LND may be conceptually subdivided into those in which SMB is produced by the administration of a purine derivative, such as, caffeine, and those in which SMB is induced by drugs that act directly or indirectly as catecholamine agonists. Most of the drugs tested act both centrally and peripherally, and many increase the general level of arousal. It has been difficult to demonstrate behavioral specificity with any model that relies solely on drug administration.

4.1. Methylxanthines

Self-biting was first reported in rats and rabbits following acute or chronic injections of very large doses (170–200 mg/kg/d) of caffeine (Boyd et al., 1965; Peters, 1967; Morgan et al., 1970; Sakata and Fuchimoto, 1973). These regimens do produce self-biting in individual animals, but the incidence of the behavior is low (Morgan et al., 1970; Seegmiller, 1976; Palmour and Pearce, 1979). In our own acute studies, all animals showed elevated motor activity and stereotypy at lower doses and earlier time points than those required for SMB (Palmour and Pearce, 1979). However, in rats treated repeatedly with a slightly lower dose of caffeine (140 mg/kg/d for 10 d), 40% exhibited SMB by d 10 in the absence of increased locomotion and stereotyped grooming (Mueller et al., 1982). Other investigators have administered methylxanthines (MX) either in the drinking water or as part of a liquid or pelleted diet. Using this approach, maximum doses of around 200 mg/kg/d caffeine can be achieved (Ferrer et al., 1982). In several stuties, theophylline elicited SMB more reliably and at lower doses than did caffeine, whereas theobromine produced no SMB and a minimal change in motor activity (Morgan et al., 1970; Ferrer et al., 1982).

Factors influencing the frequency and intensity of SMB after MX administration include, at a minimum, dosing regimen and route of administration, species and strain, sex, and general level of nutrition. In our hands, repeated administration has been more effective than continuous exposure, but there are differences of opinion on this subject (Ferrer et al., 1982; Mueller et al., 1982; Palmour et al., 1990a). Rats are generally the preferred species, if only because mice often die at doses of MX below the threshhold for SMB. In most studies, males are more susceptible to SMB than females (Peters, 1967; Mueller et al., 1982). Fischer 344 and Long-Evans rats generally display more SMB than Sprague-Dawley or Wistar strains, but the absolute incidence is subject to environmental manipulation (Lloyd and Stone, 1981). Many other reports note increased susceptibility to MX effects in food-deprived animals (Boyd et al., 1965; Peters, 1967; Sakata and Juchimoo, 1973). The appearance of SMB was also potentiated by providing an isocaloric liquid diet formulated to inhibit the activity or reduce the synthesis of enzymes that normally metabolize MXs (Alvares et al., 1976; Palmour and Pearce, 1976). When high doses of caffeine or theophylline are administered in the diet, rats progressively lose weight and will invariably die of malnutrition if the experiment is continued (Ferrer et al., 1982; R. M. Palmour and C. J. Lipowski, unpublished observations).

Caffeine (1,3,7-trimethylxanthine), theophylline (1,3-dimethylxanthine), and theobromine (3,7-dimethylxanthine) share the purine ring nucleus with Hx and guanine, the normal substrates of HPRT. If these compounds were competitive inhibitors of HPRT, then acute or chronic administration might effectively lower in vivo activity of the enzyme. In the absence of an authentic animal model (i.e., one with inherited HPRT deficiency), a biochemical model with reduced enzymatic activity might be quite valuable. In an early survey, Lau and Henderson (1971) found caffeine and theophylline to be weak inhibitors of HPRT activity. However, more detailed examination shows that caffeine and theophylline are competitive with respect to Mg^{2+} rather than substrate and that the maximum inhibition achievable, even at very high doses of MX, is on the order of 30% (Palmour et al., 1990a). In vivo, we found no change in HPRT activity of any rat brain region after repeated daily doses of theophylline, whereas Minana et al. (1984) found a 20–50% increase in brain HPRT activity after 10–12 d of dietary caffeine (Palmour et al., 1990a).

The hypothesis that high-dose MX administration produces hyperuricemia or increased *de novo* purine synthesis has also been considered. Brain levels of UA, purines, and purine nucleosides are not significantly altered by daily theophylline treatment (Palmour et al., 1990a); but CSF levels, which perhaps more accurately reflect turnover and release, have not been documented. Both 3',5'-(cAMP) and 3',5'-(cGMP) are elevated by some acute or chronic MX administration protocols, but this is not uniform (Kiebling et al., 1975; Sprugel et al., 1977; Stefanovich, 1979). In cultured cells, high-dose caffeine inhibits purine biosynthesis, perhaps by blocking an early step in the *de novo* pathway (Waldren and Patterson, 1979). Caffeine also reportedly inhibits pyrimidine biosynthesis in cultured cells (Rumsby et al., 1982). In the studies of Minana et al. (1984), selected brain enzymes of both *de novo* purine and pyrimidine synthesis exhibited increased activity.

The role of MXs as antagonists at cell-surface ADO receptors was not considered in most of these early studies (Smellie et al., 1979; Daly et al., 1981). However, doses of MX that elicit SMB are in excess of 100 mg/kg/d, whereas MXs act as ADO receptor antagonists at doses ranging from 10–60 mg/kg/d. MXs are now known, of course, to modulate the release of catecholamines, 5HT, acetylcholine, and excitatory amino acids, but there are no published reports with respect to brain or CSF levels of transmitters following administration of very high doses of MX (Stone, 1981; Dunwiddie, 1985; Williams, 1987). However, MX-induced SMB can be attenuated by potent ADO agonists, such as *N*-ethylcarboxamidoadenosine (R. M. Palmour, C. J. Lipowski, and T. W. Heshka, unpublished observations).

4.2. Dopaminergic Agonists

Stereotyped behavior, which sometimes proceeds to self-biting, is a classic action of drugs that act directly on DA receptors or that stimulate release of endogenous DA (Randrup and Munkvad, 1967; Lal and Sourkes, 1974). A dose relation for increased motor activity, stereotyped behavior, and SMB can be demonstrated with L-dopa, apomorphine, methylphenidate, and amphetamine (Lal and Sourkes, 1974; Taylor et al., 1974; Moore and Kelly, 1976). Again, SMB will occur in only a proportion (typically 10–40%) of animals treated with suprapharmacological doses of the drug. This type of SMB is typically accompanied by licking, obsessive self-grooming, and rearing, and is blocked by haloperidol, pimozide, and other DA antagonists. Mueller and colleagues have suggegted that both acute pemoline and continuous-release amphetamine pellets (Mueller and Hsiao, 1980; Mueller and Nyhan, 1981; Mueller and Nyhan, 1982,1983) produce persistent SMB with minor interference from motor agitation and other stereotypic behaviors. Pemoline is thought to act primarily on DA systems, both directly at the receptor and indirectly to release endogenous stores. Chronic amphetamine has been shown by some workers to be neurotoxic to DA terminals (Wagner et al., 1980a,b). Both drugs may also act to some extent on serotonergic pathways.

At very high doses, the α-adrenoceptor partial agonist clonidine, may stimulate self-biting (Razzak et al., 1975), and at lower doses, it potentiates amphetamine and apomorphine-induced stereotypy. Clonidine selectively potentiates caffeine- or amphetamine-induced SMB at the expense of locomotion and other stereotyped behaviors (Mueller and Nyhan, 1983). According to some investigators, the potentiation of SMB by clonidine is a consequence of antagonist actions at ADO receptors (Katsuragi et al., 1984). Suprapharmacological doses of several other classes of drugs—muscimol, morphine, enkephalins, and atropine—have been reported to elicit SMB in isolated instances.

With the possible exception of chronic amphetamine, none of these drugs is known to produce either neuronal DA depletion or persistent alterations of purine metabolism. After repeated or chronic amphetamine administration, DA and NE levels are reduced, DA terminals are swollen, tyrosine hydroxylase activity is diminished, and DA reuptake sites are diminished by 50 (Ellison et al., 1978; Segal et al., 1980; Wagner et al., 1980a). However, brain catecholamine levels slowly return to normal 3–6 mo after discontinuation of amphetamine administration. Repeated methamphetamine administration also produced prolonged DA and NE depletion (Wagner et al., 1980b). Recent evidence suggests that any dramatic increase in motor activity (including vigorous exercise) is likely to be accompanied by increased purine turnover and thus, by transient elevations of inosine, Hx, and UA, all of which are increased in LND patients. The possible long-term effects of catecholaminergic agonists on brain purine pathways has not yet received adequate attention.

4.3. Perspective

What is to be learned from these purely pharmacological models? First, one can infer that the potential for SMB is present in the normal brain and that this potential may be unmasked, or perhaps disinhibited when the endogenous neurochemical balance is disturbed, as it certainly is by effective regimens. Second, careful study of the details of pharmacologically induced SMB, even in the rat, reveals some important parallels with the clinical phenotype. In LND patients, there is a definite developmental progression that begins with disordered and involuntary movements and only later proceeds to SMB and other compulsive behaviors. Similarly, in pharmacological animal models of LND, there is a sequence of behaviors beginning with locomotion and progressing through stereotypy to SMB, temporally compressed by comparison to the clincal ontogeny, but nonetheless dose- and time-dependent. The incidence and severity of SMB in animal models is exacerbated by behavioral or physiological stress, just as it is in patients. Also, SMB in animal models typically can be blocked by DA antagonists or ADO agonists at doses much lower than those required for reversal. Indeed, once the most dramatic symptoms of SMB emerge, only anesthesia or immobilizing doses of drug will terminate behavioral expression. Likewise, in caring for children with HPRT deficiency, it is often possible to prevent an episode of SMB but much more difficult to interrupt one without using physical force.

Sophisticated pharmacologists would refuse to draw any conclusions with respect to neurochemical substrata of these models, given the massive doses of drug used to produce SMB. However, most of the drugs that induce SMB are (at more moderate doses) specific DA agonists or ADO antagonists. In some cases, drugs that elicit SMB are thought to influence both DA and ADO transmission. Because of what we now know about the neurochemistry of LND, perhaps this is not coincidental. More interesing is the increasing evidence from basic studies (detailed in a later section) of a complex and dynamic interrelationship between dopaminergic and purinergic pathways in the brain, that is likely to be quite relevant to the neurochemistry of LND.

5. Dopamine Depletion and Other Lesion Models

5.1. Background

Another rational approach to an animal model for LND is based on postmortem evidence that children with LND have progressive DA depletion in striatal brain regions (Lloyd et al., 1981). Surgical or chemical lesioning of brain DA pathways was originally utilized in investigations of the anatomy, biochemistry, and function of brain DA systems; and an extensive body of knowledge predated its use as a model for LND (reviewed by Moore and Kelly, 1976). In mammals, the cell bodies of DA neurons are located in several overlapping brainstem nuclei that project into striatal and limbic brain regions and onto the frontal cortex (reviewed by Roth et al., 1987). The nigrostriatal pathway contains more than 90% of brain DA in rodents and humans. Most of the DA in the nucleus accumbens and a portion of the DA in the amygdala derives from the mesolimbic pathway, which originates in the ventral tegmental area. Projections from a third group of cells in the retrorubral field of the midbrain also innervate accumbens, central amygdala, and limbic cortical regions. However, recent studies suggest that these rather tidy relationships are oversimplified and that many DA-containing regions receive input from more than one nuclear area (Fig. 4).

If mesolimbic DA pathways are interrupted surgically or chemically, relatively low doses of amphetamine or direct DA receptor agonists produce stereotyped behaviors and in 5–10% of rats, perseverative self-grooming and self-biting (Lal and Sourkes, 1973,1974; Moore and Kelly, 1976). These lesions thus reduce the threshhold for drug-induced SMB. Locomotion is

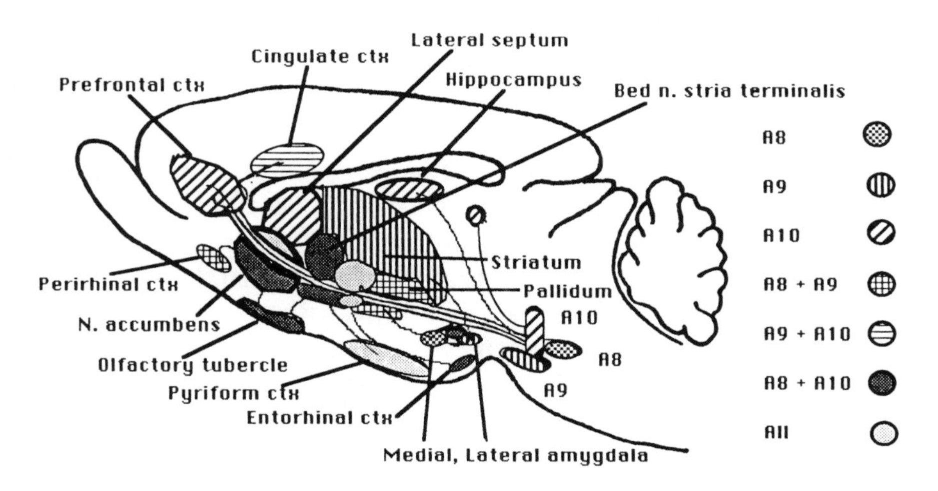

Fig. 4. Dopamine pathways in rat brain.

increased by amphetamine after selective lesioning of nigrostriatal pathways, but stereotypy is not. Also, microinjection of DA into caudate increases stereotypy, whereas microinjection into nucleus accumbens elicits locomotion. The nigrostriatal pathway is thus thought to subserve stereotypy and the mesolimbic pathway, locomotion (Moore and Kelley, 1976).

The integrated functioning of the nigrostriatal and mesolimbic pathways has also been examined. Administration of apomorphine after a unilateral lesion of the nigrostriatal pathway generates rotation ipsilateral to the lesion, whereas amphetamine stimulates contralateral rotation (Ungerstedt and Arbuthnott, 1971; Moore and Kelley, 1976). This has been interpreted in the context of denervation supersensitivity of postsynaptic receptors (Creese et al., 1977), but it has not always been possible to demonstrate receptor proliferation directly (Breese et al., 1987). In animals lesioned in both sides of the mesolimbic tract and in one side of the nigrostriatal tract, the rotational effects of apomorphine are enhanced, and those of amphetamine are suppressed, suggesting that the mesolimbic system may modulate nigrostriatal output (Moore and Kelley, 1976).

An important tool for these studies is 6-hydroxydopamine (6-OHDA), which enters the cytoplasm via the catecholamine reuptake system (Kopin, 1987). Once internalized, 6-OHDA is metabolized by amine oxidase, generating toxic oxygen radicals (Ungerstedt, 1965; Heikkela and Cohen, 1972). The effects of 6-OHDA are restricted to catecholamine neurons (Uretsky and Iversen, 1976). This can be further refined by pretreating animals with inhibitors of the NE reuptake system so that only DA neurons are destroyed, or by adjusting dosing so that NE neurons are preferentially affected (Breese and Traylor, 1960). Additional specificity may be introduced by applying 6-OHDA to restricted brain regions and by optimizing vol and concentration to minimize nonspecific tissue damage (Vaccarino and Franklin, 1984).

5.2. Dopamine Depletion in Neonatal Animals

The first deliberate attempt to utilize selective DA depletion as a model for behavioral manifestations of LND was carried out in the laboratory of Breese et al. (1984). Breese reasoned that there might be important differences in the behavioral consequences of 6-OHDA lesioning in neonatal and adult rats and so utilized a paradigm in which lesions are placed at 5 d of age. In an initial study, these investigators established that stereotypy and SMB were elicited in neonatally lesioned rats by low doses of L-dopa (30–100 mg/kg) or apomorphine (1 mg/kg), which do not effectively elicit stereotypy or SMB in adult rats (Breese et al., 1984,1985). For example, at 100 mg/kg L-dopa, 65% of neonatally treated rats showed SMB, whereas sham-operated animals or those lesioned as adults did not. The incidence of sniffing and rearing did not differ between adult and neonatally lesioned groups, whereas paw treading, head nodding, and "taffy pulling" were all enhanced in the neonatally lesioned group. Similar patterns of behavior were seen after apomorphine administration.

As used in these studies, 6-OHDA produced a selective depletion of DA and its metabolites in all brain regions in rats treated either as neonates or as adults. However, in neonates, there was greater sparing of DA of the SN (35Z vs 3.5Z). Although not noted by Breese et al. (1984), the catecholamine profile of neonatally treated animals is closer to that seen in LND patients. Rats that exhibited SMB showed more profound depeletion of DA, especially in terminal areas, than neonatally treated rats, which did not bite. Neonatal depletion of brain 5HT or NE did not elicit SMB in any animals. This model thus displays two important advantages: It has chemical similarity, albeit at a secondary level, to LND; and SMB can be stimulated with relatively low, presumably specific doses of drug.

Breese and his associates have further explored several pharmacological and biochemical characteristics of rats lesioned neonatally with 6-OHDA. Both advantages and limitations of the preparation as a model for LND have thus been revealed. Predictably, both L-dopa-induced stereotypies and increased locomotion were antagonized by haloperidol (Breese et al., 1984; Duncan et al., 1987). SCH 23390, a D1 DA antagonist, blocked SMB and stereotypies more effectively than did haloperidol or pimozide (Duncan et al., 1987). In keeping with this finding, a small dose of SKF 38393, a D1 DA agonist, enhanced locomotion in neonatally treated animals to a greater degree than it did in rats treated with 6-OHDA as adults. The reverse was true for LY 171555, a D2 selective agonist. SKF 38393 stimulated SMB in 502 of neonatally lesioned animals, whereas LY-171555 was ineffective (Breese et al., 1985). These data were taken to indicate that D1 DA receptor were sensitized in neonatally treated animals, whereas in adults, D2 receptors were preferentially affected. This finding is consistent with the report of improved behavior in one LND patient after fluphenthixol treatment (Goldstein et al., 1986). Despite behavioral evidence of supersensitivity, rats lesioned neonatally showed no change in D1 receptor number even after a 2-wk treatment with haloperidol; this treatment effectively elevated SCH 23390 binding in both unlesioned rats and rats lesioned as adults (Breese et al., 1987). Thus, functional supersensitivity in this model would appear to involve components other than postsynaptic receptors.

Another parallel with the clinical findings in LND is the report that 5HT concentrations in striatum were increased when DA was maximally depleted (Towle et al., 1989). Neonatally lesioned rats have normal levels of HPRT and purine nucleotides but are hypersensitive to the locomotor effects of theophylline, an ADO receptor antagonist (Criswell et al., 1987). Supersensitivity to theophylline was greater in rats lesioned as adults and appeared to be secondary to presynaptic release of DA. SMB stimulated by L-dopa was blocked by cerebral microinjection of *N*-ethylcarboxamideadenosine, an ADO A2 receptor agonist. In an earlier study, Goldstein et al. (1973) found that striatal HPRT activity was normal in rats after 6-OHDA lesions to the nigrostriatal pathway but declined after quinolinic lesions. To the authors, this suggested that HPRT activity was confined to DA containing intrastriatal neurons. This conclusion is unfounded, since those regions of the brain containing DA cell bodies were not examined, and HPRT is an enzyme found predominantly in cell bodies, rather than in neuronal terminals.

Neonatal 6-OHDA lesioning as a model for LND has many advantages, not the least of which is its reliability: At least one-half of the individuals in any treated group exhibit SMB. In addition, it has been characterized in far greater pharmacological and biochemical detail than other models. Among the important parallels to LND are the developmental appropriateness of the model and the apparent D1 DA specificity. Also, there is some degree of change in serotoninergic systems. However, the onset of the syndrome is very fast by comparison to the actual disease process, and spontaneous SMB is not seen. Neonatal lesioning does not allow specificity of target, so that it is not possible to lesion terminals while sparing cell bodies to the extent seen in LND. Finally, this preparation models the symptoms, not the disease and thus, can be used to test new compounds that might alleviate the symptoms but not those that might halt the progression of LND.

5.3. Dopamine Depletion in Adult Animals

There are at least two other attempts to utilize DA lesioning as a model for some aspects of LND; both pertain to neuroanatomical issues. In a single report, Hartgraves and Randall (1986) showed that rats with 6-OHDA lesions in the tail of the caudate would self-mutilate after subcutaneous administration of apomorphine or other DA agonists, in contrast to those lesioned in nucleus accumbens. SMB and stereotypy were blocked by sulpiride, halopridol, or SCH 23390. The authors proposed that these behaviors were thus a unique consequence of striatal DA depletion, rather than an exaggeration of stereotyped behavior. We also found that the incidence and severity of pemoline-induced SMB was potentiated by 6-OHDA lesions placed in caudate or amygdala and decreased (by comparison to sham-operated groups) by lesions to the nucleus accumbens (Palmour et al., 1990). Animals lesioned throughout the brain via intracerebro-ventricular application of 6-OHDA did not differ, with respect to SMB, from sham-operated controls. SMB could be blocked or reduced by pretreatment with haloperidol, SCH 23390, or *N*-ethylcarboxamidoadenosine.

In our series, SMB was exceptionally intense after 6-OHDA lesions to the medial aspect of the amygdala but was completely blocked by lesions to the lateral aspect of the amygdala (Palmour et al., 1990b). This observation may be interpreted within the context of human studies showing that electrical stimulation of the corticomedial amygdaloid nucleus is calming, whereas that of the basolateral amygdala is dysphoric and in the presence of an appropriate target, may produce attack behavior (Mark and Ervin, 1970; Adamec and Adamec, 1986). Moreover, microiontophoretic studies show that cells of the basolateral amygdala are extremely sensitive to DA inhibition (Ben-Ari and Kelly, 1976). These findings emphasize the importance of limbic contributions to SMB and might suggest that SMB is suppressed in older LND patients in parallel with the progression of DA depletion (i.e., as DA depletion goes to completion, the imbalance between functional supersensitivity and loss of DA inhibition declines).

5.4. Ventral Tegmental Lesions in Primates

Several groups of investigators have provided information regarding motor dysfunction and SMB in primates with lesions to various portions of the nigrostriatal DA pathways. A very important issue is that movement disorder but not SMB occurs without pharmacological challenge in these preparations, unlike the situation for any rodent model thus far reviewed. Similarly, rodent models for cerebral palsy simply do not exist, as anoxia and other precipitants of cerebral palsy in humans produce no spontaneous motor dysfunction in rodents.

The primate model most thoroughly explored within the context of LND is that of Goldstein et al. (1973,1986). In African green monkeys lesioned at 2–4 yr of age with a unilateral surgical cut to the ventromedial tegmental nucleus, tremors and fine motor abnormality are seen shortly after surgery. These symptoms may be reduced by L-dopa and DA agonists that in turn, elicit dyskinesias. These dyskinesias may take the form of increased restlessness and aggression or of stereotyped behaviors: lip smacking, biting of forelimbs, and chorea (Goldstein et al., 1973). After periods ranging from 10–14 yr, SMB and dyskinesias could again be revealed following administration of DA agonists, including L-dopa and apomorphine (Goldstein et al., 1986). Biting was restriced to the contralateral forelimb, and spasticity was most marked in the contralateral hindlimb. There was some specificity for D1 agonists, and substantial selectivity of blockade by D1 antagonists. Small doses of progabide or muscimol (both GABA agonists) potentiated SMB and dyskinesias, whereas larger doses of these drugs reduced them (Lloyd et al., 1985). Kanazawa et al. (1990) also showed that kainic acid lesions followed by L-dopa challenge will produce chorea and involuntary motor stereotypies in macaques. The incidence of chorea was correlated with increased activity of tyrosine hydroxylase on the lesioned side rather than with changes in postsynaptic receptors.

5.5. Denervation Syndromes

In a number of animal models of spinal injury or chronic pain, SMB occurs (reviewed by Levitt, 1985). For example, following dorsal rhizotomy or surgical lesions of spinal or peripheral nerves, there may follow a period of biting and gnawing on the deafferented part severe enough to produce autotomy (Dennis and Melzack, 1979). Levitt argues that the extent of tissue damage depends on the degree of deafferentation. In humans, intermittent periods of dysesthesia, tingling, numbness, and pain follow nerve injury; sometimes various self-injuries are inflicted upon the injured parts ostensibly in an attempt to change or interrupt the prevailing sensation. Although there may be some parallels between these phenomena and SMB in LND, there are several important differences. There is reason to believe that LND patients do not experience dysesthesias in their fingers or lips, even though we are rarely able to ask them directly. However, a small number of communicative patients have denied these sensations (Watts et al., 1982). Second, in biting or gnawing a deafferented area, nerve injury patients express no anxiety about hurting themselves; the affective component, including anticipatory fear so marked in LND patients, seems absent.

A very different type of SMB is seen in patients with congenital insensitivity to pain. In this syndrome or syndromes, which also are hereditary in some families, self-mutilation of the affected body segments is common (Pinsky and DiGeorge, 1966) and may often proceed to tissue damage. In this situation, as well as in the many reported instances of self-biting in experimental animals housed in isolation (Jones and Barraclough, 1978; Chamove and Anderson, 1981), the apparent demand is for stimulation of any kind, even if it be noxious, and for increased arousal. By contrast, in the LND patient, SMB is increased by any type of arousal, especially that which can be thought to be stressful (Nyhan, 1976; Seegmiller, 1976).

6. Metabolic Models

6.1. Intracerebral Hypoxanthine Administration

A potentially useful but inadequately explored model employs intracerebral Hx as lesioning agent (Palmour et al., 1990). Given that DA depletion is a consequence of HPRT deficiency, there should be some metabolic link between the abnormal purine profile characteristic of LND patients and DA loss. We chose Hx as a possible neurotoxic metabolite for three reasons;

1. In LND patients, Hx is known to accumulate in brain and CSF to a greater extent than in plasma or urine (Rosenbloom et al., 1967; Kelley et al., 1969);
2. Hx has been shown to promote the formation of toxic oxy-

gen radicals in animal and cellular models of ischemic neuronal damage (Demopoulos et al., 1980; Fridovich, 1983; McCord, 1985); and

3. Similar mechanisms are thought to subserve the neurotoxicicy of 6-OHDA, a chemical that causes selective depletion of the same neurotransmitters affected in the brains of LND patients (Heikkela and Cohen, 1972; Slivka and Cohen, 1985; Kopin 1987).

To evaluate this possibility, we utilized a modification of the unilateral lesion strategy of Ungerstedt and Arbuthnott (1970). Cannulae packed with solid Hx were implanted in left cautate nucleus or left lateral ventricle of rats; turning behavior was stimulated by apomorphine. Treated animals showed pronounced time- and dose-dependent ipsilateral turning, which peaked 14–18 d after cannula implantation. Consistent with the notion that rotation is mediated by alterations in DA neurons, turning was competitively blocked by haloperidol, a general DA antagonist, and more specifically by SCH 23390, a selective D1 antagonist. At 14 or 21 d, UA, xanthine, Hx, inosine, and ADO levels were elevated in caudate nuclei and other brain structures, demonstrating effective Hx delivery.

If cannulae are implanted in the head of the caudate, DA was reduced 50–65% in treated caudate and increased in contralateral caudate compared to sham-operated controls. If cannulae are implanted in the lateral ventricle, DA levels increased in both caudate over the initial 3-wk treatment phase but were reduced in animals treated for 7–12 wk. After 3 wk of treatment, DA and DOPAC levels are extensively depleted only in the amygdala, which is thoroughly bathed by ventricular fluid. NE was unchanged in any structure regardless of cannula placement, but 5HT and 5HIAA were increased 10–25%. Direct receptor studies of caudate D1 sites, using [^{3}H]-SCH 23390, suggested that during the early period, there was decreased binding to D1 receptors; at later times, binding sites were increased in number, but not in affinity (unpublished data). The observed changes in DA metabolites were quite consistent with those of laboratories that have made a detailed analysis of the time-course of metabolic changes following 6-OHDA lesioning (Hefti et al., 1980;

Heikkela et al., 1981; Fredholm et al., 1983; Altar et al., 1987). Briefly, these studies have shown that DA cells that escape lesioning overproduce DA, and hence, DA metabolites (perhaps in an attempt to normalize cellular functioning).

In about 30% of animals implanted bilaterally with Hx, spontaneous SMB appears between 4–8 d after treatment. This SMB can be exacerbated with L-dopa, amphetamine, apomorphine, or theophylline and blocked by SCH 23390 or *N*-ethylcarboxamideadenosine. The behavioral profile was produced in 83% of animals with elevated brain purine levels, and is produced much more frequently by Hx than by other purines (guanine, 20%; xanthine, 8%). If the specific effects of Hx were mediated by the catabolic generation of oxygen radicals, then allopurinol, an inhibitor of xanthine oxidase, should block the process. Preliminary studies indicate that this is not the case (Heshka, 1990).

Although the characterization of this model is still far from complete, it offers a number of potential opportunities and is complementary in many respects to the better-defined model of Breese and colleagues (reviewed in Section 5.2.). First, it is unique in that it attempts to relate the neurotransmitter depletion seen postmortem to metabolic alterations known to derive from the primary genetic lesion. Second, the time course for producing neuronal damage, although still much faster than that seen in patients, is relatively slow and progressive. Indeed, we have probably sampled only the initial portion of the time-effect curve and suspect that increasing deficit will be seen after longer Hx treatment. Because this model is progressive, it also offers an opportunity to evaluate therapies that might prevent some aspect of neuronal damage or slow disease progression. Third, rats treated with intracerebral Hx occasionally manifest spontaneous SMB. In our experience, this is unique with rodent models and suggests that further characterization is warranted.

6.2. Dopamine–Adenosine Interactions

Recent evidence from many laboratories supports the conclusion that DA and ADO interact at several levels of organization, particularly in the striatum (Stone, 1981; Williams, 1987). At the behavioral level, many investigators have shown that ADO antagonists increase the efficacy of DA agonists, and conversely, ADO agonists may block the actions of DA agonists (Fuxe and Ungerstedt, 1974; Joyce and Koob, 1981; Fredholm et al., 1983; Criswell et al., 1987). This is particularly apparent in animals lesioned in DA pathways with 6-OHDA (Criswell et al., 1987; Casag et al., 1989). At the anatomical level, ADO A2 receptors and DA D1 receptors—both positively coupled to adenylate cyclase—are heavily colocalized in caudate nucleus and in amygdala (Williams, 1987; R. M. Palmour and A. M. Babey, unpublished observations). Potent A2 ADO agonists, administered systemically, inhibit central release of DA and NE but facilitate release of 5HT (Stone, 1981; Myers and Pugsley, 1986; Wood et al., 1988). Paradoxically, there is preliminary evidence to suggest that A2 agonists increase the synthesis of DA, whereas DA autoreceptor (D2) agonists decrease it (Onali et al., 1988).

Several investigations of DA–ADO interactions have provided evidence pertinent to the behavioral phenotype of LND. Green et al. (1982) found that rats injected in the left caudate with ADO A2 agonists and systemically with low-dose apomorphine rotated vigorously to the left. With moderate doses of apomorphine, SMB was seen. One interpretation of this would be that ADO A2 receptor agonists potentiate the effects of apomorphine at postsynaptic DA receptors. However, this interpretation is not consistent with the blockade of apomorphine-induced behaviors by A2 agonists given systemically. The apparent contradiction may reflect the extremely high doses of intracaudate *N*-ethylcarboxamidoadenosine (approx 1000-fold the behaviorally active dose) used in these studies. In more detailed studies, Porter et al. (1988) administered potent A2 agonists intracerebroventricularly for 2 wk. Animals were somnolent and ataxic for much of the treatment period. Both ADO A2 and DA D1 receptors were desensitized, as measured by stimulated adenylate cyclase activity, and A2 binding sites were reduced in number. In HPRT-deficient neuroblastoma cells, we found alterations in the coupling between ADO A2 receptors and adenylate cyclase to be correlated with extracellular levels of Hx (Palmour et al., 1985). Although the isolated experiments reviewed in this section do not constitute an animal model for LND, they provide additional evidence of the functional interrelationships between DA and purines and suggest a productive venue for further investigation.

7. Gene Manipulation Models

Interest in developing an authentic, i.e., HPRT-deficient, animal model of LND dates almost from the time of the discovery of the enzymatic deficiency. Despite sometimes heroic efforts, the outcome has generally been disappointing. In the first two serious attempts, no animals with severe HRPT deficiency were produced. More recently, HPRT-deficient transgenic mice have been engineered, but they fail to express salient features of the behavioral and neurological phenotype.

7.1. Mutagenesis and Selective Breeding

In 1970, Robert DeMars suggested that an effective strategy would involve selective mating of mutagenized males with fertile XO female mice carrying X-linked coat color markers distinct from those carried by the male parent. Some female offspring of these crosses should be heterozygous for HPRT deficiency; males would all carry the single normal X from the mother. Carrier females would then serve as repositories for the mutant gene, with the expectation that one-half of their male progeny would inherit the mutant allele and thus, be HPRT-deficient.

The specific rationale for production of HPRT-deficient males was quite ingenious, utilizing both mutagenesis and chemical selection of mutated cells. DeMars reasoned that the incidence of mutation at the HPRT locus could be enhanced by irradiating male mice at an approximate LD^{50}. Spermatozoa specifically deficient in HRPT activity would then be selected by treating the irradiated males with 8-azaguanine, a toxic purine analog that is incorporated only into those cells with HPRT activity. If a few hardy males could survive these procedures, which are not unlike those used in modern cancer chemotherapy, their spermatozoa should be enriched for HPRT deficiency. In fact, when this scheme was applied, mice with orotic aciduria (an inborn error of pyrimidine metabolism) were produced rather than mice with LND. We now know that normal testicular function and spermatogenesis is highly dependent on HPRT activity; children with LND have primary azoospermia (Watts et al., 1982). Thus, it may be that HPRT-deficient spermatogonia do not give rise to viable sperm.

7.2. Teratocarcinoma and Chimeras

Another approach to devising HPRT-deficient mice utilized mutant teratocarcinoma cells injected into genetically marked blastocysts, which undergo normal development to produce allophenic mice (Dewey et al., 1977). Again, X-linked coat color markers were employed to facilitate screening of progeny. In this technically complex experiment, a blastocyst from a defined mouse strain was grafted onto a syngenic host to form a malignant teratocarcinoma. The teratocarcinoma was removed and established in tissue culture, then chemically mutagenized. HPRT-deficient clones were selected and reintroduced into the cavity of genetically marked, allogenic blastocysts. After a few mitotic divisions, the microinjected embryos were introduced into the uterus of a foster mother (in vitro fertilization), where they were allowed to develop normally. From an inspection of coat color markers, it appeared that some mosaic mice were produced.

No overt cases of HPRT-deficiency were produced either in the first or second generation offspring. Analysis of the tissue distribution of HPRT in both mosaic and nonmosaic mice showed that HPRT-deficient cell lines were plentiful in heart, kidney, spleen, liver, and pancreas. No HPRT-deficient lineages were detected in brain. Whether this indicates that HPRT-deficient cells were not repesented among brain-stem cell lineages or that HPRT-replete cells in brain simply overgrew HPRT-deficient cells is uncertain. In any event, it further confirms the importance of HPRT to normal brain function and underlines the strong positive selection for activity of this enzyme in brain.

7.3. Retroviral Vectors and Transgenic Animals

More recently, two groups have produced HPRT-deficient mice using the technique of injecting HPRT-deficient pluripotent embryonic stem cells into normal embryos at an early stage of development (Hooper et al., 1987; Keuhn et al., 1987). In one study, HPRT deficiency was produced by multiple infection with a retroviral vector, and 6-thioguanine was used to select for HPRT deficiency (Keuhn et al., 1987). Analysis of DNA digests from the four thioguanine-resistant clones suggested that at least three lines had sustained insertional mutation within the reading frame; unfortunately, HPRT activity was not measured in these clones. In the other series, spontaneous HPRT-deficient mutants were selected by growth in 6-thioguanine; seven resistant clones were found (Hooper et al., 1987). A single clone, demonstrated to be HPRT-deficient by enzymatic assay and to possess a Y chromosome, was utilized for injection into host blastocysts. After selective breeding, progeny carrying extensive contributions from injected stem cells were produced by both laboratories. Both carrier females and affected males were confirmed by a variety of techniques, including heterogeneous labeling with ^{3}H-Hx, direct enzymatic activity, DNA digestions, Southern analysis, and so on.

HPRT-deficient males produced by either technique reportedly have normal patterns of motor activity and exhibit no SMB or other behavioral abnormality (Finger et al., 1988). In addition, the mice lacked many of the metabolic features of the human disease. In particular, they did not have high levels of UA in blood or tissues, presumably because the enzymes of purine metabolism are regulated quite differently in mice than in humans. A recent investigation shows that mutant mice had 20–30% depletions of striatal and forebrain DA and compensatory increases of DA turnover (as measured by DOPAC:DA and HVA:DA ratios) of the same order of magnitude (Dunnett et al., 1989). Although these effects were statistically significant and are certainly in the right direction, they may be inadequate to produce either spontaneous or stimulated behavioral change.

Morphology of tyrosine hydroxylase-containing neurons (identified immunochemically) was normal, as were cell counts of these neurons. Serotonin was significantly depleted in HPRT-deficient mice in all areas studied, whereas epinephrine and NE levels did not differ from those seen in controls.

8. Conclusion

8.1. Summary

During the 25 years that have elapsed since the initial description of LND, a variety of animal models ranging in behavioral sophistication from mouse to monkey and in technological sophistication from acute drug administration to genetically engineered organisms have been evaluated. As noted earlier and summarized in Table 4, none of the existing models is ideal from all points of view. Ironically, those models that replicate the etiology of LND do not, because of species differences in biochemistry and behavior, mimic aspects of the disorder of particular interest to clinicians and neuroscientists. Whereas at least two of the technologically simpler models, neonatal DA and intracerebral Hx, may better reflect the behavioral phenotype and the secondary biochemical alterations of the disease state.

Despite this somewhat critical view, each class of model has, in retrospect, refined our conceptualization of the questions to be asked about SMB in the specific context of LND. Studies of acute and chronic pharmacological models have thus suggested that SMB in LND patients, though not the endpoint of classic drug-induced motor agitation, is likely to be an end point behavior reflecting the neurochemical imbalance of the disorder. Also, SMB is produced quite effectively, although not exclusively, by drugs with specificity for brain ADO and DA receptors; both systems are known to be altered in LND. Indeed, it could be argued that behavioral studies of the effects of DA agonists were instrumental, together with prior clinical investigations of the brains of patients with Parkinson's disease, in directing attention to brain DA systems in LND (Hornykiewicz, 1981; Lloyd et al., 1981).

Table 4
Characteristics of Animal Models for LND

Model	Behavioral specificity	HPRT activity reduced?	Hx, UA elevated; synthesis elevated?	*De novo* purine	DA depleted	5HT elevated?
Phamacological	Some	MX: No	?	Mx: No	Amphet: Yes	?
6-OHDA lesions						
Neonatal	Yes	No	No	No	Yes	Yes
Amygdala	Yes	No	No	No	Yes	?
VTA lesions in monkeys	Yes	No	Probably no	Probably no	Yes	?
Spontaneous movement disorder						
Intracerebral hypoxanthine	Yes Spontaneous SMB	No	Yes	Yes	Yes	Yes
Transgenic mouse	Probably no	Yes	Only slightly	?	Yes	Yes

DA depletion models were a natural outgrowth of the demonstration that caudate and accumbens DA were pathologically reduced in LND patients. The neonatal lesion model of Breese and colleagues (1984) is the most extensively characterized and perhaps the most reliable of the presented models. It is also undoubtedly the most generally accessible model and thus, has the greatest potential for testing drugs that might block or alleviate SMB in patients who have already developed these behaviors. Indeed, published studies with this model suggest that D1 and A2 agonists might have therapeutic utility; unfortunately, the most potent members of these drug classes are not available for clinical use (Criswell et al., 1987; Duncan et al., 1987). A particular strength of this model is that 6-OHDA is administered during the neonatal period, in consideration of the developmental course of symptoms in patients. In general, Breese and colleagues have not tested neonatally treated rats before 200 d of age (young adults). A more detailed characterization of the progression of neurochemical and behavioral changes in these animals might prove fruitful, as might testing of the possibility that certain drugs, administered during an early period of development, would attenuate or block the susceptibility to SMB.

The intracerebral Hx model is a unique attempt to understand the metabolic relationships between an inherited abnormality of purine metabolism and brain DA depletion (Palmour et al., 1989). A major shortcoming of this model is inadequate characterization, both with respect to detailed sequential knowledge of the consequences of prolonged exposure of the brain to pathological concentrations of Hx and with respect to the relative impact of Hx on DA receptors (both pre- and postsynaptic), DA turnover, and the enzymes of DA metabolism. Another potential problem is that Hx in normal rat brain may be metabolized differently than it would be in HPRT-deficient brain, so that an Hx-derivative might, in fact, be a more efficient antimetabolite. However, there is clearly much to be learned from this model, and if verified, it would be quite useful in testing drugs that might block or ameliorate the progression of neurodegeneration in LND. This would appear to be the only rodent model in which SMB occurs spontaneously.

The only authentic animal models of LND, with respect to HPRT deficiency, are the transgenic mouse constructs (Hooper et al., 1987; Kuehn et al., 1987). Although these models, like the Hx model, are still inadequately characterized, some disappointing problems have arisen that will likely obviate their general utility. Most particularly, purine metabolism in the mouse is regulated in a fashion that precludes the extreme hyperuricemia and hyperuricosuria seen in humans. Also, it would appear from preliminary reports that accumulation of Hx and other purines is not comparable to that seen in LND (Finger et al., 1988). It is not yet clear whether purine *de novo* synthesis is less exaggerated in the HPRT-deficient mouse or whether salvage synthesis is less critical than it is in humans. In any case, this topic deserves a more detailed investigation. On the positive side, however, is the finding of moderate DA depletion in HPRT-deficient mice (Dunnett et al., 1989). Unfortunately, the meager extent of DA depletion suggests that this model will not be particularly useful for screening possible therapeutic drugs.

Finally, although neither designed nor adequately exploited as an optimal model for LND, the ventral tegmental surgical lesion model is behaviorally the most appropriate in that movement disorder occurs without pharmacological challenge and that the natural behavioral repertoire is appropriately complex (Goldstein et al., 1986). From a biochemical point of view, a surgical cut is probably not the ideal way to interrupt DA pathways. However, this model may be unique in having predicted a therapy that is effective in some patients (Goldstein et al., 1985; Jankovic et al., 1988).

8.2. Lessons for the Clinician from Animal Models of LND

Have we learned anything from animal models of LND that we did not already know? I believe the answer is clearly yes on one count and possibly yes on two others.

Arguably the best defined contribution of animal models of LND is to allow a tentative answer to the question *Does DA depletion explain the behavioral phenotype of LND?* Earlier in this chapter, we reviewed the evidence supporting the contention that severe HPRT deficiency subserves the behavioral phenotype of LND. This evidence is derived solely from clinical biochemical studies. Clinical data that would persuasively link the movement disorder and SMB characteristic of LND patients to DA depletion does not exist. However, integration of experimental animal data with clinical evidence from other disorders, such as, Parkinson's disease, does suggest that the motor abnormalities of LND are largely a consequence of terminal DA depletion and of other neurochemical changes that compensate for DA loss. Does DA depletion also account for SMB? This question cannot be answered so clearly, but several lines of evidence suggest that at a minimum, animals with DA depletion have a reduced threshhold for SMB. However, the fact that DA-depleted animals do not spontaneously self-mutilate argues that additional chemical triggers, either endogenous (i.e., those elicited by behavioral stress) or exogenous, are required for overt SMB—at least in rodent models.

Second, the inference that DA might be specifically depleted in caudate and in accumbens is derived more from animal studies of DA depletion than from clinical parallels to Parkinson's or Huntington's disease. The specific movement disorders seen in LND patients do not resemble those found in Parkinson's patients with marked DA depletion but rather those of Huntington's patients in whom DA is not reduced. An important differentiation, and one that may explain in part the distinct symptom-atology, is that DA depletion reportedly appears first in the SN of Parkinson's patients, but is most apparent in nerve terminal regions, such as caudate and putamen, in LND patients (Hornykiewicz, 1981; Lloyd et al., 1981). Neither Huntington's nor Parkinson's patients exhibit SMB. By contrast, experimental animals given very large or repeated doses of DA agonists, or depleted of brain DA and challenged with DA agonists, often show persistent and sometimes profound self-biting and SMB.

Finally, there are a number of possible therapeutic avenues suggested by experimental animal models of LND. The most obvious are ADO A2 agonists and specific DA antagonists. Indeed, fluphenazine, a selective DA D1 antagonist reportedly blocks

SMB in some but not all LND patients (Goldstein et al., 1985; Jankovic et al., 1988). Although potent A2 agonists have not yet come to clinical trial, they are under active investigation for a variety of more common conditions and have been shown to block or ameliorate oxygen radical-induced neuronal damage in experimental models for stroke and ischemia (Evans et al., 1987; von Lubitz et al., 1988). In addition, an appreciation of the significance to LND of the multiple interactions between brain DA and ADO systems is unlikely to have arisen in the absence of animal model experimentation.

8.3. Future Directions

Although there are many possible directions for this research, it seems appropriate to call attention to three general areas that have been neglected in experimental animal model studies and in clinical investigation of LND patients. Quite broadly, these three areas concern the integration of information from multiple neurochemical systems, a detailed consideration of interactions between limbic and striatal circuits of the brain, and the level of behavioral analysis that is appropriate to a progressive neurodegenerative disorder of metabolism.

In LND, there are substantial alterations in at least three neuroregulatory metabolites—DA, ADO, and 5HT—and lesser changes in several others, including neurotransmitter amino acids and the enzymes of acetylcholine synthesis. Although there is now much information about the isolated action of each of the classical neurotransmitters, an understanding of the integrated pattern of function in the mammalian brain is far from complete. Several recent reviews cite evidence showing that at least DA, NE, acetylcholine, 5HT, and GABA contribute to the regulation of neostriatal and mesolimbic DA pathways (Cools, 1980; Penney and Young, 1983). ADO is known to modulate the release of DA, NE, 5HT, acetylcholine, glutamate, and GABA, as well as that of many hormones and neuropeptides (*see* Section 6.2. for refs.). There is an immense body of data showing that the first response of any perturbed neuronal system, whether injured, drugged, or lesioned, is to initiate compensatory changes. Now that we have identified some of the key neurochemical players in HPRT deficiency, we must begin to study their altered interactions in more detail.

Classic studies of Papez (1937) and many others allege that the striatopallitum is the motor output system for the limbic system in general and the amygdala in particular. The profound affective distress that accompanies SMB in patients with LND reveals significant limbic dysfunction, just as the disturbed motor activity indicates that basal ganglionic function is impaired. The particular pattern of irritability and self- and other-directed aggression calls attention most dramatically to the amygdala, but hippocampal involvement is also suggested by two symptoms: hydrophobia and the ability in some patients to call for restraint when SMB impends. In humans and other mammals, the amygdala is a complex structure derived in part from the older, reptilian "fight or flight" system (medial regions) and in part from the neostriatum (basolateral nuclei), which also gives rise to the head of the caudate (McLean, 1949). In general, these two subdivisions of the amygdala have opposing functional attributes: Electrical or cholinergic stimulation of medial regions activates "fight or flight," whereas electrical stimulation of lateral amygdala facilitates social and affiliative behaviors (Mark and Ervin, 1970; Adamec and Adamec, 1986). Although there is considerable evidence suggesting that lateral amygdala directly inhibits medial amygdala, there is also reason to believe that the caudate is inhibitory to medial amygdala (Nauta, 1986). Since DA is predominantly an inhibitory neurotransmitter, the progressive loss of DA from caudate and, by inference, lateral amygdala implies that excitation of medial amygdala is unopposed by normal inhibitory inputs (Penney and Young, 1983; Roth et al., 1989). Increased attention to the striatal–limbic interactions, in clinical and in basic studies is almost certain to result in improved management strategies for patients with LND.

Finally, the need for more sophisticated approaches to behavioral issues, in clinical testing of LND patients and in the conceptualizaton of animal models is obvious. In neither humans nor animals has enough emphasis been placed on the developmental progression of the disorder. Among the many dozens of published clinical investigations of LND, there is a single predominantly longitudinal clinical review and four longitudinal biochemical investigations (Mizuno, 1986). There is as yet no direct documentation of the progression of DA impairment. It is quite likely, however, that conflicting reports with respect to therapeutic trials—e.g., Watts et al., 1982 vs Jankovic et al., 1988 and Goldstein et al., 1985 vs Jankovic et al., 1988—reflect different developmental stages of neuronal degeneration rather than intrinsic differences between patients or investigators. A first step might be to initate systematic longitudinal neuropsychological testing as part of the routine care of LND patients. Another approach (in progress as part of a multicenter, collaborative study) is periodic magnetic resonance imaging (MRI) and positron emission tomography (PET) to track the extent and timing of DA deterioration. A variety of other questions in this area are perhaps best approached at the present time by systematic study of the behavioral and pharmacological consequences of neonatal DA depletion, elicited by 6-OHDA or perhaps Hx, in primates.

As is often the case, the investigation of LND patients and of animal models for the disease has raised new and challenging questions for basic science. Identification of the enzymatic basis for this disorder in the 1960s led to a complete revision of the understanding of basic human purine biochemistry and metabolism. If we are eventually able to understand the behavioral phenotype of LND in the context of developmental biochemistry—a goal not yet achieved for any metabolic disorder of the brain—our conceptualization of normal human developmental neurochemistry will likewise be revolutionized. Integration of clinical and experimental animal model studies will certainly be required for this undertaking.

References

Adamec R. E. and Stark-Adamec C. (1986) Limbic hyperfunction, limbic epilepsy and interictal behavior: Models and methods of detection, in *The Limbic System: Functional Organization and Clinical Disorders* (Doane B. W. and Livingston K. E., eds.), Raven Press, New York, pp. 129–146.

Allsop J. and Watts R. W. E. (1980) Activities of amidophosphoribosyltransferase and purine phospho-ribosyltransferases and the phosphoribosylpyro-phosphate content of rat central nervous system at different stages of development. *J. Neurol. Sci.* **46,** 221–238.

Altar C. A., Marien M. R., and Marshall J. F. (1987) Time course of adaptations in dopamine biosynthesis,metabolism and release following nigrostriatal lesions: Implications for behavioral recovery from brain injury. *J. Neurochem.* **48,** 390–399.

Alvares A. P., Anderson K. E., Conney A. H., and Kappas A. (1976) Interactions between nutritional factors and drug biotransformations in man. *Proc. Natl. Acad. Sci. USA* **73,** 2501–2504.

Anders T. F., Cann H. M., Ciaranello R. D., Barchas J. D., and Berger P. A. (1978) Further observations on the use of 5-hydroxytryptophan in a child with Lesch-Nyhan syndrome. *Neuropaediatrie* **76,** 351–355.

Barshop B. A., Alberts A. S., Laikind P., and Gruber H. E. (1989) Studies of mutant human adenylosuccinate lyase. *Adv. Exp. Biol. Med.* **253A,** 23–30.

Ben-Ari Y. and Kelly J. S. (1976) Dopamine evoked inhibition of single cells of the feline putamen and basolateral amygdala. *J. Physiol.* **256,** 1–21.

Berman P. H., Balis M. E., and Dancis J. (1969) Congenital hyperuricemia, an inborn error of purine metabolism associated with psychomotor retardation, athetosis and self-mutilation. *Arch. Neurol.* **20,** 44–52.

Boyd E. M., Dolman M., Knight L. M., and Sheppard E. P. (1965) The chronic oral toxicity of caffeine. *Can. J. Physiol. Pharmacol.* **43,** 995–1007.

Breakefield X. O., Castiglione C. M., and Edelstein S. B. (1976) Monoamine oxidase activity decreased in cells lacking hypoxanthine phosphoribosyltransferase activity. *Science* **192,** 1018–1020.

Breese G. R., Baumeister A. A., McCown T. J., Emerick S. G., Frye G. D., Crotty K., and Mueller R. A. (1984) Behavioral differences between neonatal and adult 6-hydroxydopamine-treated rats to dopamine agonists: Relevance to neurological symptoms in clinical syndromes with reduced brain dopamine. *J. Pharmacol. Exp. Ther.* **231,** 343–354.

Breese G. R., Baumeister A., Napier T. C., Frye G. D., and Mueller R. A. (1985a) Evidence that D1 dopamine receptors contribute to the supersensitive behavioral responses induced by L-dihydroxyphenylalanine in rats treated neonatally with 6-hydroxydopamine. *J. Pharmacol. Exp. Ther.* **234,** 287–295.

Breese G. R., Duncan G. E., Napier T. C., Bondy S. C., Iorio L. C., and Mueller R. A. (1987) 6-hydroxydopamine treatments enhance behavioral responses to intracerebral microinjection of D1- and D2-dopamine agonists to nucleus accumbens and striatum without changing dopamine receptor binding. *J. Pharmacol. Exp. Ther.* **240,** 167–176.

Breese G. R., Napier T. C., and Mueller R. A. (1985b) Dopamine agonist-induced locomotor activity in rats treated with 6-hydroxydopamine at differing ages: Functional supersensitivity of D1 dopamine receptors in neonatally lesioned rats. *J. Pharmacol. Exp. Ther.* **174,** 447–455.

Breese G. R. and Traylor T. D. (1960) Effects of 6-hydroxydopamine on brain norepinephrine and dopamine: Evidence for selective degeneration of catacholamine neurons. *J. Pharmacol. Exp. Ther.* **174,** 413–420.

Brennand J., Chinault A. C., Konecki D. S., Melton D. W., and Caskey C. T. (1982) Cloned cDNA sequences of the hypoxanthine guanine phosphoribosyltransferase gene from a mouse neuroblastoma cell line found to have amplified genomic sequences. *Proc. Natl. Acad. Sci. USA* **79,** 1950–1954.

Brenton D. P., Astrin K. H., Cruikshank M. E., and Seegmiller J. E. (1977) Measurement of free nucleotides in cultured human lymphoid cells using HPLC. *Biochem. Med.* **17,** 231–247.

Casas M., Ferre S., Cadafalch J., Grau J. M., and Jane F. (1989) Rotational behaviour induced by theophylline in 6-OHDA nigrostriatal denervated rats is dependent on the supersensitivity of striatal dopaminergic receptors. *Pharmacol. Biochem. Behav.* **9,** 609–613.

Casas-Bruge M., Almenar C., Grau J. M., Jane F., Herrera-Marschwitz M., and Ungerstedt U. (1985) Dopaminergic receptor supersensitivity in self-mutilatory behaviour of Lesch-Nyhan disease. *Lancet* **1,** 991,992.

Castells S., Chakrabati C., Winsberg B. G., Hurwic M., Perel J. M., and Nyhan W. L. (1979) Effects of L-5-hydroxytryptophan on monoamine and amino acid turnover in the Lesch-Nyhan syndrome. *J. Autism Dev. Disord.* **9,** 95–103.

Chamove A. S. and Anderson J. R. (1981) Self-aggression, stereotypy and self-injurious behaviour in man and monkeys. *Curr. Psychol. Rev.* **1,** 245–256.

Chang S. M., Wager-Smith R., Tsao T. Y., Henkel-Tigges J., Vaishn M., and Caskey C. T. (1987) Construction of a defective retrovirus encoding the human hypoxanthine phosphoribosyltransferase cDNA and its expression in cultured cells in mouse bone marrow. *Mol. Cell. Biol.* **7,** 854–861.

Cools A. R. and van Rossum J. M. (1980) Multiple receptors for brain dopamine in behaviour regulation. *Life Sci.* **27,** 1237–1253.

Corradetti R., Lo Conte G., Moroni F., Passani M. B., and Pepeu G. (1984) Adenosine decreases aspartate and glutamate release from rat hippocampal slices. *Eur. J. Pharmacol.* **104,** 19–26.

Creese I., Burt D. R., and Snyder S. H. (1977) Dopamine receptor binding enhancement accompanies lesion-induced behavioral supersensitivity. *Science* **197,** 596–598.

Criswell, H, Mueller R. A., and Breese G. R. (1987) Assessment of purine-dopamine interactions in 6-hydroxydopamine-lesioned rats: Evidence for pre- and postsynaptic influences by adenosine. *J. Pharmacol. Exp. Ther.* **244,** 493–500.

Crussi F. G., Robertson D. M., and Hiscox I. L. (1969) The pathological condition of the Lesch-Nyhan syndrome. *Am. J. Dis. Child.* **118,** 501–506.

Daly J. W., Bruns R. F., and Snyder S. H. (1981) Adenosine receptors in the central nervous system: Relationship to the central actions of methylxanthines. *Life Sci.* **28,** 2083–2097.

De Mars R. (1971) Genetic studies of HG-PRT Deficiency and the Lesch-Nyhan syndrome with cultured human cells. *Fed. Proc.* **30,** 944–955.

Demopoulos H. B., Flamm E. S., Pietronigro D. D., and Seligman M. L. (1980) The free radical pathology and the microcirculation in the major central nervous system disorders. *Acta Physiol. Scand.* **(Suppl.), 492** 91–119.

Dennis S. G. and Melzack R. (1979) Self-mutilation after dorsal rhizotomy in rats: Effects of prior pain and pattern of root lesions. *Exp. Neurol.* **65,** 412–421.

Dewey M. J., Martin D. W., Jr., Martin G. R., and Mintz B. (1977) Mosaic mice with teratocarcinoma-derived mutant cells deficient in hypoxanthine phosphoribosyltransferase. *Proc. Natl. Acad. Sci. USA* **74,** 5564–5568.

Duncan G. E., Criswell H. E., McCown T. J., Paul I. A., Mueller R. A., and Breese G. R. (1987) Behavioral and neurochemical responses to haloperidol and SCH-23390 in rats treated neonatally or as adults with 6-hydroxydopamine. *J. Pharmacol. Exp. Ther.* **243,** 1027–1034.

Dunnett S. B., Sirinathsinghji D. J. S., Heavens R., Rogers D. C., and Kuehn M. R. (1989) Monoamine deficiency in a transgenic HPRT mouse model of Lesch-Nyhan syndrome. *Brain Res.* **501,** 401–406.

Dunwiddie T. V. (1985) The physiological role of adenosine in the central nervous system. *Int. Rev. Neurobiol.* **27,** 63–82.

Edwards N. L., Johnston M. V., and Silverstein F. S. (1989) Cerebrospinal fluid cyclic nucleotide alterations in the Lesch-Nyhan syndrome. *Adv. Exp. Biol. Med.* **253A,** 181–188.

Ellison G., Eison M. S., Huberman H. S., and Daniel F. (1978) Long-term changes in dopaminergic innervation of caudate nucleus after continuous amphetamine administration. *Science* **201,** 276–278.

Evans M. C., Swan J. H., and Meldrum B. S. (1987) An adenosine analogue protects against long term development of ischemic cell loss in the rat hippocampus. *Neurosci. Lett.* **83,** 287–292.

Ferre S., Casas M., Cobos A., Garcia C., Jane F., and Grau J. M. (1987) L-dopa causes an acute, partial and reversible reversal of denervation-induced supersensitivity of striatal dopaminergic receptors. *Psychopharmacology* **91,** 254–256.

Ferrer I., Costell M., and Grisolia S. (1982) Lesch-Nyhan syndrome-like behavior in rats from caffeine ingestion. *FEBS Lett.* **141,** 275–278.

Finger S., Heavens R. P., Sirinathsinghji D. J. S., Kuehn M. R., and Dunnett S. B. (1988) Behavioral and neurochemical evaluation of a transgenic mouse model of Lesch-Nyhan syndrome. *J. Neurol. Sci.* **86,** 459–461.

Fox I. H. and Kelley W. N. (1971) Phosphoribosylpyrophosphate in man: Biochemical and clinical significance. *Ann. Intern. Med.* **74,** 424–436.

Fredholm B. B., Herrera-Marschitz M., Jonzon B., Lindstrom K., and Ungerstedt U. (1983) On the mechanism by which methylxanthines enhance apomorphine-induced rotation behavior in the rat. *Pharmacol. Biochem. Behav.* **19,** 535–541.

Fridovich I. (1983) Superoxide radical: An endogenous toxicant. *Ann. Rev. Toxicol.* **23,** 239–257.

Frith C. D., Johnstone E. C., Joseph M. H., Powell R. J., and Watts R. W. E. (1976) Double-blind clinical trial of 5-hydroxytryptophan in a case of Lesch-Nyhan syndrome. *J. Neurol. Neurosurg. Psychiatr.* **39,** 656–662.

Fuxe K. and Ungerstedt U. (1974) Action of caffeine and theophylline on supersensitive dopamine receptors. *Med. Biol.* **52,** 48–54.

Garrod A. E. (1908) Inborn errors of metabolism (Croonian Lectures). *Lancet* **1,** 73, 142, 214.

Goldstein M., Anderson L. T., Reuben R., and Dancis J. (1985) Self-mutilation in Lesch-Nyhan disease is caused by dopaminergic denervation. *Lancet* **1,** 338,339.

Goldstein M., Battista A. H., Ohmoto T., Anagoste B., and Fuxe K. (1973) Tremor and involuntary movements in monkeys: Effect of L-dopa and of a dopamine receptor stimulating agent. *Science* **179,** 816,817.

Goldstein M, Kuga S., Kusano N., Meller E., Dancis J., and Schwarcz R. (1986) Dopamine agonist induced self-mutilative biting behavior in monkeys with unilateral ventromedial tegmental lesions of the brainstem: Possible pharmacological model for Lesch-Nyhan syndrome. *Brain Res.* **367,** 114–129.

Gordon R. B., Stout J. T., Emmerson B. Y., and Caskey C. T. (1987) Molecular studies of hypoxanthine-guanine phosphoribosyltransferase mutations in 6 New Zealand families. *Ann. N. Z. J. Med.* **17,** 424–433.

Green R. D., Proudfit H. K., and Yeung S. H. (1982) Modulation of striatal dopaminergic function by local inection of 5'-N-ethylcarboxamide adenosine. *Science* **218,** 58–61.

Greene M. L., Boyle J. A., and Seegmiller J. E. (1970) Substrate stabilization: Genetically controlled reciprocal relationship of two human enzymes. *Science* **167,** 337–340.

Hagberg H., Andersson P., Butcher S., Sandberg M., Lehmann A., and Hamberger A. (1986) Blockade of NMDA sensitive receptors inhibits ischaemia-induced accumulation of purine catabolites in the rat striatum. *Neurosci. Lett.* **68,** 311–316.

Harkness R. A. (1989) Lesch-Nyhan syndrome: Reduced amino acid concentrations in CSF and brain. *Adv. Exp. Biol. Med.* **253A,** 159–163.

Harkness R. A., McCreanor G. M., and Watts R. W. E. (1988) Lesch-Nyhan syndrome and its pathogenesis: Purine concentrations in plasma and urine with metabolite profiles in CSF. *J. Inherited Metab. Dis.* **11,** 239–252.

Hartgraves S. L and Randall P. K. (1986) Dopamine agonist-induced stereotypic grooming and self-mutilation following striatal dopamine depletion. *Psychopharmacology* **90,** 358–363 .

Hefti F., Melamed E., and Wurtman R. J. (1980) Partial lesions of the dopaminergic nigrostriatal system in rat brain. *Brain Res.* **195,** 123–137.

Heikkila R., and Cohen G. (1972) Further studies on the generation of hydrogen peroxide by 6OHDA. *Mol. Pharmacology* **8,** 241–248.

Heikkila R, Shapiro B. S., and Duvoisin R. C.3(1981) The relationship between loss of dopamine nerve terminals, striatal [^{3}H] spiroperidol binding and rotational behavior in unilaterally 6-hydroxydopamine-lesioned rats. *Brain Res.* **211,** 285–292.

Hershfield M. S. and Seegmiller J. E. (1977) Regulation of de novo purine synthesis in human lymphoblasts. *J. Biol. Chem.* **252,** 6002–6009.

Heshka T. W. (1989) Effects of hypoxanthine on dopamine neurons. Unpublished Masters thesis, McGill University, Montreal, Canada.

Hirschhorn R., Papageorgiou P. S., Kesarwala H. H., and Taft L. T. (1980) Amelioration of neurologic abnormalities after "ezyme replacement" in adenosine deaminase deficiency. *N. Engl. J. Med.* **303,** 377–830.

Hoefnagel D., Andrew E. D., Mireault N. G., and Berndt W. O. (1965) Hereditary choreoathetosis, self-mutilation and hyperuricemia in young males. *N. Eng. J. Med.* **273,** 130–134.

Hooper M., Hardy K., Handyside A., Hunter S., and Monk M. (1987) HPRT deficient (Lesch-Nyhan) mouse embryos derived from germline colonization by cultured cells. *Nature* **326,** 292–295.

Hornykiewicz O. (1981) Brain neurotransmitter changes in Parkinson's disease, in *Neurology 2: Movement Disorders* (Marsden C. D. and Fahn S., eds.), Butterworths, London, pp. 41–58.

Jankovic J., Caskey C. T., Stout J. T., and Butler I. J. (1988) Lesch-Nyhan syndrome: A study of motor behavior and cerebrospinal fluid neurotransmitters. *Ann. Neurol.* **23,** 466–475.

Jolly D. J., Esty A. C., Bernard H. V., and Friedmann T. (1982) Isolation of a genomic clone partially encoding human hypoxanthine phosphoribosyltransferase. *Proc. Natl. Acad. Sci. USA* **79,** 5038–5042.

Jolly D. J., Okayama H., Berg P., Esty A. C., Pilpula D., Bohlen P., Johnson G. G., Scively J. E., Hunkapillar T. J. and Friedmann T. (1983) Isolation and characterization of a full-length expression cDNA for human hypoxanthine phosphoribosyltransferase. *Proc. Natl. Acad. Sci. USA* **80,** 477–481.

Jones I. H. and Barraclough B. M. (1978) Auto-mutilation in animals and its relevance to self-injury in man. *Acta. Psychiatr. Scand.* **58,** 40–47.

Joyce E. M. and Koob G. F. (1981) Amphetamine–, scopolamine- and caffeine-induced locomotor activity following 6-hydroxydopamine lesions of the mesolimbic dopamine system. *Psychopharmacology* **73,** 311–313.

Kanazawa I., Kimura M., Murata M., Tanaka Y., and Cho F. (1990) Choreic movements in the macaque monkey induced by kainic acid lesions of the striatum combined with L-dopa. *Brain* **113,** 509–535.

Katsuragi T., Ushijima I., and Furukawa T. (1983) The clonidine-induced self-injurious behavior of mice involves purinergic mechanisms. *Pharmacol. Biochem. Behav.* **20,** 943–946.

Kelley W. N., Greene M. L., Rosenbloom F. M., Henderson J. F., and Seegmiller J. E. (1969) Hypoxanthine-guanine phosphoribosyltransferase deficiency in gout. *Ann. Intern. Med.* **70,** 155–206.

Kelley W. N., Rosenbloom F. M., Henderson J. R., and Seegmiller J. E. (1967)

A specific enzyme defect in gout associated with overproducton of uric acid. *Proc. Natl. Acad. Sci. USA* **64,** 1735–1738.
Kelley W. N. and Wyngaarden J. B. (1983) Clinical syndromes associated with hypoxanthine-guanine phosphoribosyltransferase deficiency, in *The Metabolic Basis of Inherited Disease,* 5th Ed. (Stanbury J. B., Wyngaarden J. B., Frederickson D. S., Goldstein J. L., and Brown M. S., eds.), McGraw-Hill, New York, pp. 1115–1143.
Kiebling M., Lindl T., and Cramer H. (1975) Cyclic adenosinemonophosphate in cerebrospinal fluid. *Arch. Psychiatr. Nervenkr.* **220,** 325–333.
Kish S. J., Fox I. H., Kapur B. M., Lloyd K., and Hornykiewicz O. (1985) Brain benzodiazepine receptor binding and purine concentration in Lesch-Nyhan syndrome. *Brain Res.* **336,**117–123.
Kopin I. (1987) Neurotoxins affecting biogenic aminergic neurons. *Psychopharmacoloy: The Third Generation of Progress,* Raven Press, New York, pp. 351–358.
Kuehn M. R., Bradley A., Robertson E. J., and Evans M. J. (1987) A potential animal model for Lesch-Nyhan syndrome through introduction of HPRT mutations into mice. *Nature* **326,** 295–297.
Lake C. R. and Ziegler M. G. (1977) Lesch-Nyhan syndrome: Low dopamine-β-hydroxylase activity and diminished sympathetic response to stress and posture. *Science* **196,**905,906.
Lal S. and Sourkes T. L. (1973) Ontogeny of stereotyped behavior induced by apomorphine and amphetamine in the rat. *Arch. Int. Pharmacodyn. Ther.* **202,** 171–182.
Lal S. and Sourkes T. L. (1974) Apomorphine derivatives and dopaminergic activity. *Adv. Neurol.* **5,** 307,308.
Lau K. F. and Henderson J. F. (1971) Inhibitors of hypoxanthine-guanine phosphoribosyltransferase. *Cancer Chemother. Rep. (Part 2),* **3,** 87–94.
Lesch M. and Nyhan W. L. (1964) A familial disorder of uric acid metabolism and central nervous function. *Am. J. Med.* **36,** 561–570.
Levitt M. (1985) Dysesthesias and self-mutilation in humans and subhumans: A review of clinical and experimental studies. *Brain Res. Rev.* **10,** 247–290.
Lloyd H. G. E. and Stone T. W. (1981) Chronic methylxanthine treatment in rats: A comparison of Wistar and Fischer 344 strains. *Pharmacol. Biochem. Behav.* **14,** 827–830.
Lloyd K. G., Hornykiewicz O., Davidson L., Shannak K., Farley I., Goldstein M., Shibuya M., Kelley W. N., and Fox I. H. (1981) Biochemical evidence of dysfunction of brain neurotransmitters in the Lesch-Nyhan syndrome. *N. Engl. J. Med.* **305,** 1106–1111.
Lloyd K. G., Willigens M. T., and Goldstein M. (1985) Induction and reversal of dopamine dyskinesia in rat, cat and monkey. *Psychopharmacology* **89 (Suppl. 2),** 200–210.
Mark V. H. and Ervin F. R. (1970) *Violence and the Brain* Harper and Row, New York.
McCord J. M. (1985) Oxygen-derived free radicals in post-ischemic tissue injury. *N. Engl. J. Med.* **312,** 159–163.

McKeran R. O. and Watts R. W. E (1976) Use of phytohaemagglutinin stimulated lymphocytes to study effects of HGPRT deficiency on polynucleotide and protein synthesis in the Lesch-Nyhan syndrome. *J. Med. Genet.* **13,** 91–95.

McLean P. D. (1958) The limbic system with respect to self-preservation and the preservation of the species. *J. Nerv. Ment. Dis.* **127,** 1–11.

Mehta S., Mitchell D., Roitman D., Kattan A. K., Palmour R. M., and Goodyer P. (1990) Nephrogenic diabetes insipidus as the early renal manfestation of Lesch-Nyhan disease. *J. Pediatr.*

Minana M. D., Portoles M., Jorda A., and Grisolia S. (1984) Lesch-Nyhan syndrome, caffeine model: Increase of purine and pyrimidine enzymes in rat brain. *J. Neurochem.* **43,** 1556–1560.

Mizuno T. (1986) Long-term followup of two patients with Lesch-Nyhan syndrome. *Neuropediatrics* **17,** 158–169.

Mizuno T.-I. and Yugari Y. (1974) Self-mutilation in Lesch-Nyhan syndrome. *Lancet* **1,**761.

Moore K. E. and Kelly P. H. (1976) Biochemical pharmacology of mesolimbic and mesocortical dopaminergic neurons, in *Psychopharmacology: A Generation of Progress* (Lipton M. A., DiMascio A., and Killa K. F., eds.), Raven Press, New York, pp. 221–234.

Morgan L. L., Schneiderman N., and Nyhan W. L. (1970) Theophylline: Induction of self-biting in rabbits. *Psychon. Sci.* **19,** 37,38.

Mueller K. and Hsiao S. (1980) Pemoline-induced self-biting in rats and self-mutilation in the deLange syndrome. *Pharmacol. Biochem. Behav.* **13,** 627–631.

Mueller K. and Nyhan W. L. (1982) Pharmacological control of self-injurious behavior in rats. *Pharmacol. Biochem. Behav.* **16,** 957–963.

Mueller K. and Nyhan W. L. (1983) Clonidine potentiates drug induced self-injurious behavior in rats. *Pharmacol. Biochem. Behav.* **18,** 891–894.

Mueller K., Saboda S., Palmour R., and Nyhan W. L. (1982) Self-injurious behavior produced in rats by daily caffeine and continuous amphetamine. *Pharmacol. Biochem. Behav.* **17,** 613–617.

Myers S. and Pugsley T. A. (1986) Decrease in rat striatal dopamine synthesis and metabolism in vivo by metabolically stable adenosine receptor agonists. *Brain Res.* **375,** 193–197.

Nauta W. J. H. (1986) Circuitous connections linking cerebral cortex, limbic system and corpus striatum, in *The Limbic System: Functional Oranization and Clinical Disorders* (Doane B. W. and Livingston K. F., eds.), Raven Press, New York, pp. 43–54

Nyhan W. L. (1976) Behavior in the Lesch-Nyhan syndrome. *J. Autism Child. Schizophr.* **6,** 235–252.

Nyhan W. L., Johnson H. G., Kaufman I. A., and Jones R. L. (1980) Serotonergic approaches to the modification of behavior in the Lesch-Nyhan syndrome. *Appl. Res. Ment. Retard.* **1,** 25–34.

Nyhan W. L., Oliver W. J., and Lesch M. A. (1965) A familial disorder of uric acid metabolism and central nervous system function. *J. Pediatr.* **67,** 257–266.

Nyhan W. L., Parkman R., Page T., Gruber H. E., Pyatt J., Jolly D.and Friedmann T. (1986) Bone marrow transplantation in Lesch-Nyhan disease. *Adv. Med. Biol.* **195A,** 167–172.

Onali P., Olianas M. C and Bunse B. (1988) Evidence that adenosine A2 and dopamine autoreceptors antagonistically regulate tyrosine hydroxylase activity in rat striatal synaptosomes. *Brain Res.* **456,** 302–309.

Page T. and Nyhan W. L. (1989) The spectrum of HPRT deficiency: An update. *Adv. Exp. Biol. Med.* **253A,** 129–133.

Page T., Bakay B., Nissinen E., and Nyhan W. L. (1981) Hypoxanthine-guanine phosphoribosyltransferase variants: correlation of clinical phenotype with enzyme activity. *J. Inher. Metab. Dis.* **4,** 203–218.

Palmour R. M., Dyer C., and Seegmiller J. E. (1985) Kinetic alterations of adenosine receptors in HPRT deficient neuroblastoma cells. *Adv. Exp. Biol. Med.* **195A,** 231–236.

Palmour R. M., Dyer C., Pearce C., Goldman N. J., and Seegmiller J. E. (1990a) Caffeine-induced self-mutilation in rats: A critical appraisal. *Brain Res.*

Palmour R. M., Schucher K., Pacheco P., and Ervin F. R. (1990b) Dopamine depletion and self-mutilation in the rat: A model for Lesch-Nyhan disease. *Pharmacol. Biochem. Behav.*

Palmour R. M., Goodyer P., Reade T., and Chang T. M. S. (1989) Microencapsulated xanthine oxidase as an experimental therapy in Lesch-Nyhan disease. *Lancet* **2,** 687,688.

Palmour R. M., Heshka T. W., and Ervin F. R. (1989) Hypoxanthine accumulation and dopamine depletion in Lesch-Nyhan disease: An animal model. *Adv. Exptl. Biol. Med.* **253A,** 165–172.

Palmour R. M. and Pearce C. J. (1979) Behavioral and neurochemical studies of Lesch-Nyhan's syndrome. *Dev. Psychiatr.* **2,** 155–160.

Papez J. W. (1937) A proposed mechanism of emotion. *Arch. Neurol. Psychiatr.* **38,** 725–743.

Partington M. W. and Hennen B. K. E. (1967) The Lesch-Nyhan syndrome: Self-destructive biting, mental retardation, neurological disorder and hyperuricemia. *Develop. Med. Child. Neurol.* **9,** 563–572.

Penney J. B. Jr. and Young A. B. (1983) Speculations on the functional anatomy of basal ganglia disorders. *Ann. Rev. Neurosci.* **6,** 73–94.

Peters J. M. (1967) Caffeine-induced hemorrhagic automutilation. *Arch. Int. Pharmacodyn.* **169,** 139–146.

Pinsky L. and DiGeorge A. M. (1966) Congenital familial sensory neuropathy with anhidrosis. *J. Pediatr.* **68,** 1–8.

Porter N. M., Radulovacki M., and Green R. D. (1988) Desensitization of adenosine and dopamine receptors in rat brain after treatment with adenosine analogs. *J. Pharmacol. Exp. Therap.* **244,** 220–225.

Randrup A. and Munkvad I. (1967) Stereotyped activities produced by amphetamine in several animal species and man. *Psychopharmacologia* **11,** 300–310.

Rassin D. K., Lloyd K. G., Kelley W. N., and Fox I. (1982) Decreased amino acids in various brain areas of patients with Lesch-Nyhan syndrome. *Neuropediatrics* **13,** 130–134.

Razzak A., Fujiwara M., and Ueki S. (1975) Automutilation induced by clonidine in mice. *Eur. J. Pharmacol.* **30,** 356–359.

Rockson S., Stone R., van der Weyden M., and Kelley W. N. (1974) Lesch-Nyhan syndrome: Evidence for abnormal adrenergic function. *Science* **186,** 934,935.

Rosenbloom F. J., Henderson J. F., Caldwell I. C., Kelley W. N., and Seegmiller J. E. (1968) Biochemical basis of accelerated purine biosynthesis de novo in fibroblasts lacking HPRT. *J. Biol. Chem.* **243,** 1166–1173.

Rosenbloom F. M., Kelley W. N., Miller J., Henderson J. F., and Seegmiller J. E. (1967) Inherited disorder of purine metabolism: correlation between central nervous system dysfunction and biochemical defects. *J. Amer. Med. Assoc.* **202,** 175–177.

Roth R. H., Wolf M. E., and Deutch A. Y. (1987) Neurochemistry of midbrain dopamine systems, in *Pschopharmacology: The Third Generation of Progress* (Meltzer H. Y., ed.), Raven Press, New York, pp. 81–94.

Rothman S. M. and Olney J. W. (1986) Glutamate and the pathophysiology of hypoxic-ischemic brain damage. *Ann. Neurol.* **19,** 105–111.

Rumsby P. C., Kato H., Waldren C. A., and Patterson D. (1982) Effects of caffeine on pyrimidine biosynthesis and 5-phosphoribosyl 1-pyrophosphate metabolism in Chinese hamster cells. *J. Biol. Chem.* **257,** 11,364–11,367.

Sakata T. and Fuchimoto H. (1973) Stereotyped and aggressive behavior induced by sustained high dose of theophylline in rats. *Japan. J. Pharmacol.* **23,** 781–785.

Sass J. K., Itabashi H. H., and Dexter R.A. (1965) Juvenile gout with brain involvement. *Arch. Neurol.* **13,** 639–655.

Schlosberg A. J., Fernstrom J. D., Kopzynski C., Cusack B. M., and Gillis M. A. (1981) Acute effects of caffeine injection on neutral amino acids and brain monoamine levels in rats. *Life Sci.* **29,** 173–183.

Seegmiller J. E. (1976) Inherited deficiency of hypoxanthine guanine phosphoribosyltransferase in X-linked uric aciduria (the lesch-nyhan syndrome and its variants). *Adv. Human Genet.* **6,** 75–163.

Seegmiller J. E. (1980) Diseases of purine and pyrimidine metabolism, in *Metabolic Control and Disease* (Bondy P. K. and Rosenberg L. E., eds.), W. G. Saunders, Philadelphia, pp. 777–937.

Seegmiller J. E., Rosenbloom F. M., and Kelley W. N. (1967) Enzyme defect associated with a sex-linked human neurological disorder and excessive purine synthesis. *Science* **155,** 1682–1684.

Segal D. S., Weinberger S. B., Cahill J., and McCunney S. J. (1980) Multiple daily amphetamine administration: Behavioral and neurochemical alterations. *Science* **207,** 904–907.

Silverstein F. S., Johnson M. V., Hutchinson R. J.and Edwards N. L. (1985) Lesch-Nyhan syndrome: CSF neurotransmitter abnormalities. *Neurology* **35,** 907–911.

Simmonds H. A., Fairbanks L. D., Morris G. S., Morgan G., Watson A. R., Timms P., and Singh B. (1987) Central nervous system dysfunction and erythrocyte guanosine triphosphate depletion in purine nucleoside phosphorylase deficiency. *Arch. Dis. Childhood* **62,** 385–391.

Skaper S. and Seegmiller J. E. (1976) Hypoxanthine-guanine phosphoribosyltransferase mutant glioma cells: Diminished monoamine oxidase activity. *Science* **194,** 1171–1173.

Slivka A. and Cohen G. (1985) Hydroxyl radical attack on dopamine. *J. Biol. Chem.* **260,** 15,466–15,472.

Smellie F. W., Davis C.W., Daly J. W., and Wells J. N. (1979) Alkylxanthines: Inhibition of adenosine-elicited accumulation of cyclic AMP in brain slices and of phosphodiesterase activity. *Life Sci.* **24,** 2475–2482.

Snyder F. F., Cruikshank M. K., and Seegmiller J. E. (1978) Comparison of purine metabolism and nucleotide pools in normal and HPRT- neuroblastoma *Biochim. Biophys. Acta* **543,** 556–569.

Snyder S. H. (1985) Adenosine as a neuromodulator. *Ann. Rev. Neurosci.* **8,** 103–124.

Sprugel W., Mitznegg P., and Heim F. (1977) The influence of caffeine and theobromine on locomotive activity and the brain cGMP/cAMP ratio in white mice. *Biochem. Pharmacol.* **26,** 1723,1724.

Stefanovich V. (1979) Influence of theophylline on concentrations of cyclic 3',5'-adenosine monophosphate and cyclic 3',5'-guanosine monophosphate of rat brain. *Neurochem. Res.* **4,** 587–594.

Stone T. W. (1981) Physiological roles for adenosine and adenosine 5'-triphosphate in the nervous system. *Neuroscience* **6,** 523–555.

Stout J. T. and Caskey C. T. (1989) Hypoxanthine phosphoribosyltransferase deficiency: The Lesch Nyhan syndrome and gouty arthritis, in *The Metabolic Basis of Inherited Disease,* (Scriver C. R., Beaudet A. L., Sly W. S., McKusick V. A., Stanbury J. B., Wyngaarden J. B., Frederickson D. S., Goldstein J. L., and Brown M. S., eds.), McGraw Hill, New York, pp. 1007–1028.

Sweetman L. (1968) Urinary and cerebrospinal fluid oxypurine levels and allopurinol metabolism in the Lesch-Nyhan syndrome. *Fed. Proc.* **27,** 1053–1059.

Sweetman L., Borden M., Kulovich S., Kaufmn I., and Nyhan W. L. (1977) Altered excretion of 5-hydroxyindoleacetic acid and glycine in patients with the Lesch-Nyhan disease, in *Purine Metabolism in Man II: Regulation of Pathways and Enzyme Defects* (Muller M. M., Kaiser E., and Seegmiller J. E., eds.), Plenum, New York, pp. 398–404.

Sweetman L. and Nyhan W. L. (1970) Detailed comparison of urinary excretion of purines in a patient with Lesch-Nyhan syndrome. *Biochem. Med.* **4,** 121–134.

Taylor M., Goudie A. J., Mortimore S., and Wheeler T. J. (1974) Comparison between behaviours elicited by high doses of amphetamine and fenfluramine: Implications on the concept of stereotypy. *Psychopharmacology* **40,** 249–258.

Towle A. C., Criswell H. E., Maynard E. H., Lauder J. M., Joh T. H., Mueller R. A., and Breese G. R. (1989) Serotonergic innervation of the rat caudate following a neonatal 6-hydroxydopamine lesion: An anatomical, biochemical and pharmacological study. *Pharmacol. Biochem. Behav.* **34,** 367–374.

Ungerstedt U. (1965) 6-hydroxydopamine induced degeneration of central monoamine neurons. *Eur. J. Pharmacol.* **5,** 107–110.

Ungerstedt U. and Arbuthnott G. (1970) Quantitative recording of rotational behavior in rats after 6-OHDA lesions of the rat nigrostriatal dopamine system. *Brain Res.* **24,** 485–493.

Uretsky N. J. and Iversen L. L. (1970) Effects of 6-hydroxydopamine on catecholamine neurons in the rat brain. *J. Neurochem.* **17,** 269–278.

Vaccarino F. J. and Franklin K. B. J. (1984) Opposite locomotor asymmetries elicited from the medial and lateral substantia nigra by modulation of substantia nigra dopamine receptors. *Pharmacol. Biochem. Behav.* **21,** 73–77.

Van Heeswijk P. J., Blank C. H., Seegmiller J. E., and Jacobson C. B. (1972) Preventive control of the Lesch-Nyhan syndrome. *Obstetr. Gyn.* **40,** 109–113.

von Lubitz D. K. J. E., Dambrosia J. M., Kempski O., and Redmond D. J. (1988) Cyclohexyl adenosine protects against neuronal death following ischemia. *Stroke* **19,** 1133–1139.

Wagner G. C., Ricaurte G. A., Johanson C. E., Schuster C. R., and Seiden L. S. (1980a) Amphetamine induces depletion of dopamine and loss of dopamine uptake sites in caudate. *Neurology* **10,** 547–550.

Wagner G. C., Ricaurte G. A., Seiden L. S., Schuster C. R., Miller R. J., and Westley J. (1980b) Long-lasting depletions of striatal dopamine and loss of dopamine uptake sites following repeated administration of methamphetamine. *Brain Res.* **181,** 151–160.

Waldren C. A. and Patterson D. (1979) Effects of caffeine on purine metabolism and ultraviolet light-induced lethality in cultured mammalian cells. *Cancer Res.* **39,** 4975–4982.

Watts R. W. E. (1983) Some regulatory and integrative aspects of purine nucleotide biosynthesis and its control: An overview. *Adv. Enzyme Reg.* **21,** 33–48.

Watts R. W. E., McKeran R. O., Brown E., Andrews T. M., and Griffiths M. L. (1974) Clinical and biochemical studies on treatment of Lesch-Nyhan syndrome. *Arch. Dis. Childhood* **49,** 693–702.

Watts R. W. E., Spellacy E., Gibbs D. A., Allsop J., McKeran R. O.,and Slavin G. E. (1982) Clinical, post-mortem, biochemical and therapeutic observations on the Lesch-Nyhan syndrome. *Q. J. Med.* **201,** 43–78.

Williams M. (1987) Purine receptors in mammalian tissues: Pharmacology and functional significance. *Ann. Rev. Pharmacol. Toxicol.* **17,** 315–345.

Willis R., Jolly D. I., Miller A. D., Plent M., Esty A., Anderson P., Jones O., Seegmiller J. E., and Friedman T. (1984) Partial phenotypic characterization of human Lesch-Nyhan (HPRT deficient) lymphoblasts with a transmissible retroviral vector. *J. Biol. Chem.* **259,** 7842–7846.

Wilson J. M., Stout J. T., Pallela T. D., Davidson B. L., Kelley W. N., and Caskey C. T. (1986) A molecular survey of HPRT deficiency in man. *J. Clin. Invest.* **77,** 188–95.

Wilson J. M., Tarr G. E., Mahoney W. C., and Kelley W. N. (1982) Human hypoxanthine guanine phosphoribosyltransferase: Complete amino acid sequence of the erythrocyte enzyme. *J. Biol. Chem.* **257,** 10978–10,985; 14,830–14,834

Wilson J. M., Young A. B., and Kelley W. N. (1983) Hypoxanthine-guanine phosphoribosyl transferase deficiency: The molecular basis of the clinical syndromes. *N. Engl. J. Med.* **309,** 900–910.

Wood A. W., Becker M. A., Minna J. D., and Seegmiller J. E. (1973) Purine metabolism in normal and thioguanine-resistant neuroblastoma. *Proc. Natl. Acad. Sci. USA* **70,** 3880–3883.

Wood P. L., Kim H. S., Boyar W. C., and Hutchison A. (1988) Inhibition of nigrostriatal release of dopamine in the rat by adenosine receptor agonists: A1 receptor mediation. *Neuropharmacology* **28,** 21–25.

Wyngaarden J. B. and Ashton D. M. (1959) The regulation of activity of phosphoribosylpyrophosphate amidotransferase by purine ribonucleotides: A potential feedback control of purine biosynthesis. *J. Biol. Chem.* **234,** 1492–1505.

Wyngaarden J. B. and Kelley W. N. (1983) Gout, in *The Metabolic Basis of Inherited Disease* (Stanbury J. B., Wyngaarden J. B., Frederickson D. S., Goldstein J. L., and Brown M. S., eds.), McGraw Hill, New York, 1101–1114.

Yang T. P., Patel P. I., Chinault A. C., Stout J. T., Jackson L. G., Hildebrand S. M., and Caskey C. T. (1984) Molecular evidence for new mutation in the HPRT locus in Lesch-Nyhan patients. *Nature* **310,** 412–414.

Appendix 1

Sample Protocol: Neonatal Dopamine Depletion in the Rat

Because the goal of this experiment is to produce an experimental model of SMB, note must first be taken of the guidelines pertaining to ethical treatment of experimental animals. At McGill, experiments of this type would be classed as "C" under the Guidelines of the Canadian Council on Animal Care and would thus require review by both the Facility Committee and the University Ethics Committee. Care must be taken first to ensure that treated rats are handfed and handwatered during the immediate postsurgical period, since 6-OHDA often but not always stimulates aphagia and adipsia. A strategy for terminating SMB when it arises is described. A final caveat is that all animal experiments, especially those that potentially cause distress or suffering to the animal, should be carefully planned with respect to execution and with respect to cost/benefit ratio (Breese et al., 1984,1985; Duncan et al., 1987).

Pregnant female rats (Sprague-Dawley or Long-Evans) are obtained from an appropriate source 1–2 wk before delivery and are housed singly in plastic cages with woodchip bedding. Food and water are given *ad libitum*, and lighting is regulated to a 12 n:12 n light:dark, cycle. Two d after delivery, litters are culled to 10 pups. Breese et al. (1984) report no differences for sex, and so try to maintain equal numbers of males and females. In our lab, we would preferentially keep males.

The supplies needed for 6-OHDA lesioning of neonatal rats include:

1. a Hamilton syringe with a short needle for intracisternal injection;
2. 6-OHDA HBr from Regis Chemical (Chicago, IL), Sigma Chemicals (St. Louis, MO), or another reliable source;
3. desmethylimipramine (DMI) from Sigma (St. Louis, MO) or another reliable source; and
4. anesthesia (Breese et al., 1984) used ether, but we prefer ketamine (0.07 mg/g body wt, given im) as being more reliable and less stressful. The toxic threshold for pentobarbital is reduced by DMI, so that dosages should be adjusted downward if it is to be used.

On the fifth d of life, 20 mg/kg DMI (dissolved in 0.15*M* saline containing 0.05% ascorbic acid) is injected im 30 min before lesioning. Animals are anesthetized 25 min later. A syringe containing 10 μL 6-OHDA (100 μg free base dissolved in 0.05% ascorbic acid) is inserted into the cisterna without making any incision. The drug is slowly delivered over 2 min. This is ideally done with a micropump but can be controlled manually after a bit of practice. Pups are returned to the home cage but checked periodically over 2 h for recovery. Control animals should receive DMI and intracisternal injections of vehicle. *Note*: Both drugs must be dissolved immediately before use and should be kept sealed and wrapped in aluminum foil if multiple injections are to be done from a single vial.

Rats must be checked daily for aphagia and adipsia. This is best done by keeping a log of body wt and manually feeding any pup that is not gaining wt in accordance with the normal curve exhibited by controls. Behavioral evaluation may commence after several days of recovery. Breese et al. (1984,1985) reported testing at 23, 30, and 85 d of life. However, rats treated with 6-OHDA are generally quite healthy 10 d postsurgery and could reasonably be challenged any time thereafter. Behavioral challenge is with either apomorphine or L-dopa, preceeded by a dopa decarboxylase inhibitor. The usual dopa decarboxylase inhibitor is RO 4-4602, which may ordinarily be requested from Hoffman-La Roche (Nutley, NJ or Basel). Apomorphine HCl (Merck Sharpe and Dohme, Rahway, NJ or Sigma, St. Loius, MO) is dissolved in 0.05% ascorbic acid at concentrations sufficient to administer 1–10 mg/kg; L-dopa (Hoffman-LaRoche or Sigma) is suspended in 0.5% methylcellulose at concentrations sufficient to administer 10–100 mg/kg. RO 4-4602 is administered at a constant dose (50 mg/kg im) 1 h before L-dopa. For blockade, any DA antagonist will be effective. Haloperidol (McNeil Laboratories, Fort Washington, PA or Sigma) should be used at 0.3–1 mg/kg, ip, given 45 min before DA agonist challenge.

The behavioral challenge is as follows:

1. RO 4-4602 (50 mg/kg im), followed 1 h later by 10, 30, or 100 mg/kg L-dopa, given ip; or
2. 1, 3, or 10 mg/kg apomorphine given sc.

Motor abnormalities will begin quite promptly and should be well developed at 20 min after drug administration. In neonatally treated rats, most motor behaviors peak at 30–40 min; SMB does not peak until 80 min. Behaviors that have been generally useful include locomotion (operationally defined by crossing the bars of a grid, if one does not have an activity monitor), licking, rearing, sniffing, head nodding, paw treading, "taffy pulling," perseverative self-grooming, self-biting (no break in skin), and SMB (skin broken). Standard operational definitions may be used or as an exercise, students may learn to define behaviors operationally.

Two different types of scoring strategies have been used. In one, each animal is observed uninterruptedly, and each behavior is scored as it occurs by 15 s segments. This will produce a very detailed behavioral map for each animal. In the other, four animals are observed simultaneously, with each receiving the observer's attention 15 s of every min (or 1 min of every 5, with 1 min out). This strategy provides very complete data once observers develop good concentration habits and facility with the instruments. There should always be some practice sessions, followed by interobserver reliability testing, prior to data collection. Data of this sort generally should be analyzed with analysis of variance (ANOVA) for repeated measures.

If an animal exhibits SMB, i.e., bites hard enough to break the skin and draw blood, it should be promptly anesthetized. Breese et al. (1984) recommend phentobarbital (40 mg/kg ip). We typically use ketamine, as mentioned previously. In some cases, a few s exposure to 5% carbon dioxide is sufficient to interrupt SMB, particularly if it occurs more than 60 min after drug administration. If animals are too agitated to handle easily, a brief exposure to carbon dioxide followed by ketamine is advised.

Repeated testing of the same animal can be done at 1-wk intervals. It is generally considered to be appropriate to test each animal up to five times, if it remains in good physical and psychological health. Several questions that require further study are mentioned in Section 8. of this chapter. A detailed investigation of the early developmental patterns of this model would be a simple and useful experimental exercise, requiring no sophisticated equipment or detailed chemical knowledge.

Index